Meiosis. IV: Current Research

Papers by
M. Callebaut, Kwen-Sheng Chiang, Yu. F. Bogdanov, N. O. Bianchi, N. Odartchenko, Anil B. Mukherjee, Monna Crone, Herbert Stern, Yasuo Hotta, Stephen H. Howell, Spencer W. Brown, Meredith C. Gould, Kathleen Church, E. N. Antropova, G. A. Morrill, Angela Rocchi Brasiello, L. G. Parchman, M. Muramatsu, H. G. Dickinson, R. G. Kessel, et al.

MSS Information Corporation
655 Madison Avenue, New York, N.Y. 10021

Library of Congress Cataloging in Publication Data

Main entry under title:

Meiosis: current research. IV.

The 3 earlier volumes of this 4 vol. collection issued under title: The meiotic process.

Includes bibliographies.

1. Meiosis. I. Melnyk, John.

QH605.M428 574.8'7623 72-6751

ISBN 0-8422-7041-8

TABLE OF CONTENTS

CREDITS & ACKNOWLEDGEMENTS

Antropova, E. N.; and Yu. F. Bogdanov, "Cytophotometry of DNA and Histone in Meiosis of *Pyrrhocoris apterus*," *Experimental Cell Research*, 1970, 60:40-44.

Bianchi, N. O.; and Martha S. De Bianchi, "Y Chromosome Replication and Chromosome Arrangement in Germ Line Cells and Sperm of the Rat," *Chromosoma*, 1969, 28:370-378.

Bogdanov, Yu. F.; N. A. Liapunova; A. I. Sherudilo; and E. N. Antropova, "Uncoupling of DNA and Histone Synthesis Prior to Prophase I of Meiosis in the Cricket *Gryllus* (*Acheta*) *domesticus* L.," *Experimental Cell Research*, 1968, 52:59-70.

Brasiello, Angela Rocchi, "Autoradiographic Study of Ribonucleic Acid Synthesis during Spermatogenesis of *Asellus aquaticus* (Crust. Isopoda)," *Experimental Cell Research*, 1968, 53:252-260.

Brown, Spencer W., "Developmental Control of Heterochromatization in Coccids," *Genetics*, 1969, 61:191-198.

Callebaut, M.; and J. L. Bernheim, "Feulgen-cytophotometric Determination of DNA Content in the Germ Cell Nuclei of the Female Chicken Embryo during Premeiosis," *Experientia*, 1969, 25:978-980.

Chiang, Kwen-Sheng; Joseph R. Kates; Raymond F. Jones; and Noboru Sueoka, "On the Formation of a Homogeneous Zygotic Population in *Chlamydomonas reinhardtii*," *Developmental Biology*, 1970, 22:655-669.

Church, Kathleen; and D. E. Wimber, "Meiosis in *Ornithogalum virens* (Liliaceae). II. Univalent Production by Preprophase Cold Treatment," *Experimental Cell Research*, 1971, 64:119-124.

Crone, Monna; and Hannah Peters, "Unusual Incorporation of Tritiated Thymidine into Early Diplotene Oocytes of Mice," *Experimental Cell Research*, 1968, 50:664-668.

Dickinson, H. G.; and P. R. Bell, "Nucleocytoplasmic Interaction at the Nuclear Envelope in Post Meiotic Microspores of *Pinus banksiana*," *Journal of Ultrastructure Research*, 1970, 33:356-259.

Gould, Meredith C.; and Paul C. Schroeder, "Studies on Oogenesis in the Polychaete Annelid *Nereis grubei* Kinberg. I. Some Aspects of RNA Synthesis," *The Biological Bulletin*, 1969, 136:216-225.

Hotta, Yasuo; and Norman Hecht, "Methylation of *Lilium* DNA during the Meiotic Cycle," *Biochimica et Biophysica Acta*, 1971, 238:50-59.

Hotta, Yasuo; L. G. Parchman; and Herbert Stern, "Protein Synthesis during Meiosis," *Proceedings of the National Academy of Sciences*, 1968, 60:575-582.

Hotta, Yasuo; and Herbert Stern, "Analysis of DNA Synthesis during Meiotic Prophase in *Lilium*," *Journal of Molecular Biology*, 1971, 55:337-355.

Howell, Stephen H.; and Herbert Stern, "The Appearance of DNA Breakage and Repair Activities in the Synchronous Meiotic Cycle of *Lilium*," *Journal of Molecular Biology*, 1971, 55:357-378.

Kessel, R. G.; and W. R. Panje, "Organization and Activity in the Pre- and Postovulatory Follicle of *Necturus maculosus*," *The Journal of Cell Biology*, 1968, 39:1-34.

Morrill, G. A.; Adele B. Kostellow; and Janet B. Murphy, "Sequential Forms of ATPase Activity Correlated with Changes in Cation Binding and Membrane Potential from Meiosis to First Cleavage in *R. pipiens*," *Experimental Cell Research*, 1971, 66:289-298.

Mukherjee, Anil B.; and Maimon M. Cohen, "DNA Synthesis during Meiotic Prophase in Male Mice," *Nature*, 1968, 219:489-490.

Muramatsu, M.; T. Utakoji; and H. Sugano, "Rapidly-labeled Nuclear RNA in Chinese Hamster Testis," *Experimental Cell Research*, 1968, 53:278-283.

Odartchenko, N.; and M. Pavillard, "Late DNA Replication in Male Mouse Meiotic Chromosomes," *Science*, 1970, 167:1133-1134.

Parchman, L. G.; and Herbert Stern, "The Inhibition of Protein Synthesis in Meiotic Cells and Its Effect on Chromosome Behavior," *Chromosoma*, 1969, 26:298-311.

Stern, Herbert; and Yasuo Hotta, "DNA Synthesis in Relation to Chromosome Pairing and Chiasma Formation," *Genetics*, 1969, 61:27-39.

PREFACE

This multi-volume collection includes major experimental results in the field of meiosis published in English in the period 1968–1971.

Areas in which current research has developed rapidly in this field are meiotic DNA synthesis and the ultrastructure of the meiotic apparatus, particularly the synaptinemal complex. As elsewhere in contemporary biology, an intensive effort is being made to determine the basis of specificity, here the specificity of homologous chromosome recognition. The answer to this lies somewhere in the molecular structure of the synaptinemal complex, as papers in these volumes indicate. Meiotic DNA synthesis and repair have been convincingly associated with exchange of genetic material through cross-over. A great many variants of the meiotic process are now known and variants such as pairing, recombination and chromosome movement are described in these volumes.

Other papers presented deal with the study of meiosis as it occurs in man and other mammals, the radiation and drug sensitivity of meiotic events, the biochemical and physiological aspects of meiosis, and the effects on meiosis of particular genes, mainly in *Drosophila.*

Premeiotic and Meiotic DNA Synthesis

Feulgen-Cytophotometric Determination of DNA Content in the Germ Cell Nuclei of the Female Chicken Embryo During Premeiosis

M. CALLEBAUT and J. L. BERNHEIM

In a previous work[1,2], one of us investigated DNA synthesis during premeiosis in the ovarian germ cells of the chicken embryo both in vitro and in vivo, using the autoradiografic technique, following the incorporation of ^{3}H-thymidine. The successive developmental germ cell stages found during this period in the cortex of the ovary of the chicken embryo are represented in Figure 1. Only during the preleptotene stage of the germ cells, chiefly occurring in the central part of the ovarian cortex of 15- to 17-day-old embryos, does nuclear incorporation of ^{3}H-thymidine take place[3]. Germ cells with a reticulated nucleus, characterized by a more regular and much finer chromatin distribution than the oogonia at interphase, do not incorporate the DNA-precursor. Large numbers of these cells are found in the central part of the ovarian cortex of 14- and 15-day-old embryos. Since there is morphological evidence of a transition between cells with a reticulated nucleus and cells in the early preleptotene stage[4], we consider the former to be the precursors of the latter. The meiotic divisions of the female germ cells in the chicken are concluded months or years after the observed ^{3}H-thymidine incorporation wave. Hence, this DNA synthesis may represent:

(1)A premeiotic reduplication of the germ cell nuclear DNA; (2) a metabolic DNA synthesis[5] ('specific gene amplication')[6] or DNA turnover; (3) a combination of these two hypotheses. The present study was undertaken to check these explanations by comparing the DNA content of reticulated nuclei and leptotenes.

Fig. 1. Drawings of the successive developmental stages of the germ cells found in the ovarian cortex of a 15-day-old female chicken embryo, exactly as seen under the microscope. From left to right: oogonium, oocyte with reticulated nucleus, oocyte in early pre-leptotene stage, oocyte in late preleptotene stage and oocyte in leptotene stage of meiosis (fixation of the ovary with acetic-alcohol and coloration with Feulgen-fast green).

Material and methods. Fertile eggs of White Leghorn chickens were incubated at 38.5 °C for 15 and 17 days. 4 ovaries of each age were fixed in A.F.A. (alcohol 96%: 70 parts, formaline: 20 parts and chilled acetic acid: 5 parts), dehydrated in graded alcohols, embedded in paraffin and cut at 10 μ thickness.

In order to enable a uniform processing of the material from the 2 age groups, sections from 15- and 17-day-old embryos were placed on the same slide. After hydrolysis in 1 *N* HCl during 12 minutes at a constant temperature of 60 ± 1 °C, the sections were stained with Schiff's reagent during 45 min and rinsed in 3 successive baths of a freshly prepared So_2 solution. For the Feulgen-DNA determinations we used the Lison cytophotometer in visible light[7]. The nuclei were projected on a screen and their light-absorption was measured through a diaphragm of about $^1/_4$ of their projected surface area. Nuclei which on manipulating the fine adjustment appeared cut were not measured. 50 nuclei of each cell group were investigated, 5 measurements are made over each nucleus and their arithmetic mean was computed as the nuclear extinction. The Feulgen-DNA content was expressed in arbitrary units (AU) as the product of the nuclear extinction and the projected nuclear surface area. Since the Feulgen-DNA contents of the nuclei may be assumed to follow a logarithmic distribution within a given ploidy class, the mean values were calculated after logarithmic transformation of the distribution-curve. The absolute nuclear surface area of nuclei in both groups is irrelevant to the present study. Hence it was not computed separately but also expressed in arbitrary units (AU).

Results. The results of the Feulgen-DNA measurements are summarized in Figure 2 and in the Table. The mean value of Feulgen-DNA found for the reticulated nuclei is 1364 AU and for the leptotene nuclei it is 3645 AU. The average surface area of the reticulated nuclei is only slightly larger than that of the leptotenes (Figure 3). This confirms the impression gained by microscopical observation that the 2 groups can hardly be differentiated on the basis of their nuclear size.

Discussion. If the mean nuclear Feulgen-DNA content of reticulated nuclei is taken as 100%, the leptotene group has a mean of 267%. In other words, if we consider the DNA-content of the reticulated nuclei to be at the diploid level, then the leptotenes would be largely hypertetraploid, suggesting aneuploidism or DNA of combined origin. If, alternatively, we assume that the leptotenes are tetraploid, the reticulated nuclei would be on a hypodiploid level. This second hypothesis seems the most likely when the conditions of the measurements are taken into account.

(1) The particular asymmetrical shape of the histogram of reticulated nuclei, with its numerous very low Feulgen-DNA values (Figure 2), is in sharp contrast with the symmetry of the histogram of leptotene nuclei. This

asymmetry can most readily be explained by an almost unavoidable inaccuracy in the selection for measurement of the reticulated nuclei. Indeed because these cells are

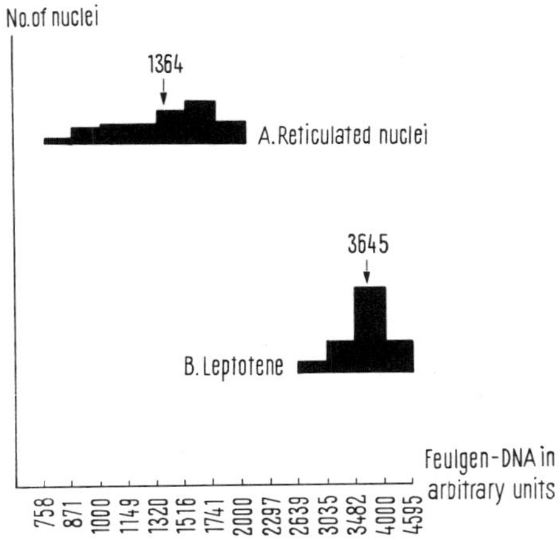

Fig. 2. Feulgen-DNA content in arbitrary units of reticulated and leptotene nuclei.

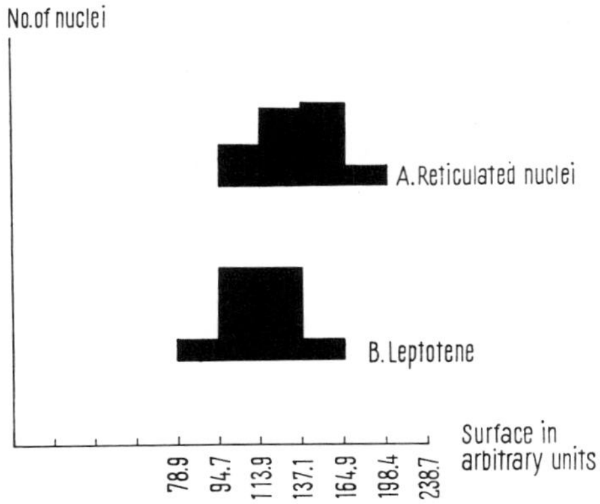

Fig. 3. Surface area in arbitrary units of reticulated and leptotene nuclei.

Feulgen-DNA content of reticulated and leptotene nuclei. $\overline{x}$, logarithmic mean; s, standard deviation; s log $\overline{x}$, standard error.

	$\overline{x}$	log $\overline{x}$	s	s log $\overline{x}$
Reticulated nuclei	1364	3.13469557	0.10344	0.014798
Leptotene nuclei	3645	3.57034185	0.045825	0.0065038

very pale, it is difficult to ascertain by manipulating the fine adjustment of the microscope whether or not a more or less important part of the nucleus has been cut. This difficulty therefore may have led to the measurement of incomplete reticulated nuclei. As suggested by the symmetry of their histogram, the selection of the leptotenes, which are much darker, was more precise.

(2) The extinction values in the reticulated group oscillate around 10–12% and in the leptotene group around 30%. Since the minute extinctions of the reticulated nuclei often fall just beyond the range of the linear section of the cytophotometer's absorption curve, these measurements are less accurate and, moreover, bring about an underestimation of the Feulgen-DNA content[8].

(3) In reticulated nuclei, many of the fine chromatin granules are studded against the inner surface of the nuclear membrane and may have escaped measurement since we had to keep the projection screen diaphragm clear of the nuclear membrane. Thus this particular distribution of DNA brings about another underestimation of extinction in the reticulated nuclei. Summarizing, error (1) causes the insertion of measurements over incomplete nuclei into the calculation of the mean Feulgen-DNA content, thus lowering the mean of the reticulated nuclei. It also accounts for the asymmetrical shape of their histogram. Errors (2) and (3) provoke an overall underestimation of the Feulgen-DNA content in the reticulated nuclei.

These limitations of the technique in this material may suffice to explain that the mean Feulgen-DNA content of the reticulated nuclei is less than half that of the leptotenes.

Conclusion. The results of the present quantitative study show that the germ cells at the leptotene stage contain at least twice as much nuclear Feulgen-DNA as the germ cells with reticulated nuclei. Considering the sources of error in this material, a simple 2/1 (tetraploid/diploid) rate is suggested, but could not be ascertained. Our data, however, indicate that during the DNA synthesis at the preleptotene stage, the Feulgen-DNA content in the germ cells of the female chicken is at least doubled. The germ cells appear to go through a period of premeiotic doubling of their DNA content, analogous to the S-phase of a regular cell cycle. Our data neither support nor rule out the possibility of a concomitant DNA synthesis of other origin at this stage of premeiosis[9].

[1] M. CALLEBAUT et R. DUBOIS, C. r. hebd. Séanc. Acad. Sci., Paris *261*, 12 (1965).
[2] M. CALLEBAUT, Experientia *23*, 419 (1967).
[3] M. CALLEBAUT, J. Embryol. exp. Morph. *18*, 299 (1967).
[4] G. C. HUGHES, J. Embryol. exp. Morph. *11*, 513 (1963).
[5] H. ROELS, Int. Rev. Cytol. *19*, 1 (1966).
[6] D. BROWN and I. B. DAWID, Science *160*, 272 (1968).
[7] L. LISON, *Histochemie et cytochimie animale* (Gauthier-Villars, Paris 1960).

[8] H. ROELS, Acta Histochem., Suppl. VI (1963).
[9] Acknowledgments. The authors are gratefull to Professor L. VAKAET (University of Antwerp) and to Professor H. ROELS (University of Ghent) for their valuable suggestions.

On the Formation of a Homogeneous Zygotic Population in *Chlamydomonas reinhardtii*[1]

KWEN-SHENG CHIANG, JOSEPH R. KATES, RAYMOND F. JONES, AND NOBORU SUEOKA

INTRODUCTION

Meiosis and accompanying events such as gametogenesis and fertilization can be studied under controllable experimental conditions in the unicellular green alga *Chlamydomonas*, which provides many synchronous meiotic cells at various developmental stages. *Chlamydomonas* is not only unicellular and heterothallic, but it also undergoes a zygotic, or initial, type of meiosis. Vegetative growth, gametogenesis, and fertilization can be manipulated in the laboratory according to standard, microbiological techniques in chemically defined liquid or agar media. Large vegetative cultures of the two mating type strains and their respective gametogenic differentiated states can be handled separately and controllably. To initiate fertilization, equal numbers of two opposite mating type gametes are mixed; virtually every gamete in a large population fuses with the other mating type gamete to give rise to zygotes.

Historically, small quantities of gametes of *Chlamydomonas* were obtained for genetic or fusion studies by flooding aged vegetative cultures on agar plates with a dilute medium or distilled water (Bold, 1949; Sager and Granick, 1954; Lewin, 1956; Tsubo, 1956; Forster and Wiese, 1954; Bernstein and Jahn, 1955; Ebersold, 1956; Trainor, 1958, 1959). The fusion process of gametes in *Chlamydomonas* has been extensively studied (Lewin, 1954; Wiese and Jones, 1963; Wiese, 1965; Friedmann *et al.*, 1968; Brown *et al.*, 1968). The study

[1] This investigation was supported by U.S. Public Health Service Research Grant GM 15114, National Science Foundation grants G7025, G15080, GB3445, GB3064, and GB5702, and Atomic Energy Commission Contract AT (30-1)-3475.

of gametogenesis per se in *Chlamydomonas* has only recently been initiated (Kates and Jones, 1964 a and b; Kates, 1966; Kates *et al.*, 1968; Jones *et al.*, 1968). A method for inducing synchronous gametogenesis from a light-dark synchronized, vegetative culture has been developed. Under certain experimental conditions, gametic populations arise that possess 100% mating efficiency with the opposite mating type gametes (Kates and Jones, 1964a). Following this basic technique with some modifications, large, homogeneous populations of freshly formed zygotes (conjugated gamete-pairs) of *C. reinhardtii* devoid of vegetative cells and unmated gametes have been obtained for meiotic DNA replication studies (Chiang *et al.*, 1965; Sueoka *et al.*, 1967; Chiang, 1968).

This paper presents further studies on the method for obtaining large populations of zygotes in a liquid medium. The following basic characteristics of the mating behavior of *C. reinhardtii* gametes, heretofore undocumented, will be demonstrated: (1) the correlation of the mating efficiency of gametes with the vegetative cell growth stage in the light-dark synchronized culture; (2) the independence of the mating efficiency of gametes and the vegetative growth stage in the continuous light nonsynchronized culture; (3) the unstable mating behavior of gametes, which varies among clones and sometimes within clones.

With an understanding of these basic properties of gametic mating behavior, homogeneous populations of young zygotes derived from 100% fusion of gametes can be repeatedly obtained under different experimental conditions and with different strains.

MATERIALS AND METHODS

Organism and strains. Plus (mt^+ female) and minus (mt^- male) mating types (137F) of wild-type *C. reinhardtii* (Dangeard) used in this investigation were kindly provided by Dr. R. P. Levine. Wild type mt^+ and mt^- strains of *C. reinhardtii* (Dangeard) obtained from the culture collection of algae at Indiana University (strains 89, 90) were also used. In the course of the present study, identical results were obtained from the two wild-type strains. Except when specified, the results presented here pertain to the strains obtained from Dr. Levine.

Culture media. Minimal liquid media described previously (Sueoka *et al.*, 1967) were used at different phases of growth. HSM and 3/10 HSM were used for the vegetative growth of wild-type plus and minus vegetative cells. A nitrogen-free (N-free) medium was used for the induction of gametogenesis and for the subsequent mating of mt^+ and mt^- gametes (Kates and Jones, 1964 ; Sueoka *et al.*, 1967).

Strain maintenance and inoculation. Stock clones of both mating types were maintained separately on HSM 1.5% agar slants grown under dim light (less than 100 ft-c). For optimum synchrony and mating conditions, cell clones were subjected to "single colony isolation" (see Experiments and Results) every 4–5 weeks.

To start each experiment with a cell population of maximum homogeneity, single colonies from stock clones were reisolated and grown in fresh 250 ml HSM liquid "pilot cultures" under a 12-hour light-12-hour dark cycle until the cell concentration reached 1 to 2 $\times$ 10^6/ml. To obtain a vegetative culture, cells from the "pilot cultures" were inoculated into 1- or 2-liter Erlenmeyer flasks to give an initial cell concentration of 1 to 2 $\times$ 10^4/ml.

Vegetative culture. Vegetative cells were grown in HSM or 3/10 HSM at 21° $\pm$ 1°C in 1- or 2-liter Erlenmeyer flasks. The cultures were aerated with a mixture of filtered air and CO_2 (3%), and stirred constantly with magnetic stirrers. Overhead illumination was provided by cool white fluorescent lamps at an intensity of about 360 ft-c as measured (using Weston Illumination Meter, Model 756) on the surface where the flasks stood. For synchronized vegetative growth an alternating light-dark cycle of 12-hour intervals was imposed upon the culture (Bernstein, 1960; Kates and Jones, 1964). For nonsynchronized growth, continuous illumination of about 360 ft-c was provided.

Induction of gametogenesis and mating of gametes. For induction of gametogenesis in liquid culture a nitrogen-withdrawal technique was used (Kates and Jones, 1964). Prior to gametic induction and during vegetative growth, cells were harvested by centrifugation at (10,500 g_{max}) at 25°C for 10–15 minutes, washed once with nitrogen-free (N-free) medium, and resuspended into fresh N-free medium at a cell concentration of 3 to 6 $\times$ 10^6/ml.

The gametogenic culture (10–20 ml) was placed in a large, 75-ml culture tube and aerated with a mixture of filtered air and CO_2 (3%). Aeration from a glass tube extending to the bottom of the culture tube provided continuous mixing of the culture. Overhead illumination was supplied by cool white fluorescent lamps, used continuously for 16–24 hours. Although the same light source and intensity were employed for both vegetative and gametogenic cultures, the effective light intensity for the gametogenic cultures was higher, due to their less dense composition (10–20 ml in 75-ml culture tubes). Approximately 16 hours after resuspension into the N-free medium, gametes were obtained from the gametogenic cell division. Their smaller sizes and more active movements (compared with vegetative cells) indicated the full differentiation of the gametes. Mating tests were performed between opposite mating type gametes to determine their mating efficiency. Usually 2–3 ml of one mating type gamete culture of a known cell concentration were mixed in a 20-ml culture tube with a calculated volume of the other culture containing an equivalent number of opposite mating type gametes. The mixture was aerated according to the method for the gametogenic culture. The mating process was observed regularly by examining small portions of the cell mixture. Large-scale gametogenesis and mating to obtain considerable quantities of meiotic cells were performed exactly according to the above procedures, except for the use of a 2-liter low-form Fernbach flask to obtain relatively high intensity illumination and more effective stirring.

Determination of mating efficiency. The efficiency of mating in the gametic population was determined from the cell number in the mating mixture containing equal numbers of two opposite mating type gametes with the following equation (Chiang, 1965; Sueoka *et al.*, 1967):

$$2(1 - N/N_0) \times 100\%$$

where N_0 is cell number per milliliter immediately after the mixing of the two mating types, and N is the cell number per milliliter (a conjugated pair is counted as one) at any given time after mixing.

For convenience, all the progeny of the cell division during gametogenic differentiation in N-free medium are designated as "gametes," whether or not they are capable of mating. A gametic population of 100% mating efficiency (determined by the above equation) is sometimes referred to as a quantitative-mating gametic population.

EXPERIMENTS AND RESULTS

Single-Clone Selection and Mating Behavior of Gametes

The mating efficiencies of a number of plus and minus isogamous strains of *C. reinhardtii* obtained from different sources were examined without any pretreatment such as single clone selection. Liquid vegetative cultures were inoculated directly from the original agar

TABLE 1
TYPICAL DATA OF A SINGLE-CLONE SELECTION EXPERIMENT[a]

Cross No.	Crosses between clones	Mating efficiency (%)
I	89-1-a × 90-1-c	60
II	89-1-a × 90-1-g	15
III	89-1-a × 90-2-a	70
IV	89-2-e × 90-1-c	102
V	89-2-e × 90-1-g	80
VI	89-2-e × 90-2-a	50
VII	89-2-b × 90-1-c	97
VIII	89-2-b × 90-1-g	64
IX	89-2-b × 90-2-a	76
X	89-2-b × 90-2-b	60

[a] The vegetative cells of two pairs of freshly tested quantitative mating cones 89-1, 90-1, and 89-2, 90-2 were plated on agar plates. After incubation, a number of single colonies were picked up and a specific number was given to each of them, indicating their origin (e.g. 89-1-a, 90-2-b, etc.). These clones were then inoculated into synchronized liquid culture for multiplication. At the end of linear phase of this vegetative growth (see text), gametogenic induction and the subsequent mating test were carried out as described in Materials and Methods.

slants. Vegetative culture, gametogenic induction, and small-scale mating tests were carried out according to the standard techniques described in the Materials and Methods. Mating efficiency fluctuated to a large extent in different pairs of strains tested under identical experimental conditions. There were also indications of noncompatibility between strains of different origin. The mating precentages were quite different among a number of different crosses, even though all these strains originated from the same Indiana Culture Collection of algae. This fact suggests that the mating behavior of a particular pair of strains is unlikely to be very stable and may vary even during their maintenance. This point was further substantiated by routine single clone isolations to ensure 100% mating efficiency. Table 1 presents typical data for such a single clone selection and mating efficiency test. In addition the mating behavior appears to be an intrinsic characteristic of a particular clone. Table 1 demonstrates the instability of mating behavior and a variation of 15–100% mating efficiency among clones from strains that originally showed quantitative mating. Furthermore, the stability of the mating property for any given clone and the transmittance of the mating property to its progeny were different in different clones. As shown in Table 2,

TABLE 2

STABILITY OF MATING EFFICIENCY IN DIFFERENT CLONES[a]

Origin of clone	Crosses	Mating efficiency	Crosses	Mating efficiency (%)
Parental clones	89-4 × 90-4	~100	89-7 × 90-7	~100
	89-3 × 90-3	~100		
	89-3 × 90-4	~100		
	89-4 × 90-3	~100		
Progeny subclones from parental clones above	89-4-b × 90-3-e	15	89-7-c × 90-7-c	91
	89-3-a × 90-3-e	0	89-7-c × 90-7-e	85
	89-3-a × 90-4-a	50	89-7-c × 90-7-h	105
	89-3-a × 90-4	24	89-7-e × 90-7-c	99
	89-3-b × 90-4-a	16	89-7-e × 90-7-e	82
	89-3-b × 90-4	0	89-7-e × 90-7-d	106
			89-7-e × 90-7-h	85

[a] For experimental details see Table 1 footnote.

parental clones 89–7 and 90–7 appear more stable than 89–3, 90–3, and 89–4, 90–4.

The above data indicate wide variations in mating efficiency both among and within clones. It was important, therefore, to carry out a small-scale single-clone selection procedure to preselect a 100% mating clone just before conducting a large-scale meiotic experiment that required developing zygotes in quantity. For this reason preselected 100% mating clones were used in all experiments below.

Gametogenic Differentiation in Synchronized Vegetative Cultures

Under an alternating 12-hour dark-12-hour light cycle, the vegetative cell division takes place toward the end of the dark period (Fig. 1). It was previously found that the best results for the synchronous control of gametogenesis were obtained when vegetative cells were harvested and resuspended into N-free medium after 6 hours in the light period (Kates and Jones, 1964). Following this procedure, vegetative cells at different growth stages in the synchronized culture (see Fig. 1) were transferred to N-free medium for gametogenesis. After 18 hours under continuous illumination, the mating efficiencies of the resultant gametes were tested by mixing them with equal numbers of opposite mating type gametes obtained by the same procedure.

Analogous to the exponential phase in the continuous light, non-

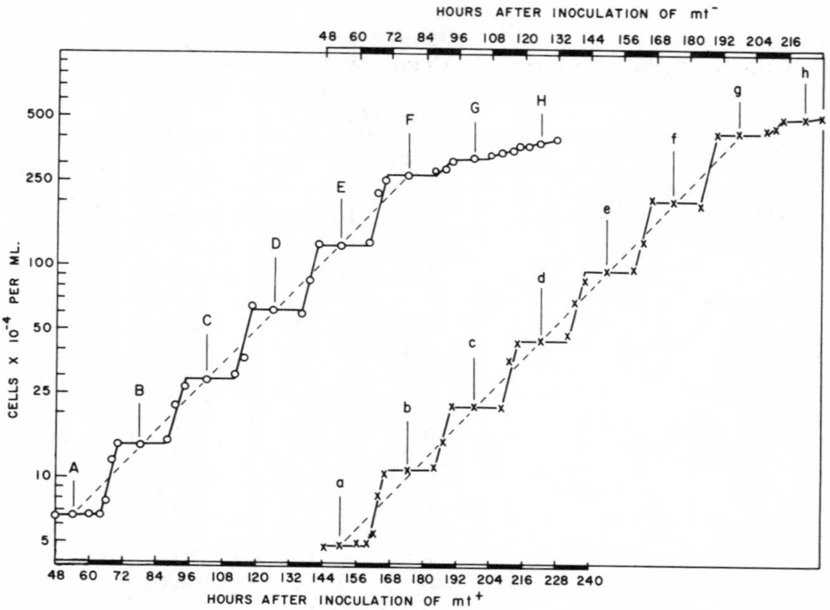

FIG. 1. Vegetative growth of synchronized cultures and gametogenic induction. The plus (mt^+) and minus (mt^-) mating types of *Chlamydomonas reinhardtii* were grown separately in HSM medium (see Materials and Methods) under an alternating light-dark cycle as indicated by open and closed horizontal bars, respectively, in the abscissa. At different growth stages, as indicated by different letters (capital letter for mt^+ and small letter for mt^-) on growth curves, vegetative cells from each culture were harvested and resuspended in the nitrogen-free medium for the induction of gametogenesis. After 18 hours under continuous illumination at 21°C equal numbers of the plus and minus gametes were mixed for the mating test. Ninety minutes after the mixing, the mating efficiency was scored using the formula described in the Material and Methods section. The results of the gametogenic cell division are given in Table 3.

synchronized culture, the early growth stage in the synchronized culture is here called "linear" phase (see Fig. 1), while the subsequent nonlinear phase is referred to as the "stationary" phase. Figure 1 and Table 3 show that only those gametes induced from vegetative cells toward the end of the "linear" growth phase gave nearly 100% mating. Those induced at the middle of the "linear" or at the "stationary" phase failed to give quantitative mating. Gametes induced from vegetative cells during the very early part of linear growth did not mate at all (Table 3). Furthermore, this correlation between the vegetative growth stage and the mating efficiency of gametes existed in both plus and minus mating type synchronized cultures. When a

TABLE 3

GAMETOGENIC CELL DIVISIONS AND MATING EFFICIENCY OF GAMETES INDUCED FROM A SYNCHRONIZED VEGETATIVE CULTURE AT VARIOUS GROWTH STAGES

Point on Fig. 1	Mating efficiency of vegetative cells (%)	Cells × 10^{-4} per ml			Mating efficiency of gametes (%)
		Vegetative culture	Gametogenic culture		
			Starting	Finishing	
A	0	6.6	584	3270	0
a		4.9	602	6200	
B	1	14.2	606	2995	11
b		11.1	591	6470	
C	−4	28.9	576	3465	38
c		22.0	641	6750	
D	0	61.8	455	2370	42
d		45.1	606	7250	
E	0	125	548	3275	47
e		96.2	687	7800	
F	5	270	467	3425	100
f		210	610	8150	
G	−3	322	613	3263	99
g		431	573	3531	
H	0	383	589	2065	43
h		507	596	2705	

pair of quantitative mating gamete populations (F, G, f, and g in Fig. 1) were mated with opposite mating type gametes induced from two separate vegetative cultures, both at their early linear growth and with low vegetative cell concentrations, low mating efficiencies resulted (Table 4).

No mating between the vegetative cells of opposite mating types from any growth stage in the synchronized culture was observed (Table 3). The plus and minus numbers appearing in Table 3 reflect small counting errors using the mating efficiency formula (see Materials and Methods).

Thus, the mating efficiency of gametes is apparently correlated with the growth stage of the synchronized culture in which gametogenesis is induced; quantitative mating occurs only in gametes produced at the end of the "linear" growth phase.

TABLE 4

Mating Efficiency of Gametes Induced from Synchronized Vegetative Cultures at Different Growth Stages[a]

Point on Fig. 1	Cells $\times$ 10^{-4} per ml			Mating efficiency of gametes (%)
	Vegetative culture	Gametogenic culture		
		Starting	Finishing	
F	270	467	3425	0
—[b]	13.8	605	6475	
f	210	610	8150	−4
—[c]	22.6	600	4255	
G	322	613	3263	1
—[b]	28.5	587	6943	
g	431	573	3531	17
—[c]	49.0	592	4340	

[a] See text (Experiments and Results: Gametogenic Differentiation in Synchronized Vegetative Cultures) for experimental conditions.

[b, c] From two separate vegetative cultures, both at early "linear" growth (b = mt^-, c = mt^+; not shown in Fig. 1).

Gametogenic Differentiation in a Continuous Light, Nonsynchronized Vegetative Culture

At different growth stages (see Fig. 2), plus and minus mating type vegetative cells in both nonsynchronized cultures were harvested and resuspended in N-free medium for the induction of gametogenesis. After 16–18 hours in continuous light, following a round of cell division, gametogenic differentiation was completed. Gametes so obtained at different vegetative growth stages were mixed with approximately the same number of opposite mating type gametes obtained at about the same vegetative growth stage. Mating efficiency of 100% was obtained with gametes induced from the vegetative cultures at a wide range of growth stages, from the very early exponential stage to the early stationary phase (Fig. 2). Mating efficiency of 100% was also obtained when gametes were induced from two nonparallel vegetative cultures at different growth stages, i.e., early linear with later linear or early stationary growth stages. No

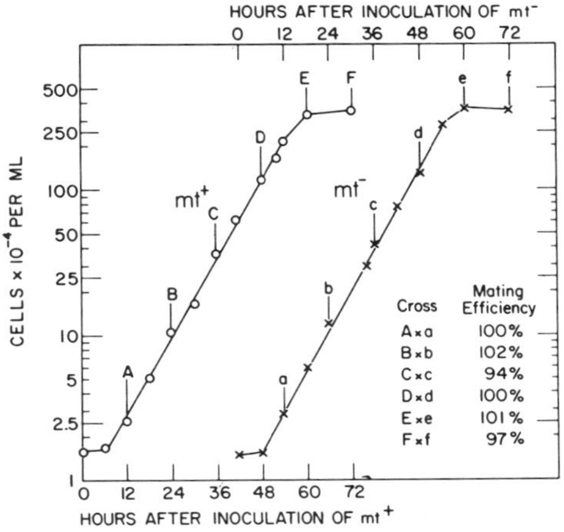

FIG. 2. Vegetative growth of nonsynchronized cultures and mating efficiency after gametogenic induction. The mt^+ and mt^- vegetative cells of *Chla reinhardtii* were grown separately in HSM medium under continuous illumination. At different growth stages, as indicated by different letters on growth curves, vegetative cells from each culture were harvested and resuspended in the nitrogen-free medium for gametogenic induction. After 18 hours under continuous illumination at 21°C equal numbers of the mt^+ and mt^- gametes were mixed and the mating efficiency was determined 90 minutes later.

detectable mating between the vegetative cells at any growth stage was observed. It is thus clear that the mating efficiency of gametes induced from a continuous light, nonsynchronized vegetative culture was independent of the growth stages at which gametogenesis was induced.

DISCUSSION

Correlation between Mating Efficiency and Growth Stage in Synchronized Culture

Under lower light intensity and lower CO_2 concentration than the present conditions, Kates and Jones (1964) have shown that a gametic population with 100% mating efficiency could be induced from a light-dark synchronized vegetative culture after the cell concentration reached 1×10^6 cells per milliliter. No attempt was made at that time to study the mating efficiency of gametes induced from the vegetative culture at an earlier growth stage. Apparently the "linear" growth phase of the synchronized vegetative culture terminated at about 1×10^6 cells per milliliter (Fig. 1 in Kates and Jones, 1964).

The results presented here demonstrate the striking differences in the mating efficiency (0 to 100%) of gametes induced at different vegetative growth stages in the synchronized culture. Evidently the mating behavior of gametes was directly related to the growth stage or age of synchronized vegetative cultures at the time of gametogenic induction. Gametogenesis that yields gametes with 100% mating efficiency was normally obtained only from the synchronized vegetative culture at the end of the "linear" growth stage.

Independence of Mating Efficiency and Growth Stage in Nonsynchronized Culture

No correlation between the mating efficiency and growth stages was observed in the continuous light, nonsynchronized vegetative culture. In contrast to the synchronized culture, the mating efficiency of gametes appeared independent of the growth stage at the time of gametogenic induction. Gametes with 100% mating efficiency could be induced from virtually all growth stages except late stationary phase, when the actual decline in cell population started. With photosynthetic organisms satisfactory synchrony in cell division could often be obtained under an alternating light-dark regime. Such synchrony is characterized by the occurrence of cell division in the dark period, together with cell growth in preparation for division in the light period. This kind of synchrony probably resulted from the manipulation of biosynthetic pools; the pools were built up during the light period but depleted during the dark period, just after cell division. In a continuous light, nonsynchronized culture, on the other hand, the fluctuation of such light-dark regulated pool sizes was probably nonexistent. The nonsynchronized culture probably also contained a pool that did not fluctuate between cell divisions as greatly as the pool in the synchronized culture (whether or not the pools under these two different growth conditions were identical in composition.) In the synchronized culture the pools probably diminished to sizes similar to those of the nonsynchronized culture toward the end of the "linear" growth phase in the synchronized culture. Only at this time could gametes with 100% mating efficiency be induced from the synchronized culture.

It was particularly under continuous illumination that the nonsynchronized culture grew much more rapidly than the synchronized culture under a light-dark cycle (compare Figs. 1 and 2). This drastic difference in growth time between synchronized and nonsynchronized cultures may suggest that the vegetative cell under continuous illumination utilized all its biosynthetic resources for cell division, rather than building up light-dark regulated pools as in the synchronized culture.

In a separate experiment gametes from a continuous light culture were mixed with gametes from a light-dark synchronized culture to test their cross-mating efficiency. Mating efficiency of 100% was obtainable from gametes derived from asynchronized and synchronized cultures as long as the latter was induced at the end of its "linear" growth phase. If the gametes from the asynchronized culture were mated with gametes derived from the synchronized culture at the early "linear" growth phase, no quantitative mating was observed. The mating efficiency of such crosses was about the same as the efficiency of the latter gametic population when mated with opposite mating type gametes obtained also at an early "linear" growth phase.

The Unstable Characteristics of Mating Behavior

The foregoing results demonstrate that: (1) mating behavior of gametes could be altered considerably during the routine maintenance of the original culture on the agar slants; some strains of 100% mating efficiency lost their high mating efficiency over a period of about two months; (2) not all the progeny of a particular strain inherited the identical mating efficiency of their parents (Table 1). The inheritance of this quantitative-mating behavior therefore did not seem uniform among the progeny. The mating behavior thus appeared to be a rather unstable, complex expression of different factors. The frequency with which the mating efficiency altered was much too high to be due to genetic changes involving a single factor or a small number of factors. Mating of gametes required at least the following known steps: mating type reaction (i.e., the flagella agglutination reaction) and the formation of the protoplasmic bridge (Wiese and Jones, 1963; Friedmann *et al.*, 1968) and the eventual fusion of two complete gametes into one common cytoplasmic zygote. Each of these steps may in turn have required the coordination of a number of physiological and biochemical processes. The failure of the quantitative-mating process in the gametic population could have been due to a failure in any of the above essential processes, some of them purely physiological. It was therefore not totally surprising to find very unstable characteristics in the mating behavior of *C. reinhardtii* strains.

Precautions in Obtaining a Large Homogeneous Population of Young Zygotes

To obtain a large quantity of zygotes from two gametic populations with 100% mating efficiency in *C. reinhardtii* and possibly other *Chlamydomonas* species, three prerequisites must be satisfied.

1. High mating efficiency clones: it is necessary to obtain a relatively stable and high mating efficiency clone, either from an established clone or by single clone selection. Routine single cloning and mating tests are recommended every few months to maintain the high efficiency of the clone.

2. Characterization of the growth stages of the culture, when a synchronized vegetative culture is used: even if the same clone and same medium are used, the cell concentration at which the "linear" growth phase terminates in the light-dark synchronized culture still varies considerably under different conditions, such as effective light intensity, CO_2 content in the aeration mixture, temperature, etc. The induction of gametogenesis must be carried out at the end of the linear growth phase.

3. High light intensity and/or relatively dilute cell concentration for the gametogenic culture: it is critical that a sufficiently high effective light intensity be provided for the gametogenic culture so that all the vegetative cells in the culture can divide to form gametes in the absence of an exogenous nitrogen source. Insufficient light intensity for the gametogenic culture invariably results in incomplete gametogenic differentiation and low mating efficiency of gametes. Lower cell concentration (2 to 3 $\times$ 10^6/ml) in the gametogenic culture will compensate for high light intensity, if the latter is not available.

SUMMARY

Upon transfer into a liquid medium devoid of a nitrogen source, vegetative cells of *Chlamydomonas reinhardtii* undergo a gametogenic differentiation process to give rise to sexually active gametes that are capable of mating with opposite mating type gametes. Striking differences have been demonstrated in the mating efficiency (0 to 100%) of gametes induced at different vegetative growth stages in the light-dark synchronized culture. This difference in mating efficiency of gametes is independent of mating-type but directly related to the growth stage of the synchronized vegetative cultures at the time of gametogenic induction. In contrast to the synchronized culture, the mating efficiency of gametes in the continuous light, non-synchronized vegetative culture is independent of the growth stage at the time of gametogenic induction.

The mating behavior of gametes appears to be rather unstable since the mating efficiency of a given clone can be altered considerably during routine maintenance of the culture. Not all the progeny of a particular strain inherit the identical mating efficiency of their parents. The stability of the mating behavior and the transmittance of the mating behavior to the vegetative progeny have been found to vary among different clones. The high frequency at which the mating behavior alters suggests that the mating behavior is a rather complex expression of a number of different factors.

The authors are indebted to Mrs. Kazuko Tsuji, Mrs. Brenda Mihan, and Mrs. Regina Milasius for their competent technical assistance. The assistance given by Mrs. Tina Chiang in preparing the manuscript is gratefully acknowledged.

The authors also wish to thank Messrs. G. Gibson, G. Grofman, J. Hanacek, and P. Kwiatkowski for their mechanical, electronic, and photographic work.

REFERENCES

BERNSTEIN, E. (1960). Synchronous division in *Chlamydomonas moewusii*. *Science* **131,** 1528–1529.

BERNSTEIN, E., and JAHN, T. L. (1955). Certain aspects of the sexuality of two species of *Chlamydomonas*. *J. Protozool.* **2,** 81–85.

BOLD, H. C. (1949). The morphology of *Chylamydomonas chlamydogama*. *Bull. Torrey Bot.* Club Sp. Nov. **76,** 101–108.

BROWN, R. U., JOHNSON, C., and BOLD, H. C. (1968). Electron and phase-contrast microscopy of sexual reproduction in *Chlamydomonas moewusii*. *J. Phycol.* **4,** 100–120.

CHIANG, K. S. (1965). Meiotic DNA replication mechanism in *Chlamydomonas reinhardi*. Ph.D. thesis, Princeton University, University Microfilms, Ph.D. Dissertation 65–13, 130, Ann Arbor, Michigan.

CHIANG, K. S. (1968). Physical conservation of parental cytoplasmic DNA through meiosis in *Chlamydomonas reinhardi*. *Proc. Nat. Acad. Sci. U. S.* **60,** 194–200.

CHIANG, K. S., and SUEOKA, N. (1967). Replication of chromosomal and cytoplasmic DNA during mitosis and meiosis in the eucaryote *Chlamydomonas reinhardi*. *J. Cell. Physiol.* **70,** Suppl. 1, 89–112.

CHIANG, K. S., KATES, J. R., and SUEOKA, N. (1965). Meiotic DNA replication mechanism in *Chlamydomonas reinhardi*. *Genetics* **52,** 434–435.

EBERSOLD, W. T. (1956). Crossing over in *Chlamydomonas reinhardi*. *Amer. J. Bot.* **43,** 408–410.

FORSTER, H., and WIESE, L. (1954). Gamoniwirkungen bei *Chlamydomonas eugamentos*. *Z. Naturforsch. B* **9,** 548–550.

FRIEDMANN, I., COLWIN, A. L., and COLWIN, L. H. (1968). Fine-structural aspects of fertilization in *Chlamydomonas reinhardi*. *J. Cell. Sci.* **3,** 115–128.

JONES, R. F., KATES, J. R., and KELLER, S. J. (1968). Protein turnover and macromolecular synthesis during growth and gametic differentiation in *Chlamydomonas reinhardtii*. *Biochim. Biophys. Acta* **157,** 589–598.

KATES, J. R. (1966). Biochemical aspects of synchronized growth and differentiation in *Chlamydomonas reinhardtii*. Ph.D. thesis, Princeton University, University Microfilms, Ph.D. Dissertation, 66–9619, Ann Arbor, Michigan.

KATES, J. R., and JONES, R. F. (1964a). The control of gametic differentiation in liquid cultures of *Chlamydomonas*. *J. Cell. Comp. Physiol.* **63,** 157–164.

KATES, J. R., and JONES, R. F. (1964b). Variation in alanine dehydrogenase and glutamate dehydrogenase during synchronous development of *Chlamydomonas*. *Biochim. Biophys. Acta* **86,** 438–447.

KATES, J. R., CHIANG, K. S., and JONES, R. F. (1968). Studies on DNA replication during synchronized vegetative growth and gametic differentiation in *Chlamydomonas reinhardtii*. *Exp. Cell Res.* **49,** 121–135.

LEWIN, R. A. (1954). "Sex in Unicellular Organisms," pp. 100–133. Amer. Assoc. Advan. Sci., Washington, D.C.

LEWIN, R. A. (1956). Control of sexual activity in *Chlamydomonas* by light. *J. Gen. Microbiol.* **15,** 170–185.

SAGER, A., and GRANICK, S. (1954). Nutritional control of sexuality in *Chlamydomonas reinhardti*. *J. Gen. Physiol.* **37,** 729–742.

SUEOKA, N., CHIANG, K. S., and KATES, J. R. (1967). DNA replication in meiosis of *Chlamydomonas reinhardi*. I. Isotopic transfer experiments with a strain producing eight zoospores. *J. Mol. Biol.* **25,** 47–66.

TRAINOR, E. R. (1958). Control of sexuality in *Chlamydomonas chlamydogama*. *Amer. J. Bot.* **45,** 621–629.

TRAINOR, E. R. (1959). A comparative study of sexual reproduction in four species of *Chlamydomonas*. *Amer. J. Bot.* **46,** 65–70.

TSUBO, Y. (1956). Observations on sexual reproduction in a *Chlamydomonas*. *Shokubutsugaku Zasshi* **69,** 1–6.

WIESE, L. (1965). On sexual agglutination and mating-type substances (gamones) in isogamous heterothallic Chlamydomonads. I. Evidence of the identity of the gamones with the surface components responsible for sexual flagellar contact. *J. Phycol.* **1,** 46–54.

WIESE, L., and JONES, R. F. (1963). Studies on gamete copulation in heterothallic *Chlamydomonas*. *J. Cell. Comp. Physiol.* **61,** 265–274.

UNCOUPLING OF DNA AND HISTONE SYNTHESIS PRIOR TO PROPHASE I OF MEIOSIS IN THE CRICKET *GRYLLUS* (*ACHETA*) *DOMESTICUS* L.

Yu. F. BOGDANOV, N. A. LIAPUNOVA, A. I. SHERUDILO and E. N. ANTROPOVA

The problem of meiosis, i.e. the question why cells undergo meiosis and any question on the mechanism of homologous pairing of chromosomes, is still the most intriguing problem in cytogenetics. These questions have arisen with the origination of cytogenetics, but are still unanswered [22, 35]. It is, nevertheless, evident that the solution of these problems should necessarily involve the cytochemical study of cells undergoing transition from mitotic state to meiosis [22].

It is known that the principal chemical component of a chromosome is the DNA-histone complex (DNP) which accounts for 90 per cent of nuclear chromatine (for review see [21]). Hence, all essential quantitative and conformational rearrangements in the DNP-complex should cause changes in the "architecture" of chromosomes. Thus trypsin digestion of chromosomal protein including histone in formalin-fixed metaphase chromosomes leads to despiralization of chromosomes [34]. This and other facts [15] suggested that it is the histone that plays an important role supporting a certain spiralization (condensation) of chromosomes.

During the mitotic cycle DNA and histone syntheses occur simultaneously in the S-period of the interphase. Cells which undergo mitosis have already a double content of DNA and histone [5, 36].

The information on the synthesis of DNA and histone in meiosis is scanty. A first report along this line was made by Ris [24] who studied meiotic cells of the grasshopper and demonstrated the doubling of the histone content during phophase I. His technique was based on differential cytophotometry of tissues treated according to Milllon before and after histone extraction with sulphuric acid. Photometry was made only for a small number of cells on sections and that paper belongs to early cytospectrophotometric studies of cytological protein staining.

In 1954–58 Ansley published several papers on quantitative investigations of DNA and histone during spermatogenesis of the centipede *Scutigera forceps* and the bug *Loxa flavicolis* [2–4]. Based on the data of single wavelength cytospectrophotometry [20] of DNA stained according to Feulgen, histone according to Alfert and Geschwind [1] and total protein accoding to Millon before and after histone extraction, this author came to the conclusion that meiotic cells have a low histone/DNA ratio. While this coefficients is equal to 3:2 for somatic cells, it changes to 1:1 in spermatogonia and during meiosis. According to Ansley [2, 3] and Ris [24] DNA and histone syntheses take place in early prophase I (zygotene) and a double content (4C) of DNA and histone per diploid set of chromosomes is found only in pachytene.

This conclusion, however, contradicts the results presented by other authors on premeiotic DNA synthesis. Using various techniques, e.g. quantitative cytophotometry [31, 32], ^{32}P incorporation into DNA [33], ^{3}H-thymidine incorporation [12, 17] and different objects (Lilium and Tradescantia, mice and Orthoptera), it was shown that meiotic DNA synthesis terminates prior to leptotene or during leptotene. As was found recently, some DNA synthesis is observed during zygotene and pachytene [14], but the amount of DNA thus synthesized is only 0.3–2 per cent of that synthesized during interphase prior to meiosis.

The purpose of this work was to investigate the timing of DNA and histone synthesis at different stages of meiosis on the same object and particularly during the transition of cells from mitosis in gonia to prophase I of meiosis and during this prophase. (The results of this investigation were published in Russian [8].)

MATERIALS AND METHODS

As an object we chose the cricket *Gryllus* (*Acheta*) *domesticus* L. which can easily be supported in laboratory culture during the whole year.

Fixation and staining.—Testes isolated from larvae during the 4th to 5th stage were fixed for 12 h in 10 per cent neutral formalin and washed in tap water over night. Separate tubes of testes were squashed in a drop of 45 per cent acetic acid, coverslips were removed on dry ice and preparations were dried in air. Preparations were stained according to Feulgen either with fast green FCF at pH 8.2, according to Alfert and Geschwind [1] or with bromphenol blue (BPB) at pH 8.2 in borate buffer after extraction of nucleic acids with hot 5 per cent TCA [6, 23]. Some slides were specially treated for deamination according to Van-Slyke in two portions of mixture containing equal amounts of 5 per cent TCA and 5 per cent $NaNO_2$ during 15 min in each portion at 4°C [7] and then stained with fast green FCF at pH 8.2. We could thus subtract the fraction of dye bound to ε-amino groups of lysine in histone [1]. To remove water from preparations stained with fast green and BPB, ethanol was substituted for less polar tertiary butanol. The latter lowers the amount of bound dye washed from cells [10].

Selectivity of DNA and histone staining and morphology of some stages of spermatogenesis after fixation of the testes in formalin are shown in Figs 1–6.

Cytospectrophotometry.—The amount of bound dye (basic fuchsine, fast green and BPB) was determined by the double wavelength technique [19, 18]. Measurements were taken with a cytospectrophotometer belonging to the Institute of Cytology and Genetics in Novosibirsk [26]. The details of the methods have been published previously [27]. A set of 7 exchangeable plugs with a diameter ranging from 4.5 μ to 18.5 μ permitted a suitable plug to be chosen for a nuclei of certain diameter.

Calculations were made according to the method of Sherudilo [27]. The amount of light-absorbing material was expressed in arbitrary units. The full random error of methods, including the errors of staining and of the measurements, was determined as coefficient of variation (C.V.) of measurement representation in those cells which do not synthesize DNA or histone [28]. Telophases of spermatogonia and metaphases I (C.V. = 11.5 per cent) stained with fast green pH 8.2 were used as cells which do not synthesize histone, and metaphase I cells stained according to Feulgen with C.V. = 11.9 per cent for cells with no DNA synthesis.

RESULTS

Data on DNA (Feulgen) and histone (fast green pH 8.2) content at various stages of cricket spermatogenesis are presented in Table 1 and as histograms (Fig. 7). One can easily find spermatogonial cells with 2C and 4C of DNA (Fig. 7A-1). Mathematical treatment of histograms (for methods see [28]) shows that spermatogonia with an intermediate content of DNA, i.e. cells in the state of DNA synthesis, do exist (Fig. 7A-1, slope shading). But this fraction of spermatogonia is small. Cells at the preleptotene stage of meiosis (chromatin represented by fine fibres, X-chromosome is not seen) contain 4C of DNA (Fig. 7A-1, black shading). At the leptotene-zygotene stage (it is not possible to differentiate between these stages of meiosis in cricket after formalin fixation) the DNA content of the nuclei is also 4C (Fig. 7A-2). Therefore premeiotic synthesis of DNA takes place at the interphase prior to meiosis.

By histone content of nuclei spermatogonial cells can be divided into 3 groups containing 2C, 4C and approx. 3C of histone (Fig. 7B-1). The average content of total histone at leptotene-zygotene stages corresponds to 3C (Fig. 7B-2) and becomes 4C only by the end of pachytene (Fig. 7B-3). The histone/DNA ratio becomes 1:1 at pachytene and does not change during other stages of meiosis as well as at early spermatides (Fig. 7, A-3, 4, 5; B-3, 4, 5).

If histone is deaminated (ε-amino groups of lysine are blocked) and fast

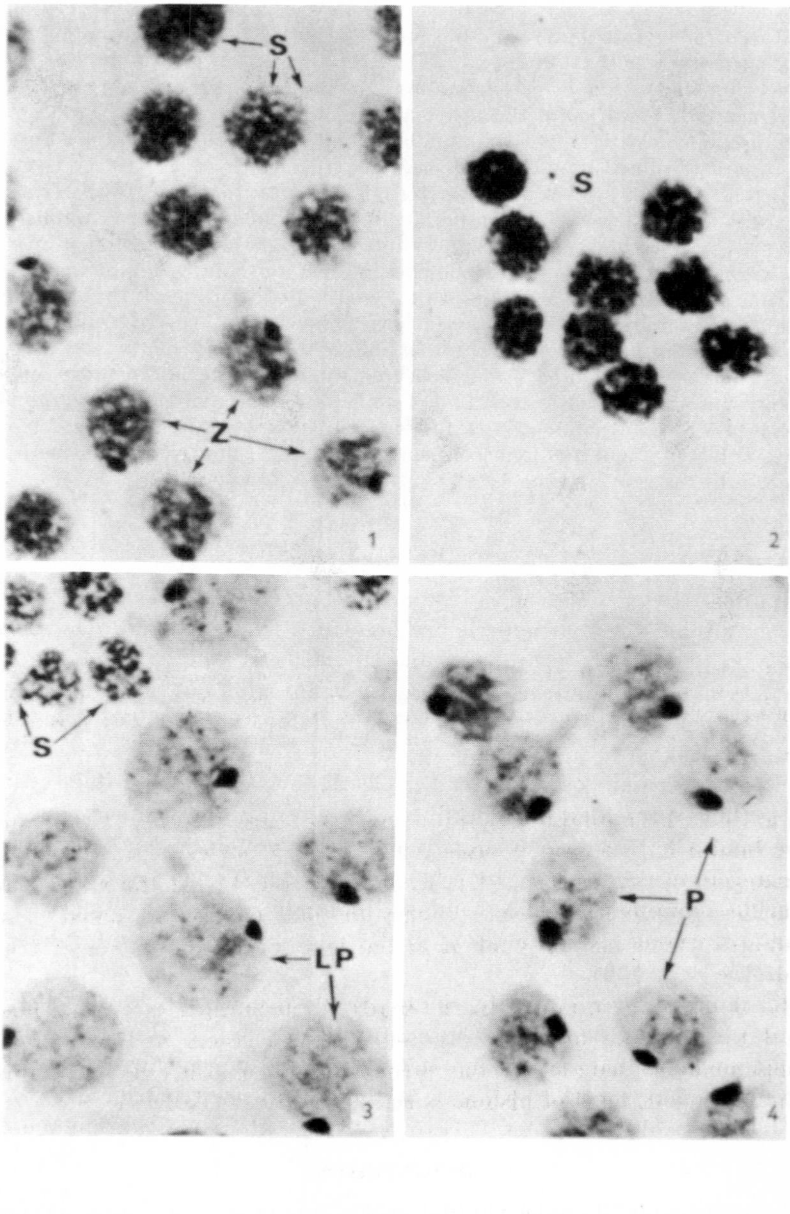
S
Z
1
S
2
S
LP
3
P
4

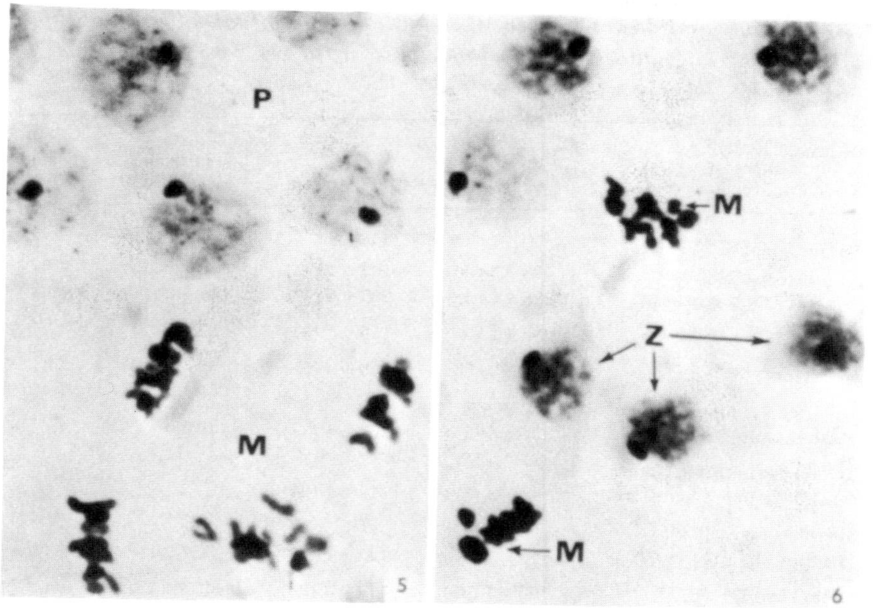

Figs 1–6.—Morphology of nuclei squashed at some stages of spermatogenesis in the cricket *Grillus (Acheta) domesticus* after formalin fixation (pH 7.0) and selective staining according to Feulgen for DNA or with bromphenol blue (BPB) at pH 8.2 for histone. *Figs 1, 3, 5:* Feulgen reaction. *Figs 2, 4, 6:* BPB. *S*, Spermatogonia; *Z*, lepto-zygotene; *P*, pachytene; *LP*, late pachytene; *M*, metaphase I.

green at pH 8.2 is bound to guanidine groups of arginine only [1], the amount of dye bound by nuclei decreases 2-fold (49 ± 5 per cent in average) at all spermatogenesis stages studied (Fig. 7, C; Table 1). At the same time all regularities characterizing the changes in total (non-deaminated) histone, including lowered histone content at the leptotene-zygotene stage, are preserved (Fig. 7C, 1–5).

Data obtained by quantitative BPB pH 8.2 technique (see Materials and Methods) are similar to those obtained with fast green. At the preleptotene and leptotene-zygotene stages the chromosomes contain only 3C of histone (Table 2). The 4C level of histone is reached only at pachytene.

DISCUSSION

Our experiments demonstrate that at the premeiotic interphase of spermatogonia the uncoupling of DNA and histone synthesis is observed. By the

TABLE 1. *Cytophotometrical measurements (in arbitrary units) of mean of DNA, histone, and histone after deamination content of nuclei at different stages of spermatogenesis in the cricket*

Number of cells measured for each stage is given in parentheses

N/N	Stages of spermatogenesis	A DNA	B Histone	C Histone deaminated	Difference (B–C) in %
1	Spermatogonia				
	prior to DNA synthesis	108 ± 5 (36)	103 ± 4 (28)	58 ± 2 (14)	43.7 ± 4 9
	after DNA synthesis	225 ± 6 (52)	205 ± 5 (29)	111 ± 6 (15)	45.8 ± 3.9
2	Leptotene–zygotene	233 ± 2 (55)	156 ± 3 (61)	82 ± 2 (37)	47.4 ± 2.6
3	Pachytene	260 ± 3 (55)	196 ± 3 (67)	90 ± 5 (39)	54.1 ± 3.1
4	Diplotene, diakinesis Metaphase I	257 ± 8 (25)	234 ± 4 (37)	119 ± 4 (30)	49.2 ± 2.6
5	Telophase I (chromosomes at each pole separately)	98 ± 2 (37)	119 ± 4 (24)	—	—
6	Spermatocytes II at interkinesis	110 ± 3 (46)	123 ± 11 (9)	57 ± 3 (18)	53.6 ± 9.0
7	Metaphase II	106 ± 3 (35)	111 ± 3 (28)	62 ± 3 (31)	44.2 ± 3.6
8	Telophase II (chromosomes at each pole separately)	—	59 ± 3 (7)	26 ± 1 (32)	55.9 ± 5.1
9	Early spermatide	66 ± 2 (54)	55 ± 2 (42)	27 ± 1 (20)	50.9 ± 3.6
10	Middle spermatide	52 ± 3 (13)	41 ± 1 (13)	22 ± 1 (41)	46.3 ± 2.4

TABLE 2. *Cytophotometrical data (in arbitrary units) for the nuclei at some stages of spermatogenesis stained with bromphenol blue at pH 8.2 for histone*

n, Number of cells measured; S.E., standard error; C.V., coefficient of variation in %

N/N	Stages	*n*	Mean and S.E.	C.V.	Ploidy units
1	Spermatogonia metaphases	12	271 ± 5	6	4
2	Spermatogonia telophases	9	133 ± 3	8	2
3	Preleptotene	4	217 ± 9	8	3
4	Leptotene–zygotene	10	206 ± 9	13	3
5	Preleptotene plus leptotene–zygotene	14	209 ± 7	13	3
6	Pachytene	14	280 ± 6	8	4

beginning of meiotic prophase I DNA is completely or almost completely duplicated whereas histone synthesis is far from termination and the histone/DNA ratio expressed in haploid units of histone and DNA is 3:4. This ratio was found, for instance, for zygotene, i.e. during chromosome pairing. One may think that it is the state of chromosome when the amount of DNA-bound histone has not yet reached its normal level that causes the lag in chromosome spiralization, thus providing the ability for homologous pairing. It is therefore possible that the hypothesis of chromosome pairing suggested by Darlington in 1937 and known as "precocity theory of meiosis" [9] which states that chromosomes pair in meiosis because they enter prophase I of

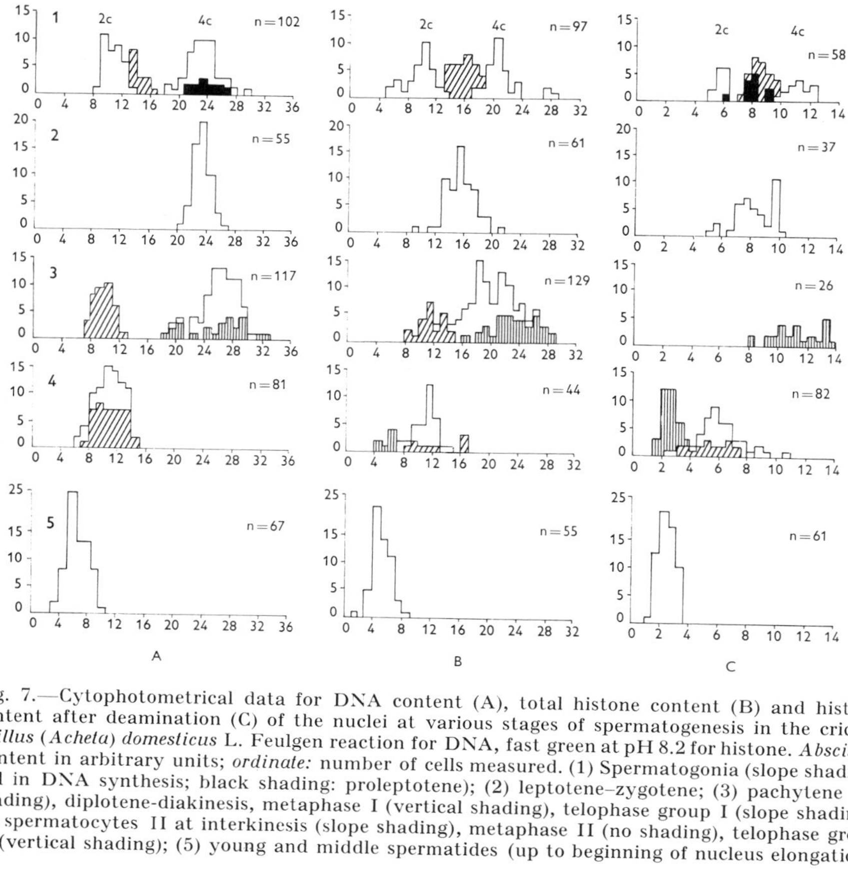

Fig. 7.—Cytophotometrical data for DNA content (A), total histone content (B) and histone content after deamination (C) of the nuclei at various stages of spermatogenesis in the cricket *Grillus* (*Acheta*) *domesticus* L. Feulgen reaction for DNA, fast green at pH 8.2 for histone. *Abscissa:* Content in arbitrary units; *ordinate:* number of cells measured. (1) Spermatogonia (slope shading: cell in DNA synthesis; black shading: proleptotene); (2) leptotene–zygotene; (3) pachytene (no shading), diplotene-diakinesis, metaphase I (vertical shading), telophase group I (slope shading); (4) spermatocytes II at interkinesis (slope shading), metaphase II (no shading), telophase group II (vertical shading); (5) young and middle spermatides (up to beginning of nucleus elongation).

meiosis without duplication is correct, though not with respect to whole chromosomes as Darlington believed but rather with a histone component as demonstrated by our experiments.

Since we suggest that the histone content might play an important and possibly decisive role in homologous chromosome pairing, it is desirable to discuss the data available in literature on qualitative changes of histones in meiosis.

The first suggestions on the occurrence of qualitative changes of histones, particularly on the appearance of histones with different amino acid composition at prophase I of meiosis, were made by Ansley [4] who demonstrated that thistone in spermatocytes I of *Loxa flaviocolis* contains less lysine than histone from somatic cells or histone of abnormal asynaptic spermatocytes I. This conclusion was based on the difference in amount of fast green bound at pH 8.2 after acetylation of normal and asynaptic spermatocytes I and somatic cells. Unfortunately, in this study Ansley did not differentiate between spermatogonia and spermatocytes I, neither did he divide spermatocytes I into meitotic stages.

As already mentioned we did not find in the cricket any differences in the relative content of fast green bound by chromosomes after deamination in spermatogonia and spermatocytes at various stages of meiosis.

Sheridan and Stern [25] have recently determined electrophoretic characteristics of histone isolated from somatic tisues and microsporogenic tissues of anthers of *Lilium longiflorum*. They observed some changes in the electrophoretic picture of total histone from microsporogenic cells which were in the process of transition from mitosis to meiosis and further to microspore formation. The first cycle of these changes is shown to occur at the premeiotic interphase and terminates before diplotena. The histone formed during this cycle was called "histone of meiosis". Electrophoretic characteristics of this histone do not change during both meiotic divisions, i.e. in the interval from metaphase I to anaphase II. Other changes of histone are noticed during formation of tetrades and microspores. The latter changes in *Lilium* are analogous to those observed during spermiogenesis of invertebrates and characterized by enrichment of histone with arginine [7]. We are concerned primarily with the first cycle of histone changes, i.e. the appearance of "histone of meiosis". In contrast to us Sheridan and Stern did not determine the histone content per chromosomal set of a single cell and their results only qualitatively characterize histone of meiosis. Nevertheless, a good agreement between their results and our data is easily noticed. First of all, these authors stated (but did not discuss) that histone synthesis continues up to the diplotene stage, since it is only the *de novo* histone synthesis that may explain the formation of "histone of meiosis" in prophase I. Further, one may suggest that the synthesis of histone of the new type and the substitution of somatic histone for this "histone of meiosis" are quite prolonged processes, not comparable with usual interphase histone synthesis. This might help to understand why in meiosis the histone content equal to 4C per nucleus is found only by the beginning of diplotene, as shown in our experiments.

One may argue that the histone/DNA ratio in zygotene (equal to 3:4) found in our work might be explained not by deficiency of histone in zygotene but rather by changes in the ability of the new "histone of meiosis" to bind fast green and bromphenol blue at pH 8.2, due to changes in conformation of histone molecules. This objection is hardly serious, however, since as was shown in our work, the amount of dye bound increases by pachytene-diplotene in such a proportion that 4C of histone is reached. It looks as if the tetraploid histone content does exist and the histone/DNA ratio in haploid units becomes typical 1:1 (as is required by our suggestion, Sheridan's and Stern's results demonstrate that histone synthesis continues up to pachytene stage). It is understood that a simple conclusion would be drawn only when it is proved in a model experiment that one weight unit of the "histone of meiosis" binds the same amount of dye as one weight unit of the "somatic histone". This would finally demonstrate that 3 arbitrary units of bound dye in zygotene corresponds to 3 arbitrary units of histone per nucleus at this stage of meiosis.

The experiments on incorporation of labelled amino acids into testis cells can serve as a useful additional control for the existence of histone synthesis in prophase. These studies are in progress at present.

What is the role that histone may play in the mechanism of homologous chromosome pairing? Using our data, one may think of two hypothetical mechanisms of chromosome pairing:

(1) Pairing of those regions in DNA which are "stripped" or "partially stripped" due to histone deficiency in zygotene.

(2) Pairing due to formation of histone bridges between homologues.

The first mechanism is preferable since in this case pairing of homologous loci might be realized through some kind of interaction between specific sequences of DNA nucleotides of two homologues. If, however, this pairing was actually brought about by this mechanism it would have occurred during premeiotic interphase when DNA content per nucleus (as it follows from our experiments) is already 4C but histone content does not exceed 3C. In recent years some authors postulated chromosome pairing in premeiotic interphase

[11]. However, as Henderson [13] has demonstrated, these conclusions were erroneous because of the low resolution of the methods used for comparison of time intervals between premeiotic DNA synthesis and crossing over. He returned to the classical idea of chomosome pairing at zygotene. One may therefore conclude that histone deficiency at interphase is not a sufficient condition for chromosomes to pair since pairing cannot proceed only through the first mechanism. Pairing due to histone bridges is also quite unlikely, for this mechanism can hardly provide a high specificity of pairing. We believe that both mechanisms may be involved in homologue pairing.

Following the sequence of events in chromosome pairing or the "histone deficiency hypothesis" may be postulated. At premeiotic interphase DNA synthesis is already over and the chromosomes show a deficiency of histone; infact the chromosomes are potentially ready to pair. The only obstacle for pairing is the condensed state of chromosome in interphase nuclei: during premeiotic interphase nuclei preserve a crude block structure of chromatin specific for psermatogonia. This type of nuclear structure is characteristic for last-generation nuclei of B type mouse spermatogonia [16] and for cricket nuclei during premeiotic DNA synthesis. The transition from premeiotic interphase to leptotene is accompanied by decondensation of chromatin. The new cycle of chromosome spiralization starts at zygotene. Chromosomes which just began meiotic spiralization undergo pairing. It is not unlikely that "histone of meiosis" which replaces "somatic" histone in the interval from the last spermatogonial mitosis to the termination of pachytene is an important factor in the appearance of a new type of chromosome spiralization in meiosis. Pairing of "stripped" DNA regions might be accompanied by the formation of histone bridges between homologues. As electron micrographs show, there exist submicroscopic bridges in a pairing space of the synaptinemal complex [30]. The elucidation of the chemical nature of these bridges as well as of other elements of the synaptinemal complex would be an important step towards the final solution of the problem of chromosome pairing.

We are still far from suggesting a working hypothesis. New facts are needed. The main obstacle in our progress here is the insufficient development of the microchemical technique for the study of nuclei and chromosomes.

We plan further to find out whether or not histone deficiency at zygotene is typical for other objects—plants and animals having typical meiosis.

SUMMARY

Using double wavelength cytospectrophotometry of Feulgen staining, fast green staining according to Alfert and Geschwind at pH 8.2 and bromphenol blue staining at pH 8.2, the authors have studied DNA and histone content per nucleus at various stages of spermatogenesis in the cricket.

Uncopuling of DNA and histone synthesis was found to occur during premeiotic interphase. While DNA synthesis is terminated or almost terminated prior to the leptotene-zygotene stage, histone synthesis is far from termination. Zygotene nuclei contain 4C DNA and only 3C of histone. The 4C level of histone is reached only at pachytene-diplotene. The histone/DNA ratio stays constant during other stages of meiosis. It is suggested that histone deficiency in chromosome during homologous pairing and also changes in chemical properties of histone may cause pairing to start. This suggestion resumes, but in much modified form, the "precocity theory of meiosis" proposed by Darlington and the new "histone deficiency hypothesis" is developed.

We are very grateful to Professor A. A. Prokofieva-Belgovskaya for support during this investigation. We thank Mr V. E. Barsky for useful advice in the cytospectrophotometrical technique.

REFERENCES

1. ALFERT, M. and GESCHWIND, J., *Proc. Natl Acad. Sci. US* **39**, 991, (1953).
2. ANSLEY, H. R., *Chromosoma* **6**, 656 (1954).
3. —— *ibid.* **8**, 380 (1957).
4. —— *J. Biophys Biochem Cytol.* **4**, 59 (1958).
5. BLOCH, D. P. and GOODMAN, G. C., *J. Biophys. Biochem Cytol.* **1**, 17 (1955).
6. BLOCH, D. P. and HEW, H., *ibid.* **4**, 593 (1958).
7. —— *ibid.* **8**, 69 (1960).
8. BOGDANOV, YU, F., LIAPUNOVA, N. A., SHERUDILO, A. I. and ANTROPOVA, E. N., *Cytologia* (Russ.) **9**, 986 (1967).
9. DARLINGTON, C. P., Recent advances in cytology, 2nd ed. Philadelphia, Blakinstone, 1937.
10. DEITCH, A. D., *Lab. Invest.*, **4**, 324 (1955).
11. GRELL, R. I. and CHANDLEY, A. C., *Proc. Natl Acad. Sci. US* **53**, 1340 (1965).
12. HENDERSON, S. A., *Chromosoma* **15**, 345 (1964).
13. —— *Nature* **211**, 1043 (1966).
14. HOTTA, Y., ITO, M. and STERN, H., *Proc. Natl Acad. Sci. US* **56**, 1185 (1966).
15. LITAU, V. C., BURDICK, C. J., ALLFREY, V. C. and MIRSKY, A. E., *Proc. Natl Acad. Sci. US* **54**, 1204 (1965).
16. MONESI, V., *J. Cell Biol.* **14**, 1 (1962).
17. MUCKENTHALER, F. A., *Exptl Cell Res.* **35**, 531 (1964).
18. ORNSTEIN, L., *Lab. Invest.* **1**, 250 (1952).
19. PATAU, K., *Chromosoma* **5**, 341 (1952).
20. POLISTER, A. W., *Lab. Invest.*, **1**, 106 (1962).
21. PROKOFIEVA-BELGOVSKAJA, A. A. and BOGDANOV, YU. F., *J. All-Union Mendeleeff Chem. Soc.* (Russ.) **8**, 33 (1963).

22. RHOADES, M. M., *in* The Cell, vol. **3**, p. 1. Academic Press, New York–London, 1961.
23. RINGERTZ, N. R. and ZETTERBERG, A., *Exptl Cell Res.* **42**, 243 (1966).
24. RIS, H., *Cold Spring Harbor Symp. Quant. Biol.* **12**, 147 (1947).
25. SHERIDAN, W. E. and STERN, H., *Exptl Cell Res.* **45**, 323 (1967).
26. SHERUDILO, A. I., *Isvestia Sibirskogo Otdelenia Akademii Nauk USSR, Ser. Med. Biol.* (Russ.) **12**, 145 (1964).
27. —— *Optika i spectroscopiya* (*Russ.*) **20**, 543 (1966).
28. —— *Cytologia* (Russ.) **8**, 120 (1966).
29. —— *Biophysika* (Russ.) In press.
30. SOTELO, J. R. and WETTSTEIN, R., *Chromosoma* **20**, 234 (1966).
31. SWIFT, H., *Proc. Natl Acad. Sci. US* **36**, 643 (1950).
32. SWIFT, H. and KLINFELD, R., *Physiol. Zool.* **26**, 301 (1953).
33. TAYLOR, H., *Exptl Cell Res* **4**, 164 (1953).
34. TROSKO, E. and WOLFF, SH., *J. Cell Biol.* **26**, 125 (1965).
35. WHITEHOUSE, H. L. K., The mechanism of heredity. Arnold, London, 1965.
36. WOODARD, J., RASH, E. and SWIFT, H., *J. Biophys. Biochem Cytol.* **9**, 445 (1961).

Y Chromosome Replication and Chromosome Arrangement in Germ Line Cells and Sperm of the Rat

N. O. BIANCHI and MARTHA S. DE BIANCHI

Introduction

The chronology of replication of the Y chromosome has been studied in several species of mammals. In somatic cells most investigators have found that this chromosome is late replicating (Bianchi, 1966). On the other hand, in germ line cells the results are scanty and controversial. In Chinese hamster Utakoji and Hsu (1965) were unable to find an out-of-phase replicating Y. Conversely, Tiepolo, Fraccaro, Hulten, Lindsten, Mannini and Ming (1967) in mice and Mukherjee and Ghosal (1968, 1969) in Golden hamster observed the presence of a late replicating Y. Moreover, the latter authors claim that the Y shifts from late to early replicating during the process of spermatogonial maturation.

This paper presents data pointing out the existence of a late replicating Y chromosome in germ line cells of the rat. Moreover, by tracing the out-of-phase replicating Y through the spermatogenic process we shall also furnish information regarding the pattern of chromosome arrangement in the sperm.

Material and Methods

Twenty μC of ^{3}HTdR (Sp.Ac. 6.9 C/mM) were injected into each testis of 84 adult male *Rattus norvegicus*. Every two hours since the ^{3}HTdR injection up to 24 hours after the injection, and every two days since the ^{3}HTdR injection up to

64 days after the injection two males were sacrificed. The testes were dissected, the tubules finely minced placed in 10 ml. of phosphate buffer solution and vigorously pipeted until obtaining a fine cellular suspension. After decanting gross debris, the suspension was centrifuged and the cellular pellet resuspended for 7 min in a hypotonic solution formed by 3 parts of distilled water and 1 part of phosphate solution. Following the hypotonic shock cells were fixed overnight in 3-to-1 ethanol-acetic acid. After a wash in fresh fixative, the cells were resuspended in a suitable amount of fixative and slides were prepared by the air drying technique. Cellular spreads were stained with carbol-fuchsine and autoradiograms prepared as described elsewhere (Bianchi, Lima-de-Faria and Jaworska, 1965). Autoradiograms were initially exposed for 7, 15, 30 and 60 days. However, after a few trials it was noticed that 15 days was the most suitable exposure time and from that moment onwards only this time was employed.

In the strain of rats employed the Y is the smallest acrocentric chromosome of the set; therefore, no difficulty in differentiating it from the autosomes was found in spermatogonial metaphases. Interphase nuclei of spermatocytes and of type A and B spermatogonia could easily be distinguished in spite of the disruption of the seminiferous tubules architecture. Type A spermatogonia showed a large nucleus faintly stained and with finely spread chromatin; type B spermatogonia had a smaller nucleus with coarse masses of heterochromatin; spermatocytes had a still smaller nucleus deeply stained and with homogeneous chromatin distribution. Since no PAS technique was employed it was impossible to distinguish the 19 stages of spermatid maturation described by Leblond and Clermont (1952). Consequently, based on nuclear morphology spermatids were arbitrarily divided into tree groups: young spermatids comprising the stages 1—7 of Leblond and Clermont (Golgi and cap phases); middle spermatids comprising the stages 8—14 (acrosome phase); sperm comprising the stages 15—19 and the mature sperm (maturation phase).

Results

Taking into account that we were working with mixed populations of spermatogonia and spermatocytes probably having different lengths of cell cycle (Monesi, 1962) no attempt of determining the duration of these cycles was made. Nevertheless, since no labeled metaphase was found in isotope treatments shorter than 6 hours it can be inferred that the G_2 period in spermatogonia and spermatocytes is longer than this time-lapse.

When the replicating behavior of the Y chromosome was analysed in spermatogonial metaphases, four different patterns of labeling could be plainly recognized. In the first one the Y chromosome was unlabeled while the remaining complement showed radioactivity (Fig. 1); this pattern was mainly found in spermatogonia labeled 16 to 22 hours before entering metaphase. In the second pattern all the complement showed radioactivity and no differential behavior of the Y was noticed; these cells chiefly appeared 12—16 hrs after the ^{3}HTdR treatment. In the third pattern some chromosomal areas or even some entire chromosomes were free from radioactivity. In these cells the Y chromosome was always labeled exhibiting a very high density of silver grains; metaphases

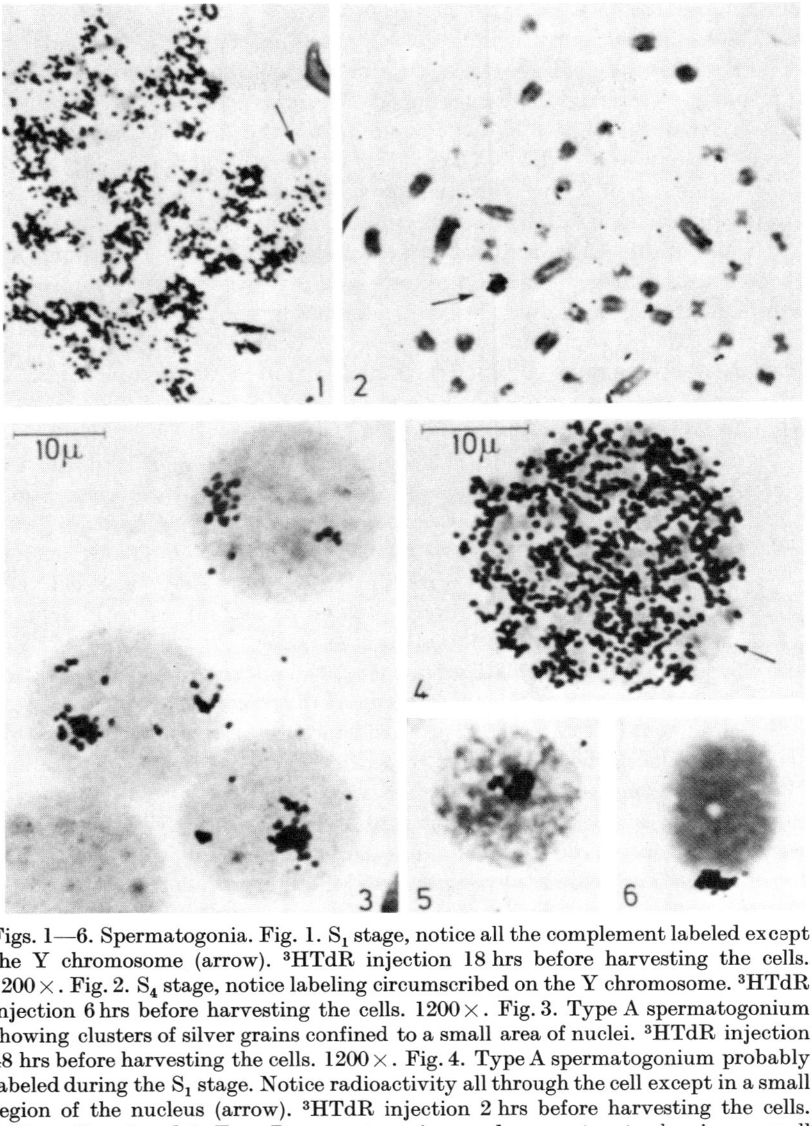

Figs. 1—6. Spermatogonia. Fig. 1. S_1 stage, notice all the complement labeled except the Y chromosome (arrow). ^{3}HTdR injection 18 hrs before harvesting the cells. 1200×. Fig. 2. S_4 stage, notice labeling circumscribed on the Y chromosome. ^{3}HTdR injection 6 hrs before harvesting the cells. 1200×. Fig. 3. Type A spermatogonium showing clusters of silver grains confined to a small area of nuclei. ^{3}HTdR injection 48 hrs before harvesting the cells. 1200×. Fig. 4. Type A spermatogonium probably labeled during the S_1 stage. Notice radioactivity all through the cell except in a small region of the nucleus (arrow). ^{3}HTdR injection 2 hrs before harvesting the cells. 1400×. Figs. 5 and 6. Type B spermatogonium and spermatocyte showing a small radioactive area. ^{3}HTdR injection 4 hrs before harvesting the cells, 1200×

with this pattern mainly stemmed from treatments performed 8—12 hrs before killing the animal. Finally, in the fourth pattern the Y chromosome remained radioactive while the other chromosomes of the set were unlabeled or slightly labeled (Fig. 2); this pattern was observed in spermatogonia labeled 6 to 10 hrs before mitosis.

Taking into account their chronology of appearance; the four patterns of labeling described above can be considered representatives of the initial, intermediate, late and final stages of the S period (S_1, S_2, S_3 and S_4 stages). Consequently, it is possible to conclude that the Y is the last chromosome to begin and end DNA synthesis in spermatogonial cells. Similar conclusions have been previously reported for somatic cells of fetuses and adult rats (Bianchi, 1966; Bianchi and Bianchi, 1966).

It is notable that during the S_1 and S_4 stages the Y chromosome showed a replicating behavior unobserved in any other chromosome of the complement. Accordingly, the absence of labeling in a discrete region of a complete labeled complement (S_1 stage), or conversely, the high concentration of silver grains confined to a small region of the chromosomal set (S_4 stage), are two autoradiographic clues indicating the Y location in cases in which this chromosome cannot be recognized.

By employing the autoradiographic criteria outlined above it was possible to detect an out-of-phase replicating Y in the interphase nuclei of A and B spermatogonia and primary spermatocytes (Figs. 3—6). In the three types of cell 1—2% of labeled nuclei showed the S_4 pattern of labeling while less than 0.5% exhibited the S_1 pattern. Labeled spermatogonia and spermatocytes were detected from 2 hrs up to 10 or more days after the ^{3}HTdR injection. However, when the cells were studied several days after the ^{3}HTdR treatment, the silver grain concentration decreased on account of the successive cell divisions and no clear pattern of labeling could be detected. The most representative patterns were observed in cells harvested from 2 to 48 hrs after the ^{3}HTdR treatment. Moreover, in the 24- and 48 hrs treatments groups of cells having the same pattern of silver grain distribution could be observed. These cells were considered daughter cells repeating the labeling pattern of the mother cell from which they stemmed (Fig. 3).

During meiosis an out-of-phase replicating Y was detected in zygotene, pachytene, diplotene and metaphase I stages. In rat testes the duration of the remaining meiotic stages is very short and they are rarely observed. Considering that the pattern of the late replicating Y is observed in a small percentage of cells it was seldom or never detected in meiotic figures other than those mentioned above.

In the fourth day after the ^{3}HTdR treatment the appearance of zygotenes and pachytenes with the sex vesicle labeled and the remaining complement unlabeled could be detected. Further after labeling the same pattern was also noticed in middle and advanced pachytenes (6th and 8th days after labeling) (Figs. 7, 8). The reverse pattern (silver grains scattered all over the complement except in the sex vesicle) was not

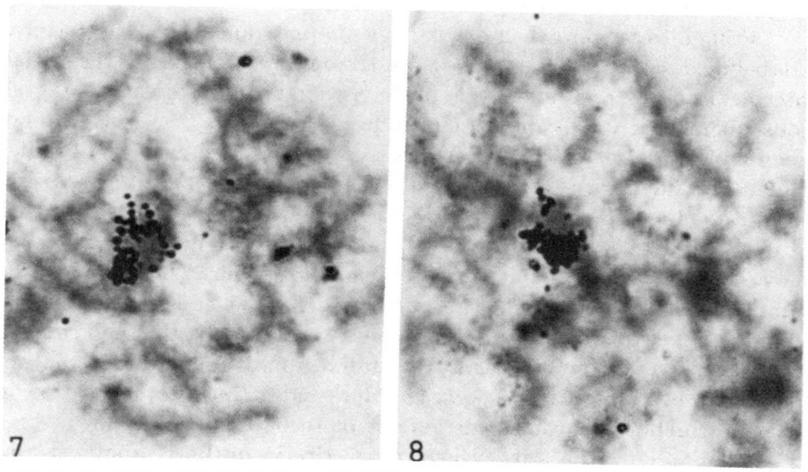

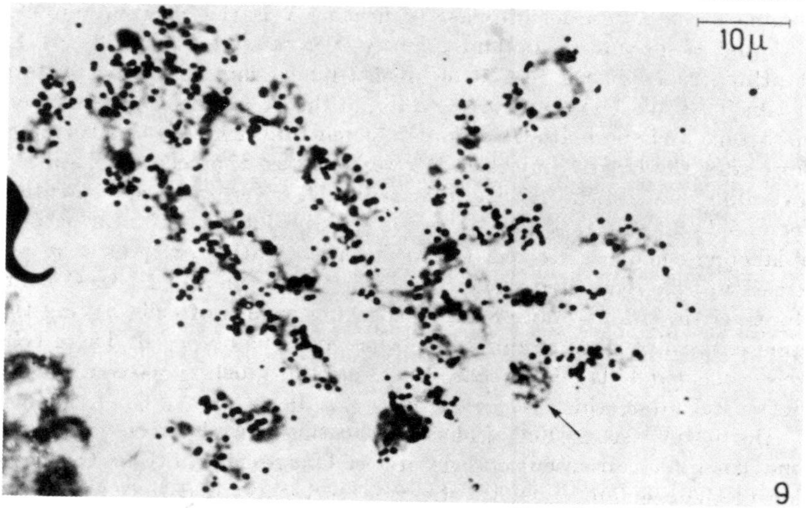

Figs. 7 and 8. Middle pachytene showing silver grains on the sex vesicle only. ^{3}HTdR injection 6 days before harvesting the cells. 1200×

Fig. 9. Pachytene-diplotene probably labeled in the S_1 stage. Notice silver grains all over the complement except in a still condensed bivalent (arrow). ^{3}HTdR injection 10 days before harvesting the cells, 1200×

observed in any case. However, in some pachytenes and early diplotenes labeling occurred throughout the complement and the sex vesicle showed slight radioactivity; such a pattern probably indicates a lack of labeling in the Y chromosome and the presence of labeling in the remaining complement including the X chromosome (Fig. 9).

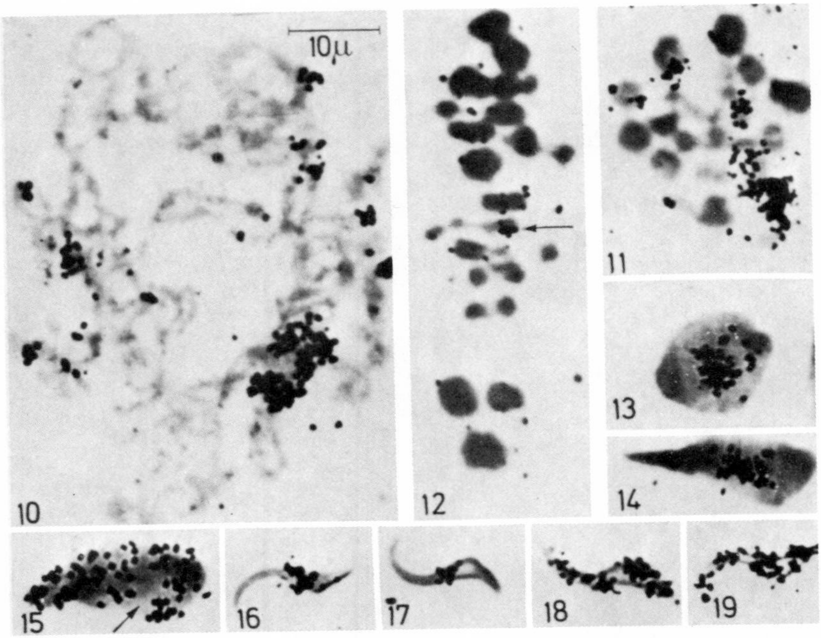

Fig. 10. Diplotene showing a high concentration of silver grains in a small region of the complement. ^{3}HTdR injection 10 days before harvesting the cells. 1200×

Fig. 11. Metaphase I showing a strongly radioactive bivalent. ^{3}HTdR injection 20 days before harvesting the cells. 1200×

Fig. 12. Metaphase I exhibiting radioactivity on part of a bivalent (arrow). ^{3}HTdR injection 20 days before harvesting the cells. 1200×

Figs. 13 and 14. Young and middle spermatids showing labeling in a small area located in the center of the nucleus. ^{3}HTdR injection 28 days before harvesting the cells. 1200×

Fig. 15. Middle spermatid showing the center of the nucleus free from silver grains. ^{3}HTdR injection 28 days before harvesting the cells. 1200×

Figs. 16 and 17. Sperms showing a band of labeling transversally crossing the middle of the nucleus. ^{3}HTdR injection 38 days before harvesting the cells. 1200×

Fig. 18. Sperm showing an unlabeled gap in the middle of the nucleus. ^{3}HTdR injection 38 days before harvesting the cells. 1200×

Fig. 19. Sperm nucleus showing four different bands of labeling. ^{3}HTdR injection 38 days before harvesting. 1200×

In diplotene and metaphase I figures showing one very radioactive bivalent and the other bivalents with slight or no radioactivity were observed (Figs. 10—12). Extreme examples of asynchronous replication were given by some metaphases I which showed silver grains bound to a part of one bivalent only (Fig. 11).

Young and middle spermatids and sperms appeared labeled 20, 26 and 30 days after the ^{3}HTdR injection respectively. Radioactivity distribution suggesting the existence of a late replicating Y were found in both types of spermatids (Figs. 13—15); however, the most informative patterns arose from sperms; 8—10% of labeled sperms showed a single band of silver grains transversally crossing the middle of the sperm head (Figs. 16, 17). On the other hand, the reverse pattern consisting of a narrow unlabeled gap located halfway in a thoroughly labeled head was observed in 4—7% of labeled sperms (Fig. 18). These two patterns of labeling probably represent Y chromosomes from germ line cells which received the isotope at initial (unlabeled gap) or final stages (labeled band) of the S period. The existence of several bands of labeling and the complete labeling of the head were two other patterns of silver grain distribution found in sperms (Fig. 19).

Discussion

Our work points out that it is possible to trace an out-of-phase replicating Y chromosome from spermatogonia to sperm in rat testes. Accordingly, we have to accept that at the moment of egg fertilization male determining sperms have Y chromosomes which replicated out-of-phase in the last DNA synthesis carried out by the cell. Since no studies on chromosome replication in early embryonic stages of the rat have been performed the existence of an out-of-phase replicating Y in preimplantation embryos cannot be ruled out. However, although possible, such a phenomenon seems improbable in the light of present knowledge.

It is well known that sex differentiation occurs early in the embryonic life of mammals and that Y chromosomes are the only carriers of male determining factors. Consequently, to trigger the mechanisms of male differentiation we have to accept that Y chromosomes are genetically active in young embryos. Since late replication, heterochromatin and genetic inactivity seem to be associated events it can be inferred that Y chromosomes are probably early replicating in young mammal embryos. Direct evidence supporting this statement has been provided by Hill and Yunis (1967) who found that in preimplantation embryos of Golden hamster the Y chromosome finishes its replication synchronously with autosomes. Furthermore, in rabbit embryos, Issa, Blank and Atherton (1969) found that the percentage of cells with late replicating Y decreased coincidentally with the decrease of age of the embryo. Thus, in 144-hrs-old embryos 65.9% of cells had a late replicating Y, while in 27-hrs-old embryos only 6.7% of cells showed a similar phenomenon. According to this it seems reasonable to assume that rabbit embryos in the stage of morula have no late replicating Y chromosomes.

In Chinese hamster (Utakoji and Hsu, 1966) and perhaps in some other species of mammals, the absence of a late replicating Y in germ line cells accounts for the lack of an out-of-phase replicating Y in young embryos. On the other hand, in mammals having a late replicating Y in germ cells we have to assume that before or immediately after fertilization the Y chromosome changes its chronology of replication. In Golden hamster Mukherjee and Ghosal (1968, 1969) demonstrated that the Y chromosome shifts from late to early replicating during the process of spermatogonial maturation. In rats, our results suggest that the replicating shift of the Y occurs in the first divisions of the fertilized egg.

In 1946 Hughes-Schrader, working with iceryne coccids having a haploid number of two, could demonstrate a linear alignment of the chromosomes in the sperm. Similar conclusions for the cave cricket were later on obtained by using polarized ultraviolet light (Inoué and Sato, 1962). More recently, Taylor (1964) has confirmed the tandem arrangement of chromosomes in the grasshopper sperm by means of autoradiography with ^{3}HTdR. It was known that the X chromosome of grasshoppers replicated out-of-phase (Lima-de-Faria, 1959); therefore, Taylor injected males of *Romalea microptera* with ^{3}HTdR during the last instar and studied the testes within about 60 days later. In sperms derived from spermatocytes labeled at the time when only the heterochromatic X was radioactive, only one segment along the mature sperm showed silver grains. Conversely, spermatocytes labeled in most chromosomes except the heterochromatic X gave rise to sperm having silver grains over all but a short length of its extent. The autoradiographic findings of Taylor are clear cut evidences of tandem chromosome arrangement in the sperm. If otherwise, chromosomes had been intermingled or ordered side by side, any of the two patterns of asynchronous X replication described above would have given diffuse labeling.

The out-of-phase replicating Y chromosome of rat spermatogonia could be identified in the sperm by the existence of nuclei showing either a narrow labeled segment or an unlabeled gap. Such a banded pattern of labeling is analogous to that described by Taylor in grasshopper and suggests an end-to-end chromosome association in rat sperm. Lately, based on the existence of a banded pattern of labeling in Chinese hamster sperm Utakoji (1966) has also suggested a tandem chromosome package for this species.

In the case of rats, the labeled and the unlabeled bands representing the Y chromosome were always in the middle part of the sperm nuclei. Therefore, it can be concluded that this chromosome, and perhaps the autosomes as well, are aligned end-to-end in a constant order. Recently, Comings (1968) has demonstrated that chromatin in interphase nuclei of human amnion cells is attached to specific sites of the nuclear membrane.

If a similar chromatin attachment appeared in spermatids, the basis for a constant arrangement of chromosomes in sperm nuclei of rats would be provided.

References

Bianchi, N. O.: Chromosomes of the rat II. DNA replication sequence of bone marrow chromosomes *in vivo*. Cytologia (Tokyo) **31**, 276—293 (1966).

— Bianchi, M. S. de: Shifting in the duplication time of sex chromosomes in the rat. Chromosoma (Berl.) **19**, 286—299 (1966).

— Lima-de-Faria, A., Jaworska, H.: A technique for removing silver grains and gelatin from tritium autoradiographs of human chromosomes. Hereditas (Lund) **31**, 207—211 (1964).

Comings, D. E.: The rationale for an ordered arrangement of chromatin in interphase nucleus. Amer. J. hum. Genet. **20**, 440—460 (1968).

Hill, R. N., Yunis, J. J.: Mammalian X-chromosomes: change in patterns of DNA replication during embryogenesis. Science **155**, 1120—1121 (1967).

Hughes-Schrader, S.: A new type of spermiogenesis in iceryne coccids with linear alignment of chromosomes in the sperm. J. Morph. **78**, 43—84 (1946).

Inoué, S., Sato, H.: Arrangement of DNA in living sperm: a biophysical analysis. Science **136**, 1222—1224 (1962).

Issa, M., Blank, C. E., Atherton, G. W.: The temporal appearance of sex chromatin and of the late replicating X chromosome in blastocysts of the domestic rabbit. Cytogenetics **88**, 219—237 (1969).

Leblond, C. P., Clermont, Y.: Definition of the stages of the cycle of the seminiferous epithelium in the rat. Ann. N. Y. Acad. Sci. **55**, 548—569 (1952).

Lima-de-Faria, A.: Differential uptake of tritiated thymidine into hetero- and euchromatin in *Melanoplus* and *Secale*. J. biophys. biochem. Cytol. **5**, 457—466 (1959).

Monesi, V.: Autoradiographic study of DNA synthesis and the cell cycle in spermatogonia and spermatocytes of mouse testis using tritiated thymidine. J. Cell Biol. **4**, 1—18 (1962).

Mukherjee, B. B., Ghosal, S. K.: Differentiation of sex chromosomes in germ line cells of male golden hamster. Proc. III rd. Intern. Congr. of Genetics **1**, 205 (1968).

— — Replicative differentiation of mammalian chromosomes during spermatogenesis. Exp. Cell Res. **54**, 101—106 (1969).

Taylor, J. H.: The arrangement of chromosomes in mature sperm of grasshopper. J. Cell Biol. **21**, 286—289 (1964).

Tiepolo, L., Fraccaro, M., Hultén, M., Lindsten, J., Mannini, A., Ming, L.: Timing of sex chromosome replication in somatic and germ-line cells of the mouse and rat. Cytogenetics **6**, 51—66 (1967).

Utakoji, T.: Chronology of nucleic acid synthesis in meiosis of the male Chinese hamster. Exp. Cell Res. **42**, 585—596 (1966).

— Hsu, T. C.: DNA replication patterns in somatic and germ-line cells of the male Chinese hamster. Cytogenetics **4**, 295—315 (1965).

Late DNA Replication in Male Mouse Meiotic Chromosomes

N. ODARTCHENKO
M. PAVILLARD

The kinetics of spermatogenesis in laboratory rodents have been studied by cell blocking (*1*) or autoradiography. Patterns of labeling of DNA with ^{3}H-thymidine (^{3}H-TdR), however, have been examined in histologic sections (*2*) or on widely time-spaced squashes (*3*) or have been limited to spermatogonial mitosis (*4*). The patterns have been studied over individual meiotic chromosomes only in plants, insects (*5*), or lower vertebrates (*6, 7*). We studied carefully timed tissue samplings containing mouse meiotic chromosome figures and describe here technically good preparations, in terms of chromosome morphology and of autoradiographic resolution.

The percentage of well-spread labeled cells in late diakinesis and early metaphase I (DMI) was recorded as a function of time after injection of ^{3}H-TdR (*8*). It takes a minimum of approximately 9 days for cells to proceed from the last premeiotic phase of DNA synthesis to the observed stages (*9*) (Fig. 1). This is shorter than the 14 to 15 days described in rats (*2*) or the 20 to 21 days observed in newts (*6*). Earlier and later phases of spermatogenesis are correspondingly shorter in the same approximate ratio. The curves of Fig. 1 cannot be taken as supporting or as excluding a low amount of DNA synthesis during the zygotene-pachytene phase (*7, 10*) because of the relative imprecision of autoradiographic data.

Labeling was continuous for at least 2 days (*11*); with time, all DMI figures became labeled, and all chromosome bivalents progressively became uniformly covered with grains. In these conditions, only the regions of chromosomes which are late labeling became apparent in partially labeled cells, and thus only slides in which less than 15 percent of DMI were labeled were selected. This corresponds approximately to less than one DNA premeiotic synthesis time (*2*) between the beginning of injection of ^{3}H-TdR and the entry of cells into nonsynthesizing phases of the mid-leptotene phase (*7*).

Suitable DMI figures (771) exhibiting a partially labeled chromosome complement were analyzed on slides taken from six different animals, according to the above criteria. These figures represent a heterogenous group of cells, because of variability in the speed of progression from the end of the last period of DNA synthesis to the DMI stage.

When only one bivalent was labeled (11 DMI), it was always the sexual XY complex, and label was consistently restricted to one end (Fig. 2a). When grains were also seen over some autosomal bivalents, the XY complex exhibited grains either over one end (Fig. 2b) or over both ends (Fig. 2c). In male mouse mitotic chromosomes the entire Y is uniformly late in replication (*12*). If this holds true for the meiotic Y, these findings confirm earlier morphological observations of an end-to-end association between the X and the Y at meiosis (*13*); the Y

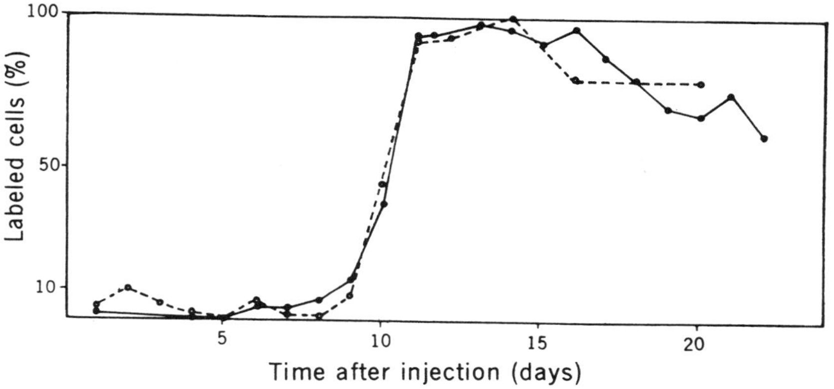

Fig. 1. Percentage of labeled cells in late diakineses and early metaphases I (DMI) as a function of time after an intratesticular injection of tritiated thymidine (0.1 μc). ●——● Slides exposed for 30 days; ○----○ exposed for 70 days.

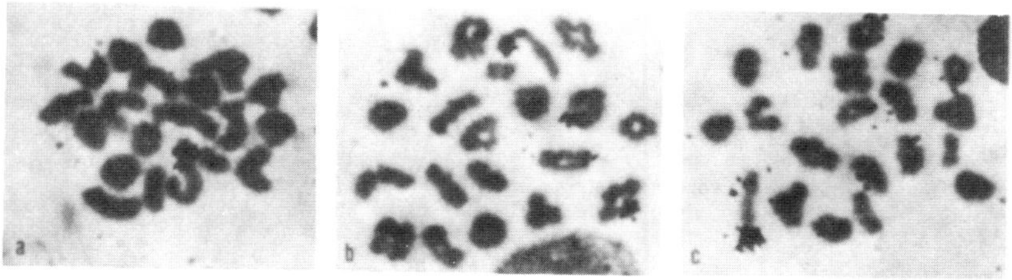

Fig. 2. Meiotic mouse chromosomes in DMI with partial labeling of heterochromosomes. (a) Late labeling of the Y only. (b) Late labeling over the Y and some parts of few autosomal bivalents. (c) Late labeling of the Y, of some autosomes, and of the distal part of the X.

can even be particularly recognized in preparations such as Fig. 2b. However, until technically good anaphase figures are obtained it is not possible to determine the actual mode of X-Y association, whether nonchiasmatic or with a chiasma formed between short arms of the X and the Y, as a result of extreme shortness of these arms in the latter case.

When more than 8 to 12 autosomal bivalents became labeled the sexual complex was entirely labeled. In fact, when only one bivalent was entirely labeled, it was always the XY complex (Fig. 3, a and b), as far as it could be morphologically defined in the conditions of heavy labeling.

In autosomal bivalents, asynchrony of DNA-replicating regions was frequently observed (Fig. 3, a and b). In fact, out of 760 DMI with partial labeling of autosomal bivalents, 13 percent of all pooled autosomes had grains localized in only one spot (zero to nine such bivalents per cell). In more than 75 percent of these bivalents, good morphology and autoradiographic resolution enabled localization of the grain spot over or very close to the distal

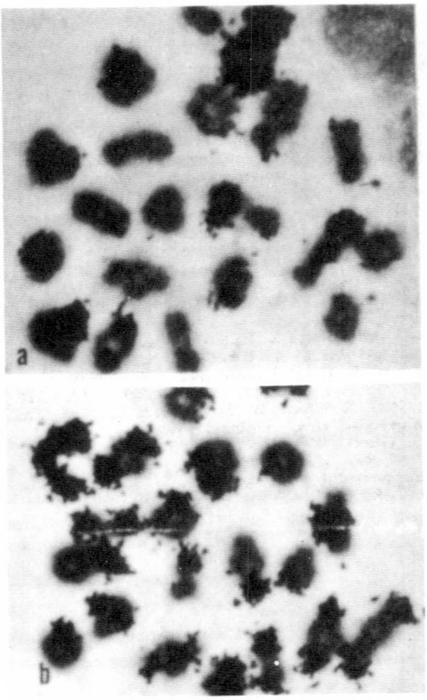

Fig. 3. Meiotic mouse chromosomes in DMI with partial labeling of autosomal bivalents and complete labeling of the sexual complex. (a) Some autosomes partially labeled. (b) All autosomes partially labeled.

part of an arm. Thirteen percent of all labeled autosomes had grains localized in two spots (zero to ten bivalents per cell) that were also close to arm-ends, with the same reservations as above. The number of grains was more frequently grossly equal in both spots than unequal. A mean of 22 percent of all autosomes were observed to be entirely labeled, whereas 52 percent were devoid of grains.

This overall pattern corresponds well with data for murine mitotic chromosomes (*12*), except for the mitotic X chromosome that has not been observed as being late labeling in males. The distal parts of meiotic bivalents are near-centromeric in mouse telocentric autosomes and generally considered as heterochromatic (*14*). Distal late labeling has also been observed in meiosis in *Triturus* (*6, 7*).

We could not determine an exact relation between termination of DNA synthesis in the two distal parts of a chromatid. Even for those X chromosomes where grains were seen at the end opposite to the Y, possible grains in the proximal end would be undistinguishable from the grain spot of the Y chromosome (Fig. 2c).

Comparison of the length and thickness of mouse meiotic DMI autosomes with the size of grain spots indicates further that synchronously late-synthesizing segments in homologous chromatids do actually pair with each other. Indeed, drawing axes of symmetry through one spot, whenever only one is present in a bivalent, or through the unit constituted by a double spot at both ends of one bivalent, shows that these axes do actually cross the center of the chromatid figures. This observation was technically possible for approximately 70 percent of all autosomal bivalents that were analyzed. In Fig. 3, a and b, this simple geometrical relation applies to an even larger proportion of autosomal chromosomes. Even though it appears quite logical that homologous chromatids pair with a gene-to-gene correspondence and that replication should also be sequentially synchronized, direct proof of the latter point has been lacking. Within the limits of autoradiographic resolution, this is actually the case for the presumably heterochromatic parts of mouse autosomes, owing to their uniformly telocentric configuration in this species and assuming that centromeres are indeed located in the close neighborhood of late-labeling regions.

References and Notes

1. E. F. Oakberg, *Amer. J. Anat.* **99**, 507 (1956).
2. W. Hilscher and B. Hilscher, *Z. Zellforsch.* **96**, 625 (1969); V. Monesi, *J. Cell Biol.* **14**, 1 (1962).

3. T. Utakoji, *Exp. Cell Res.* **42**, 585 (1966); ——— and T. C. Hsu, *Cytogenetics (Basel)* **4**, 295 (1965).
4. M. Fraccaro, I. Gustavsson, M. Hulten, J. Lindsten, L. Tiepolo, *Cytogenetics (Basel)* **8**, 263 (1969).
5. A. Lima-de-Faria, *Hereditas* **47**, 674 (1961); R. B. Nicklas and R. A. Jaqua, *Science* **147**, 1041 (1965).
6. H. G. Callan and J. H. Taylor, *J. Cell Sci.* **3**, 615 (1968).
7. D. E. Wimber and W. Prensky, *Genetics* **48**, 1731 (1963).
8. A-Swiss albino mice (8-week-old inbred males, free of pathogens) were given injections of sterile thymidine-[^{3}H-methyl] (0.1 μc; specific activity, 6.0 c/mmole) in buffered saline directly into the testicular parenchyma. Mice were killed at 8-hour intervals from the moment of injection up to day 26. Autoradiograms were prepared from air-dried smears [E. P. Evans, G. Breckon, C. E. Ford, *Cytogenetics (Basel)* 3, 289 (1964)] with Eastman-Kodak NTB-2 liquid emulsion, as described previously [N. Odartchenko, H. Cottier, L. E. Feinendegen, V. P. Bond, *Exp. Cell Res.* 35, 402 (1964)] and, after exposure times of 30, 70, and 110 days, were stained through the film with Giemsa buffered at *p*H 5.75.
9. This time is not influenced by autoradiographic exposure time (Fig. 1) and appears to be independent of major seasonal variations, since a second group of mice, treated exactly like the first at a 6-month time interval, revealed only an approximate 12-hour prolongation of this time lag.
10. Y. Hotta, M. Ito, H. Stern, *Proc. Nat. Acad. Sci. U.S.* **56**, 1184 (1966).
11. As evidenced by an evaluation of free ^{3}H-TdR in testes tissue at various time intervals after injection.
12. M. Galton and S. F. Holt, *Exp. Cell Res.* **37**, 111 (1965); H. J. Evans, C. E. Ford, M. F. Lyon, J. Gray, *Nature* **206**, 900 (1965).
13. I. Geyer-Duszynska, *Chromosoma* **13**, 521 (1963); S. Ohno, W. D. Kaplan, R. Kinosita, *Exp. Cell Res.* **18**, 282 (1959).
14. S. Ohno, W. D. Kaplan, R. Kinosita, *Exp. Cell Res.* **13**, 358 (1957).
15. Supported by the Swiss National Foundation for Scientific Research.

ANIL B. MUKHERJEE
MAIMON M. COHEN

DNA Synthesis during Meiotic Prophase in Male Mice

DNA SYNTHESIS is thought to be confined to a specific segment of interphase (S) preceding cell division and to be completed prior to the onset of either mitosis or meiosis[1]. Chromosome pairing and crossing over between homologous chromosomes in meiosis are separate events both of which involve the DNA moiety of the chromosomes. An explanation of these processes in terms of DNA synthesis has only recently been suggested[2,3]. Hotta *et al.*[4–6] have demonstrated by biochemical techniques that approximately 0·3 per cent of the total DNA complement of the cell is synthesized in the zygonema-pachynema stages of meiotic prophase in lily microsporocytes. We have cytological evidence of a little DNA synthesis during early pachytene stage in the male mouse.

Tritiated thymidine (specific activity 1·9 Ci/mmole, obtained from Schwarz BioResearch, Inc., Orangeburg, New York) in the final concentration of 5 μCi/ml. (0·2 ml.) was injected directly into the testes of four male mice (C_3H He/Ha strain). After 30 min the animals were killed by cervical dislocation and testicular preparations were made following the method of Evans *et al.*[7]. Kodak nuclear track liquid emulsion (NTB_3) was used to prepare the autoradiographs which were incubated from 15 days to 1 month at 4° C and developed in D-19 (Kodak) developer. The autoradiographs were stained for 5 min using a 1 per cent Cresyl Violet solution. Photomicrographs were obtained and actual grain counts over the premeiotic interphase cells as well as early and late pachytene stages were made. Cells with one to five overlying silver grains were considered lightly labelled, while between six and fifteen grains constituted a moderately labelled cell and cells with more than 15 grains were considered heavily labelled.

Cells in pachytene were readily identified and differentiated from the premeiotic interphases by the degree of chromosome condensation and pairing. The presence of a prominent sex bivalent (Fig. 1) was an additional criterion for identification of pachytene.

^{3}H-Thymidine is incorporated only into cells actively engaged in DNA synthesis and in the 30 min incubation period these cells do not normally pass from the premeiotic S-period to pachytene[8,9], so the silver grains over the pachytene nuclei indicate isotopic uptake at this stage only. Seventy per cent of all the premeiotic interphase nuclei incorporated ^{3}H-thymidine and 63 per cent of them

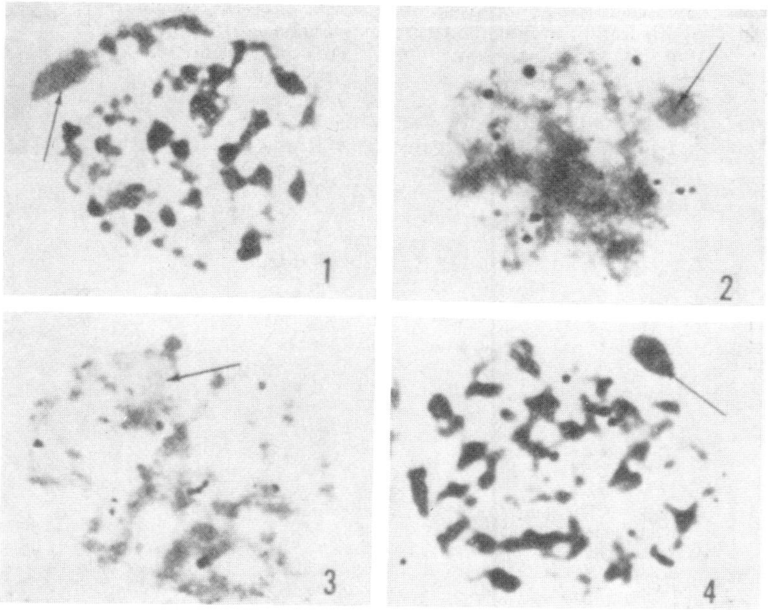

Fig. 1. Unlabelled pachytene stage indicating XY-bivalent (arrow).

Figs. 2 and 3. Early pachytene showing moderate (approximately 10 grains/cell) labelling. Arrows indicate XY-bivalent.

Fig. 4. Late pachytene showing six overlying silver grains. Condensation of chromosomes is greater than in early pachytene figures. Arrow indicates XY-bivalent.

were heavily labelled (Table 1). Only 57 per cent of the early pachytene nuclei incorporated ^{3}H-thymidine, however, and none was heavily labelled (Figs. 2–4). During the late pachytene stage only 12 per cent of the cells were labelled, all of these with less than five overlying silver grains. In the diplotene stage no incorporation of tritiated-thymidine was observed.

These results suggest that a little DNA synthesis takes place during meiotic prophase, particularly during the

Table 1. UPTAKE OF ^{3}H-THYMIDINE BY MOUSE TESTICULAR CELLS AT DIFFERENT STAGES OF MEIOSIS

Cell stage	Hours after injection of ^{3}H-thymidine	No. of each type of cells counted	No. labelled	Type of label (grain counts) Low (1–5)	Moderate (6–15)	Heavy (<15)
Premeiotic interphase	30 min	50	35	7	6	22
Early Pachytene	30 min	50	28	2	26	0
Late Pachytene	30 min	50	6	6	0	0
Diplotene	30 min	32	0	0	0	0

early pachytene stage, in the mouse. Such synthesis may be an integral component of the repair mechanism operating following the physical exchange of chromatin material during crossing over. This suggestion has been proposed by Hotta *et al.*[4] and supported by the genetic data of Whitehouse *et al.*[2] and segregational data of Kitani *et al.*[10].

Ito *et al.*[11] have shown that inhibitors of DNA synthesis interfere with the meiotic cycle if administered during zygonema and/or pachynema. The present autoradiographic evidence together with the cytological effects of inhibitors[11], correlates well with the chemical evidence[4] for a small amount of DNA synthesis during the zygonema-pachynema stages of meiosis.

1 Taylor, J. H., and McMaster, H., *Chromosoma*, **6**, 489 (1954).
2 Whitehouse, H. L. K., *Nature*, **199**, 1034 (1963).
3 Ratnayake, W. E., *Nature*, **217**, 1170 (1968).
4 Hotta, Y., Ito, M., and Stern, H., *Proc. US Nat. Acad. Sci.*, **56**, (1184 1966).
5 Hotta, Y., and Stern, H., *J. Cell Biol.*, **16**, 259 (1963).
6 Hotta, Y., Bassel, A., and Stern, H., *J. Cell Biol.*, **27**, 451 (1965).
7 Evans, E. P., Breckon, G., and Ford, C. E., *Cytogenetics*, **3**, 289 (1964).
8 Crone, M., Levy, E., and Peters, H., *Exptl. Cell Res.*, **39**, 678 (1965).
9 Utakoji, T., *Exptl. Cell Res.*, **42**, 585 (1966).
10 Kitani, Y., Oliver, L. S., and El Ani, A. S., *Amer. J. Bot.*, **49**, 697 (1962).
11 Ito, M., Hotta, Y., and Stern, H., *Dev. Biol.*, **6**, 54 (1967).

UNUSUAL INCORPORATION OF TRITIATED THYMIDINE INTO EARLY DIPLOTENE OOCYTES OF MICE

MONNA CRONE and HANNAH PETERS

WE have previously reported that the premeiotic DNA synthesis in mouse oocytes takes place during the last week of foetal life and is completed before the animal is born [9]. During this study we have, by autoradiography, observed an occasional labelled oocyte after injection of ^{3}HTdR shortly after birth [1, 7].

To investigate this apparently rare phenomenon more in detail 0d-, 1d-, 2d-, 3d-, 4d- and 5d-old female mice were injected intraperitoneally with 10 μCi/g of ^{3}HTdR and killed 1 h, 6 h, 24 h, 4 days and 7 days later. Autoradiographs of 5 μ serial sections were prepared using Ilford K2 liquid emulsion. Half of the slides were stained by the Feulgen technique before, the other half with haematoxylin and eosin after the autoradiographic procedure. The slides were exposed one, two or three weeks. Every second section was scored for labelled oocytes.

A few oocytes showed nuclear labelling (Table 1). Labelled oocytes were found only in ovaries from animals that were injected on one of the first three days after birth. The number of labelled oocytes is very low indeed, only 0.1 per cent or less of the total number of oocytes; they were all in early diplotene, surrounded by an incomplete or complete single layer of follicle cells (Fig. 1). They are located centrally in the ovary and belong to the biggest oocytes seen at that age. The grains over the oocyte nuclei are distributed diffusely over the chromosomes (Figs 1 and 2). The number of grains after one or two weeks exposure varies from 10 to about 60 grains per nucleus. There are fewer grains over the oocyte nuclei than over the labelled follicle cells. However, it is difficult to estimate the amount of ^{3}HTdR incorporated into the oocyte as the size of their nucleus is considerably larger than that of the follicle cells and the absorption of the β-particles in these two types of nuclei is different.

Each labelled oocyte can be followed in two or three neighbouring sections, which strongly indicate that the labelling is due to the incorporation of ^{3}HTdR into the cell nucleus and is not an autoradiographic artefact. Labelled oocytes are observed in sections stained by the Feulgen technique. This supports the assumption that ^{3}HTdR is incorporated into the nuclear DNA of these oocytes.

TABLE 1. *Labelling of oocytes after injection of ^{3}HTdR*

In some cases labelled oocytes were observed and shown as +. The total number could not be determined as these ovaries were not serial sectioned

Age of the mouse at injection of ^{3}HTdR in d.	Time from injection to death	Total number of labelled oocytes in the ovary	
		Right ovary	Left ovary
0	6 h	0	
0	24 h	0	
0	24 h	+	
1	1 h	0	
1	6 h	0	
1	24 h	4	
1	24 h	+	
1	24 h	+	
1	7 days	0	
2	1 h	0	0
2	1 h	7	7
2	6 h	11	8
2	24 h	4	12
2	24 h	+	
2	72 h	+	
2	96 h	0	0
2	96 h	0	0
2	7 days	0	
3	6 h	0	
3	24 h	0	
4	6 h	0	
4	24 h	0	
5	6 h	0	
5	24 h	0	

This was confirmed by performing DNase digestion on a slide with labelled oocytes. The old emulsion and the grains were removed, half of the slide was incubated in DNase (0.02 per cent solution of DNase in 0,02 *M* $MgSO_4$ at 37°C for 2 h, pH = 6.8). Afterwards a new autoradiograph was prepared and exposed one week. The new autoradiograph showed no labelling of the DNase treated previously labelled oocytes.

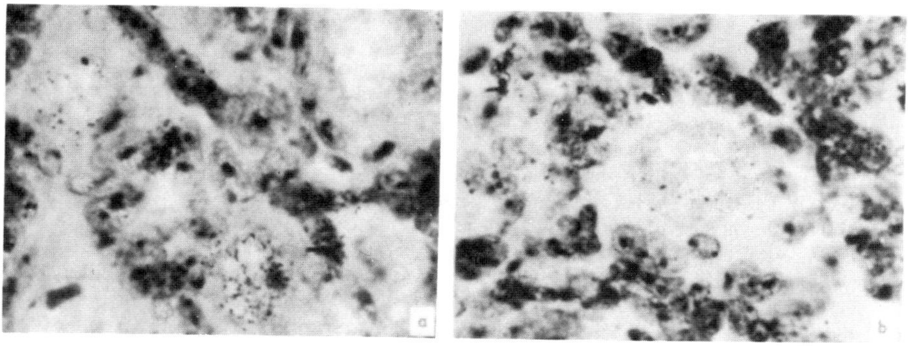

Fig. 1.—Autoradiographs of labelled oocytes in early diplotene from a mouse injected with ^{3}HTdR at the age of 2 days and killed 24 h later. *a*, Stained with haematoxylin and eosin; *b*, stained by the Feulgen technique.

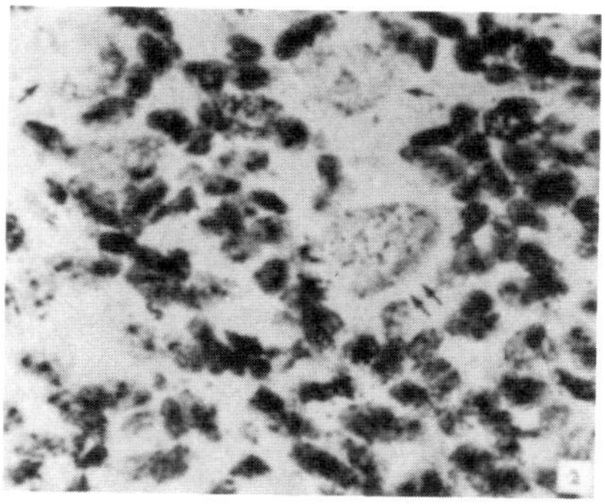

Fig. 2.—Autoradiograph from a 2-day-old mouse killed 24 h after the injection of ^{3}HTdR. Unlabelled oocytes of normal appearance (↑) and a labelled oocyte with signs of early degeneration (↑↑) stained by the Feulgen technique.

The question arises what process underlies this rare incorporation of ^{3}HTdR into the diplotene nucleus of an oocyte.

It seems most improbable that the labelled oocytes were oogonia in premeiotic S-phase at the time of injection that have passed through the first stages of meiotic prophase (leptotene, zygotene and pachytene) in the time interval from injection to sacrifice of the animal. It was previously found that it takes more than three days for an oocyte to proceed from premeiotic S-phase to the onset of diplotene [2]. In the present investigation labelled oocytes in diplotene are already seen one hour after injection of ^{3}HTdR. Furthermore the first stages of meiotic prophase, leptotene and zytogene, are only found in the foetal ovary and not seen after birth. It therefore seems unlikely that the incorporation of ^{3}HTdR into these early diplotene oocytes represents normal chromosome replication in premeiotic oogonia but that an abnormal DNA synthetic activity, perhaps a DNA repair synthesis is in process.

Recent studies on the recovery of microorganisms after irradiation or chemical treatment indicate that complex enzymic mechanisms protect DNA against a variety of damages by DNA repair synthesis [3].

Similar mechanisms may play a role in the process of genetic recombination. It has been shown, that genetic recombination in phages occurs by breakage and reunion of doublestranded DNA molecules and there is some indication that a small amount of DNA is removed and resynthesized in the formation of recombinant molecules [5].

Very little is known about genetic recombination ("crossing-over") in higher animals, neither about the mechanism itself nor at which stage of meiosis it occurs. It is accepted that crossing-over takes place during or after premeiotic DNA replication and as a general working hypothesis it is assumed that crossing-over at any level and in all organisms has as a prerequisite a precise process of homologous chromosome pairing (synapsis). The process of crossing-over is therefore supposed to occur during late zytogene and pachytene in animal cells where the homologous chromosome pairing is complete, but it cannot be excluded that it can occur also at later stages with pairing of segments of the chromosomes [10].

The observed incorporation of ^{3}HTdR into oocytes seems to be linked to a certain stage of meiotic prophase, namely early diplotene. It is, however, not necessary that this DNA synthesis is a step during the process of genetic recombination as no incorporation of ^{3}HTdR is seen in pachytene where synapsis is complete.

It seems more probable that the labelling of the oocytes is due to some abnormal DNA synthesis or DNA repair synthesis after an accidental damage to the DNA of the cell, the mechanism of DNA repair in microorganisms is partly clarified. Painter and Cleaver [6] recently observed repair replication in cultures of HeLa cells after large doses of X-irradiation. Lett *et al.* [4] have demonstrated rejoining of single strand mammalian DNA broken by ionizing radiation. These observations indicate that mammalian cells also possess means to repair damaged DNA.

If the labelled oocytes observed in our study were cells in the process of DNA repair synthesis, it seemed puzzling that only early diplotene oocytes were labelled. It has been shown, however, that oocytes in early diplotene are more resistant to X-rays than oocytes in pachytene or late diplotene [8]. The resistance of the early diplotene oocyte might partly be due to a more active repair synthesis at this particular stage of meiotic prophase. If so other kinds of damage might also initiate a more active DNA repair in oocytes in diplotene than in oocytes in other stages of meiotic prophase.

Another possibility is that the few oocytes that have incorporated the DNA precursor are in the process of degeneration and during this process have embarked upon some abnormal DNA synthesis or some unsuccessful DNA repair. Most of the labelled oocytes seem morphologically normal, however, some of them show early signs of degeneration namely "swelling" of the chromosomes and some distorsion of the nucleus (Fig. 2). A significant number of oocytes in late pachytene are degenerating shortly after birth and are seen in the histological sections as typical degenerating cells with heavily stained homogeneous nuclei, but none of these degenerating oocytes have incorporated ^{3}HTdR.

The labelled oocytes seem to disappear after three to four days. A mouse injected at the age of 2 days and killed 3 days later had labelled oocytes without any sign of degeneration, two animals killed after 4 days and 2 killed after 7 days had no labelled oocytes. This finding suggests that the most likely explanation of the labelled oocytes is some abnormal DNA synthesis during the process of degeneration.

The actual number of labelled oocytes is very low. This does not necessarily mean that the mechanism behind this phenomenon is a rarity. It might very well be, that the amount of DNA synthesized often is so small that it can hardly be detected by the autoradiographic technique used in this study.

Summary.—After a single injection of ^{3}HTdR into mice shortly after birth a few labelled oocytes were observed. These were all in early diplotene, a stage of meiotic prophase in which no normal chromosome replication takes place. Several possibilities to explain this phenomenon are discussed including abnormal DNA synthesis during degeneration and DNA repair after some kind of damage or as a step in the process of genetic recombination. The first possibility is considered to be the most likely. The mechanism underlying this DNA synthesis is not known.

REFERENCES

1. Crone, M., Levi, H. and Peters, H., Reports 2nd. Conf. Cell Res., *Lund Univ. Årsskr.* **56**, 22 (1960).
2. Crone, M., Levy, E. and Peters, H., *Exptl Cell Res.* **39**, 678 (1965).
3. Haynes, R. H., *Rad. Res.* Suppl. **6**, 1 (1966). (Literature summarized.)
4. Lett, J. T., Caldwell, I., Dean, C. J. and Alexander, P., *Nature* **214**, 790 (1967).
5. Meselson, M., *J. Mol. Biol.* **9**, 734 (1964).
6. Painter, R. B. and Cleaver, I. E., *Nature* **216**, 369 (1967).
7. Peters, H. and Crone, M., Colloque internationale sur la physiologie de la reproduction chez les mammifères. Paris 1966. In press.
8. Peters, H. and Levy, E., *J. reprod. Fertil.* **7**, 37 (1964).
9. Peters, H., Levy, E. and Crone, M., *Nature* **195**, 915 (1962).
10. Westergaard, M., *Compt. Rend. Lab. Carlsberg* **34**, 359 (1964–65).

DNA SYNTHESIS IN RELATION TO CHROMOSOME PAIRING AND CHIASMA FORMATION

HERBERT STERN AND YASUO HOTTA

THIS article is addressed to the phenomena of chromosome pairing and crossing-over in meiotic cells. Its purpose is to summarize the evidence thus far obtained which bears on developmental and biochemical mechanisms underlying these two principal events in meiosis. Since meiotic cells are products of mitotic divisions, critical biochemical events should accompany the shift from mitotic to meiotic behavior. Some of these events should precede and determine the entry of a cell into the meiotic cycle proper.

Our studies of the meiotic cells of lily and trillium lead to the general conclusion that much of the developmental programming of these cells (the "microsporocytes") is laid down several generations before the cells enter meiosis. The microsporocytes of *Trillium erectum*, for example, may be induced to undergo a mitotic rather than a meiotic division and yet, having bypassed the entire meiotic cycle, such cells ultimately grow pollen tubes. The nature of this programming and the extent to which it may represent a series of cumulative changes spanning preceding generations is beyond the scope of this article. Indeed it still appears to be beyond experimental reach. On the other hand, the terminal events accompanying the shift from mitotic to meiotic behavior are open to experimental study. These events, insofar as they relate to the chromosomes, are the subject of this article.

A SUMMARY DESCRIPTION OF CRITICAL EVENTS

The premeiotic S-phase, unlike the premitotic one, does not include total chromosome replication. A small but significant DNA component of the chromosomes, about 0.3% of the total DNA, fails to replicate by the termination of the S-phase. Its replication is delayed to the zygotene stage. This delay is, of itself, not determinative. At a point approximately midway between completion of the S-phase and leptotene, the cell becomes irreversibly committed to enter a meiotic type of division. This commitment is manifested by the subsequent entry of the cells into a prolonged prophase which includes the characteristic stages of leptotene, zygotene, and pachytene. The commitment is, however, incomplete with respect to chromosome pairing since perturbations of the cell at this stage may render subsequent pairing unstable. The basis for this instability is removed by the time the cell enters the leptotene stage. Actual pairing and formation of the associated synaptinemal complex begin in zygotene coincident with a delayed DNA replication. The pairing process is unaccompanied by chiasma formation.

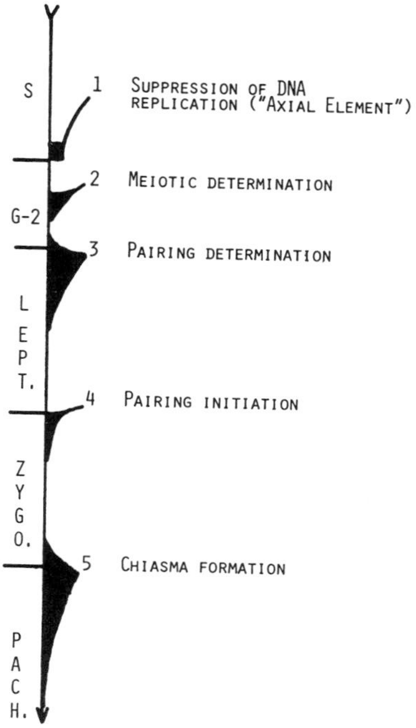

FIGURE 1.—A diagrammatic representation of the sequence of events leading to chiasma formation in meiotic cells. The scheme is based upon the experimental evidence discussed in the text.

At the end of zygotene and early in pachytene a set of events occurs which brings about chiasma formation.

The experimental basis for this sweeping picture of meiotic events is described below. Much of the evidence is circumstantial and begs further documentation. It is nevertheless clear that the two visible structural events of chromosome pairing and chiasma formation are dependent upon prior molecular events which reach as far back as the termination of the premeiotic S-phase. Moreover, this picture which is depicted graphically in Figures 1 and 3 is consistent with the evidence obtained by DAVIES and LAWRENCE (1967) and by LAWRENCE and DAVIES (1967) for the existence of two intervals in the meiotic cell cycle during which genetic recombination or chiasma formation may be modified experimentally.

EXPERIMENTAL EVIDENCE

Delay of DNA synthesis: The conclusion that the DNA synthesized during zygotene represents a delayed replication is based upon two types of experiments. In the first, purified DNA was prepared from the following groups of cells: (a) Somatic cells or presumptive meiotic cells prior to the last round of premeiotic DNA synthesis; (b) premeiotic cells in the G-2 phase and leptotene cells; (c) cells at pachytene and later meiotic stages. Each of the preparations was denatured and bound to membrane filters. Zygotene-labeled DNA was also pre-

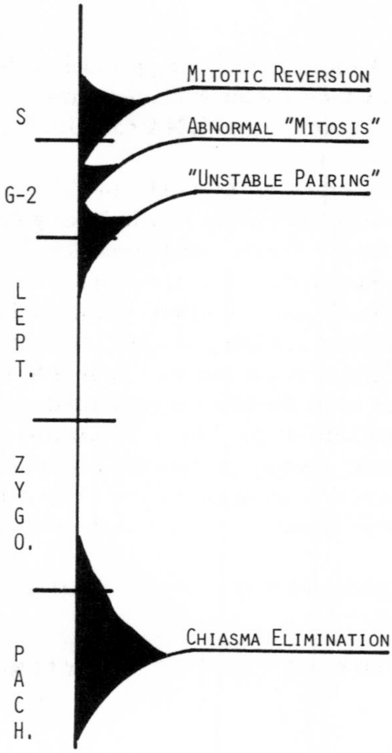

FIGURE 3.—A diagrammatic representation of various intervals in meiotic development during each of which the cells are susceptible to the transformation indicated. It should be noted that "unstable pairing" leads to achiasmatic cells just as does the elimination of chiasmata by interference with protein synthesis in late zygotene and early pachytene.

pared and the amount of such DNA which was bound to each of the unlabeled preparations under saturating conditions was measured. The results indicated that DNA prepared from group (b) bound about half as much of the labeled zygotene DNA as that from the other groups. Parallel experiments using DNA labeled during the S-phase revealed no disparity in respective amounts bound to the DNA prepared from the different groups of cells (Table 1). The most plausi-

TABLE 1

Hybridization of zygotene- and pachytene-labeled DNA with DNA from cells at various stages of the cycle

DNA on filter	Zygotene	Pachytene	Somatic
Premeiotic S-phase	0.56	0.97	1.0
G_2-Leptotene	0.54	1.00	1.0
Zygotene	0.54	0.97	1.0
Pachytene	0.91	0.96	1.0
Div. II	0.89	0.96	1.0

The percentage of each of the three types of DNA which hybridized with the unlabeled DNA derived from cells at the indicated stages was determined under saturating conditions. The values given represent the ratio of ^{32}P-zygotene or pachytene DNA to ^{32}P-somatic DNA. The data are taken from unpublished studies of Y. Hotta and H. Stern.

ble explanation of these results is that DNA replication during premeiotic S-phase is incomplete and that the regions of DNA which are unreplicated remain so until the zygotene stage. As already indicated, the amount of DNA contained in these regions is of the order of 0.3% of the total.

The second type of experiment relates to the fact that cells in the early G-2 stage may be explanted from their normal environment and induced to enter a mitotic rather than a meiotic type of division (STERN and HOTTA, 1967). If the conclusion previously drawn is correct, cells explanted at this stage should have unreplicated regions of DNA which are homologous with that synthesized during zygotene. These regions, to the extent that they are integral parts of chromosome structure, should therefore be replicated prior to the entry of cells into mitosis. Thus, a zygotene type of synthesis should be evident in the explanted cells even though the cells do not pass through a zygotene stage. The experiments confirm the expectation. Cells which are explanted during premeiotic interphase and which have terminated the S-phase synthesize a small amount of DNA virtually identical with that synthesized at zygotene (Figure 2) prior to their entry into mitosis.

These two sets of observations point to the first identifiable event associated

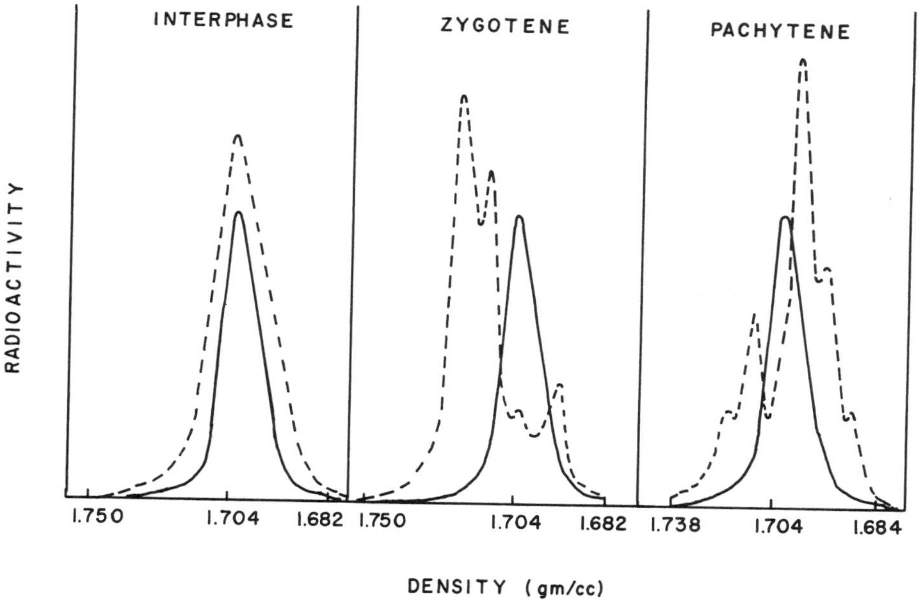

FIGURE 2.—Patterns of DNA synthesis in meiotic cells. The dotted lines trace the pattern of radioactive DNA after equilibration in a CsCl gradient. The levels of radioactivity representing the DNA synthesized respectively during premeiotic interphase, zygotene and pachytene are not comparable quantitatively. About 99.7% of the DNA is synthesized during interphase. The small amounts synthesized during zygotene and pachytene are qualitatively distinctive; their possible functions are discussed in the text.

with the shift from a mitotic to a meiotic type of division, a selective suppression of DNA replication. One may speculate that these regions of delayed replication represent specific sites of terminal DNA synthesis which in effect control the conversion of a chromosome into two chromatids. However, no information is available concerning the location of these regions except for the evidence that they are in the chromosomes. The significance which can be attached to this delay is to be found in studies of the zygotene stage.

Initiation of meiosis: The premeiotic delay in terminal DNA replication is not sufficient to determine the entry of a cell into meiosis. This conclusion follows from the possibility of inducing cells to enter mitosis after the termination of premeiotic DNA synthesis. The mitosis may be normal or abnormal. In the latter case, the chromosomes fail to undergo anaphase separation and behave as though centromere division were blocked (Stern and Hotta, 1967). In both cases, the prophase stage is typically mitotic with respect to both cytological appearance and relative brevity of duration. The capacity of premeiotic cells to revert to mitosis does not persist, however, through the entire G-2 period. Although the precise point at which this capacity is lost has been difficult to determine, the loss is evident in cells which have developed for one or two days beyond the termination of the S-period. Such cells enter a meiotic prophase and do not synthesize DNA until they reach the zygotene stage. Thus, the mechanism which delays DNA replication at the termination of premeiotic S-phase must be followed during the interphase by another set of events which initiates the meiotic cycle. No evidence is available concerning the nature of these events. Protein synthesis occurs at this time and it appears to be required for the transition since inhibition of protein synthesis by cycloheximide prevents the cells from entering meiotic prophase (Parchman and Stern, 1969).

Pairing determination: Between the time that the cells lose their capacity to revert to mitosis and their entry into leptotene, another change must occur in the cells, the effect of which becomes apparent much later, at the pachytene stage. Premeiotic G-2 cells which are explanted into culture media and subsequently enter meiosis fall into two classes. One class ultimately gives rise to achiasmatics; the other progresses to the normal diplotene configuration. Both groups pass through the zygotene and pachytene stages and although cytological observations of the achiasmatic group suggest an instability of pairing, the evidence is inadequate for an unequivocal conclusion. On the other hand, the evidence is sufficient to establish that perturbation of meiotic cells at an interval close to the beginning of leptotene leads to delayed effects which are clearly apparent at late pachytene but not earlier. The perturbations we have used are of two kinds, explantation of cells from their normal environment into artificial culture media, or exposure of cells during that interval to an elevated temperature. Meiotic cells of the lily "Cinnabar", for example, divide achiasmatically if explanted close but prior to leptotene. The frequency of the achiasmatics decreases as the cells are explanted closer to and within early leptotene. If, however, the temperature of incubation is raised from the standard 20°C to 25°C, the high frequency of achiasmatics is restored. Cells explanted at zygotene and incubated at the elevated temperature do not become achiasmatic. The evidence thus indicates that cells in late G-2 or early leptotene are susceptible to physiological disturbances which have no apparent effect on cytological progress through leptotene, zygotene, and pachytene but which render the cells visibly achiasmatic at diplotene. The conclusion may be drawn that a critical set of events occurs during late premeiotic G-2 and that these events influence subsequent chiasma formation. We have no information on the nature of these events but their occurrence would appear to be widespread among meiotic systems. The studies of DAVIES and LAWRENCE (1967) and of LAWRENCE and DAVIES (1967) on recombination frequency in Chlamydomonas all point to a sensitive premeiotic interval which, if perturbed, leads to changes in recombination or chiasma frequency.

Initiation of pairing: The evidence that chromosome pairing is initiated at the zygotene stage comes from observations through the light and electron microscopes. The cytological data have been critically reviewed by a number of investigators and our own studies add nothing that is novel to this particular issue. The very fine features of the problem, such as the exact point in development at which a cell can be said to have entered or left the zygotene stage, are probably unresolvable by the techniques of observation and may be more properly regarded as issues of nomenclature. A more substantive question is whether certain preconditions for pairing are laid down at some stage prior to initiation. The evidence for prior events determining the initiation of pairing has been discussed above. The nature of these events remains undefined except for the delay in replication of some DNA. Thus far, we have been unable to bring about pairing in the absence of DNA synthesis. Two observations lead to the conclusion that zygotene synthesis of DNA is essential to chromosome pairing. (1) The two events begin and terminate simultaneously or nearly so. (2) Inhibition of the initiation of DNA synthesis at zygotene inhibits the initiation of pairing and the arrest of DNA synthesis during zygotene correspondingly arrests pairing. The latter arrest is clearly seen by monitoring the synaptinemal complex formation through the electron microscope (ROTH and ITO, 1967).

Convincing information is lacking concerning the relationship of the DNA synthesized during the zygotene interval to chromosome structure. The evidence for this synthesis being a delayed replication has already been cited. Another relevant fact is that these replicated regions must, at the molecular level, constitute relatively long regions of DNA. This is apparent from an examination of Figure 2. Under conditions of isopycnic centrifugation zygotene-synthesized DNA appears as a discrete band with an average density that is significantly higher than the total DNA. Chemical determination of composition matches that calculated from density values (HOTTA, ITO and STERN, 1966). The discrete band thus reflects a population of discrete molecules. It would seem reasonable to infer that the discreteness also exists within the chromosomes.

No information is available on the location of these postulated regions within the chromosomes. The assumption is made that these are more or less evenly distributed along the chromosome. The only grounds for this assumption is the fate of zygotene chromosomes which are partially inhibited in their DNA replication. After 2–3 days of partial synthesis the chromosomes begin to fragment. The fragmentation occurs throughout each chromosome and in all chromosomes. Such fragmentation may be attributed to the lesions resulting from incomplete DNA synthesis (Ito, Hotta and Stern, 1967). Other interpretations could be provided but the one chosen is at least more attractive from the standpoint of speculations concerning pairing mechanisms.

The Pairing Process: The morphological data clearly establish that a structure, the synaptinemal complex, is elaborated concomitantly with chromosome pairing. The function of the complex is unclear. It may be regarded either as a mechanism for facilitation of chromosome pairing or as a stabilizer of the otherwise independent pairing process. Regardless of its function, however, the DNA synthesis which occurs during zygotene is insufficient to account for the structure associated with paired chromosomes since only a small proportion of the synaptinemal complex consists of DNA (Moses and Coleman, 1964). Much of the complex probably consists of protein and, if so, zygotene pairing should be accompanied either by the assembly of preformed proteins or by *de novo* protein synthesis.

There is no doubt that protein synthesis occurs in zygotene cells and that newly formed proteins appear in the nucleus of these cells (Hotta, Parchman and Stern, 1968). Moreover, protein synthesis is essential to the progress of meiotic cells through zygotene since inhibition of synthesis by means of cycloheximide immediately arrests meiotic development. The question which cannot yet be fully answered is whether the synaptinemal complex is formed from proteins synthesized during zygotene or whether its formation from preexisting proteins is directly dependent upon protein synthesis for the production of appropriate enzymes or accessory structural elements. The difficulty in satisfactorily answering the question is that proteins synthesized during zygotene appear throughout the cell when analyzed by radioautography and appear to be heterogeneous in composition when fractionated chemically. In the absence of additional data zygotene arrest by cycloheximide could be attributed to the failure of one or more of many processes which might be dependent upon the synthesis of protein.

Nevertheless, some evidence has already been obtained which points to, but does not prove, the direct participation of protein synthesis in the pairing process. First, certain chromatographically distinctive non-histone nuclear proteins, which are extractable at pH 8.0 in media of low ionic strength, are synthesized mainly during the zygotene stage (Hotta, Parchman and Stern, 1968). The synthesis of these proteins can be selectively inhibited by low concentrations of cycloheximide without appreciably affecting the synthesis of other proteins which are not exclusively associated with the pairing interval. Under such conditions of inhibition, zygotene development is arreseted. The relationship between this inhibition and pairing is, however, not necessarily due to a direct utilization of the proteins for the pairing structure. The proteins could play a supportive but accessory role. The likelihood of this possibility is increased by the observation that zygotene DNA synthesis is also readily inhibited by low concentrations of cycloheximide. DNA synthesis during zygotene thus appears to be tightly coupled to protein synthesis during that same interval. Even though the coupling itself is of major interest, its existence does permit the simple explanation that the efficacy of zygotene arrest by cycloheximide is due to its indirect effect on DNA synthesis.

The solution of choice to the problem of relating protein synthesis to the pairing process could be obtained by first identifying the proteins of the synaptinemal complex. Such identification is yet to be accomplished. However, some evidence has been obtained which indicates a physical association between some of the proteins synthesized during the zygotene stage and the DNA synthesized during that same interval. Zygotene-synthesized DNA is selectively extracted under the conditions used to extract these nuclear proteins. Since very little of the total DNA is extracted under these conditions, the possibility exists that the simultaneous extraction of zygotene DNA and protein reflects some physical association between them. Two types of association are, in fact, demonstrable. The protein and DNA, when fractionated on a DEAE column, elute together when a gradient of increasing salt concentration is used as eluant. Much of the association, however, appears to be ionic in nature, since centrifugation in concentrated solutions of cesium chloride releases a large fraction of the protein. The remaining fraction appears to be tightly bound to the DNA. The identification of the bound proteins and the nature of the binding are currently under study.

The experimental evidence concerning the pairing process may be summarized thus: Distinctive DNA and proteins are synthesized during zygotene and inhibition of either synthesis inhibits pairing. The DNA synthesis is dependent upon concurrent protein synthesis. Some of the protein synthesized during zygotene is bound to the DNA then synthesized. The evidence is still insufficient to say how much of this protein-DNA association consists of newly synthesized material. On the other hand, the evidence from inhibitor studies is sufficient to conclude that the association has a functional correlate in meiosis.

Chiasma formation: The evidence for the occurrence of chiasma formation *after* the completion of pairing derives from observations that achiasmatic divisions can be induced by perturbing meiotic cells at the end of zygotene. A number of reports bearing on this point are in the literature, and although the precise interval of sensitivity has been debated, there is general agreement that the interval spans some portion of meiotic prophase. Most of the reports have been based upon the use of elevated temperatures at different intervals of the meiotic cycle to induce the achiasmatic state (HENDERSON, 1966). The studies in our laboratory have been based on the use of cycloheximide. As described earlier, low concentrations of cycloheximide partially and selectively inhibit protein synthesis and simultaneously arrest meiotic development. If cells which have been exposed to the inhibitor for 2–3 days are returned to normal growth media, meiotic development is resumed. Such development, however, is not always normal. The kind of abnormality, if any, depends upon the stage at which the cells have been exposed. The consequence of arresting late zygotene or early pachytene cells for 2 days by low concentrations of cycloheximide is distinctive. Such cells resume development through pachytene but the chromosomes do not form chiasmata. The fact that induction of achiasmatics by partial inhibition of protein synthesis is restricted to a very narrow interval surrounding the completion of pairing is probably significant. It not only suggests that chiasmata are absent prior to the completion of pairing but also that the mechanism responsible for chiasma formation is irreversibly set by the mid-pachytene stage. If these broad interpretations of the data are correct, then the conclusion must be drawn that chiasma formation is a deliberately timed event in meiosis and is not a concomitant of the replication process proper.

The fact that an inhibition of protein synthesis causes a failure in chiasma formation need not be interpreted to signify that the synthesis of a specific protein is essential to the process. One possibility is that the transient inhibition of protein synthesis causes premature desynapsis and thereby prevents an interchange between homologs. This possibility begs further study but, regardless of the outcome, the principal conclusion to be examined is that chiasmata are deliberately induced at the pachytene stage. Such induction requires at least the presence of an appropriate endonuclease to produce scissions in the DNA chain and a ligase-type enzyme system to restore the breaks. Recently, both types of enzymes have been identified in the meiotic cells (HOWELL, unpublished results, this laboratory). The identification *per se*, however, only establishes the presence of an adequate mechanism to form chiasmata, but it does not clarify the temporal regulation which is a significant feature of the process. Such regulation could be achieved by activating the endonuclease at the appropriate time, or by modifying the DNA at the termination of pachytene so as to make it susceptible to nuclease action, or perhaps, by a combination of both. Measurements of nuclease activity at various intervals of meiotic prophase do show a distinct and substantial increase at pachytene (HOWELL, unpublished results, this laboratory). To the extent that these measurements faithfully reflect the *in situ* behavior of the enzyme, the results can be interpreted as indicating that the temporal control of chiasma formation is regulated via endonuclease activity. However, even if correct, this interpretation should be regarded as no more than a pointer to the nature of the regulatory mechanisms governing chiasma formation during meiosis.

The data on DNA synthesis during pachytene provide additional material for interpreting the mechanisms underlying chiasma formation. The fact that such synthesis does occur removes any question concerning the *possibility* of chromatid interchange at this particular stage. Were DNA synthesis not demonstrable, strong doubts could be entertained on molecular grounds concerning the possibility of chiasma formation. A deeper question, however, is the nature of the relationship between the observed synthesis of DNA and the mechanism of chiasma formation. The answer to this is still unclear, but the evidence on hand does provide some pointers. Pachytene synthesis, like that at zygotene, appears to be related to the integrity of chromosome structure. Partial interruption of the synthesis leads to chromosome fragmentation. However, the data obtained from mitotic revertants and from DNA-DNA hybridization measurements preclude the interpretation that pachytene synthesis represents a delayed replication. Cells in premeiotic G-2 which are induced to revert to mitosis synthesize DNA characteristic of zygotene but not of pachytene. Moreover, pachytene-labeled DNA, unlike zygotene DNA, is hybridized to the same extent with DNA prepared from any stage of the meiotic cycle (Table 1). Thus, both types of analyses provide no evidence in favor of a delayed replication but, on the contrary, provide grounds for an interpretation which is consistent with repair activity including the excision and resynthesis of some DNA regions.

If pachytene synthesis largely reflects repair activity, the data pose a problem which, for the present at least, is difficult to resolve. The composition of pachytene DNA differs markedly from that of zygotene. Inasmuch as the DNA synthesized during zygotene is essential for pairing, it would seem reasonable to suppose that the paired regions in the chromosomes contain this DNA. If, however, pachytene synthesis reflects repair activity, such activity could not occur in the regions of DNA synthesized during the zygotene stage. If repair did occur in these regions then the respective compositions of pachytene and zygotene DNA would be similar. Thus, regardless of their relationships to pairing and chiasma formation, the two syntheses must occur in different regions of the chromosomes. A variety of speculative schemes to rationalize the apparent inconsistency is possible. One might suppose that the bulk of DNA synthesis occurring during pachytene is unrelated to chiasma formation and that the repair which does occur in the regions replicated at zygotene is obscured by those unrelated syntheses. Alternatively, one might conjure up a dynamic situation in which the DNA threads initially matched at zygotene move through the synaptinemal complex by unspooling at one end and spooling at the other. In this way, regions very different from the originally paired ones would become exposed to the actions of the scission and repair enzymes. Such speculations, however useful, cannot obviate the primary need to determine whether or not pachytene synthesis does in fact represent repair activity and whether zygotene replication occurs in exclusive pairing regions.

DISCUSSION

The information we have is much too fragmentary to permit drawing a complete picture of the mechanisms underlying chromosome pairing and crossing-over. Indeed, even the elementary question of whether genetic recombination is entirely based on chiasma formation is still debatable although the evidence in favor of identifying the two events as aspects of one process is becoming increasingly compelling (Mather, 1938; Peacock, 1968). The various studies reported in this article nevertheless lend themselves to two major considerations which should contribute to our understanding of meiosis.

The events which underlie the process of pairing and crossing-over are laid out in a temporal sequence (see Figure 3). This fact, even though obvious, has important implications for meiotic regulation. One generalization which may be made on the basis of the evidence is that recombination can be affected by disturbing the sequence of events without necessarily disturbing the events themselves. For example, precocious desynapsis of chromosomes could, if it occurs early enough relative to the activation of scission and repair mechanisms, result in achiasmatic cells. Less extreme effects could be imagined by supposing slight shifts in timing of either repair or pairing events. Whether such shifts do in fact occur is a matter of conjecture. The only evidence consistent with such shifts is that reported by SANDLER *et al.* (1969) concerning meiotic mutants in Drosophila.

The individual events in the process of pairing and crossing-over, quite apart from their temporal regulation, are presumably under genetic control. However, no evidence is yet available as to whether the proteins which are specifically involved in either pairing or crossing-over represent distinctive molecules which are unique to the meiotic cells. If the proteins also function in somatic tissues then mutants of these would not be manifest as specific meiotic mutants. Such meiotic mutants would have to be attributed to regulatory genes. Regardless, however, of the ultimate genetic mechanism, it is clear from the data presented that some of the component events are experimentally dissociable from one another and genetic lesions in any one of these events would lead to meiotic aberrations. Presence of a synaptinemal complex, for example, does not by itself assure the occurrence of crossing-over. Moreover, certain events in the meiotic cycle, such as pairing, may be partly determined by events which occur much earlier and which do not appear to affect intervening processes. The production of achiasmatics by perturbing cells in late G-2 or early leptotene is an example of such a relationship and may account for the observations of others that upsets in recombination may be induced either prior to meiotic prophase or during pachytene.

A second major consideration concerns the evidence that replication of distinctive chromosomal regions is required for pairing and that this replication is not directly a part of the recombination process. The question naturally arises as to the relationship between such replication and the mechanism of pairing. At present, any answer to this question is bound to be speculative, and what follows is therefore a selected piece of speculation.

In most, if not all, considerations of pairing specificity it is presumed that such specificity resides in base pairing between complementary DNA strands. No convincing alternative to this mechanism has yet been presented. If then, the primary mechanism of pairing is in a bonding between complementary strands each associated with different but homologous chromatids, the principal problem is to explain how DNA duplexes within a chromatid become separated and thus available for interaction with those of a homologous chromatid. Separation of complementary strands may be a key question, for if annealing of single strands can be accomplished *in vitro*, it is highly probable that it can also occur *in situ*. The problem of separation is essentially one of energetics. The total energy of hydrogen bonding between successive base pairs of two long DNA chains is considerable. Stabilization is probably increased by the various protein molecules present in the chromosome. To devise a scheme whereby complementary strands in the chromatid of a fully replicated chromosome become separated is conceptually difficult. This difficulty is obviated if comparatively short intercalary regions of the chromosome remain unreplicated prior to pairing. In these regions each chromatid would contain a single strand of DNA which probably, though not necessarily, would form a duplex with the strand of the sister chromatid. Whether or not the duplex is present, the total bonding energy maintaining the duplexes between sister chromatids of the intercalary regions is very small compared to the energy maintaining the DNA duplexes within the chromatid. Initiation of replication of the intercalary regions would further reduce the forces holding sister chromatid DNA strands. Whether "breathing" of duplexes in the intercalary regions or separation due to intiation of replication make available single DNA strands, either mechanism would be sufficient to permit duplex formation between DNA strands of homologous chromatids following what may be chance collision.

To make such chance collisions effective initiators of chromosome pairing require a number of regulatory devices. Obviously, if pairing depended upon chance collision immediately following DNA replication of the intercalary regions, most regions would probably complete replication prior to an appropriate collision. It would appear reasonable to suppose that some mechanism prohibits extensive, if any, replication of the intercalary DNA prior to pairing between homologous DNA strands. Such a mechanism would account for the extended period (about 2 days) which appears to be required for zygotene DNA synthesis. However, it is also apparent that with progression of replication, conditions favorable to duplex formation between homologous strands of DNA are removed. Unless the initial pairing is stabilized by some mechanism other than hydrogen bonding, synapsis would not persist. The coupling of DNA replication with a synthesis of protein associated with the synaptinemal complex could be regarded as the mechanism for stabilization.

Although some consideration might be given to the possibility of a prealignment of homologous chromosomes at a specific point prior to zygotene, the evidence for such a mechanism is even less compelling than that for the speculative scheme just described. One major conclusion to be drawn from the scheme is that pairing of homologous chromosomes requires DNA replication, whereas crossing-over requires DNA repair and is temporally separated from the process of pairing. The conclusion is, of course, as speculative as the scheme itself, and the validity of either can only be established by future experiments.

SUMMARY

One characteristic property of meiotic cells relates to the timing of DNA replication. During premeiotic S-phase, most but not all of the DNA undergoes replication. A small but significant amount remains unreplicated until the zygotene stage. At that time, replication of that DNA and chromosome pairing begin simultaneously. If replication is inhibited, pairing does not progress and a synaptinemal complex is not formed. This DNA which accounts for about 0.3% of the total has a distinctive base composition which is virtually the same in several species tested. Completion of DNA synthesis and of pairing during the zygotene stage does not appear to involve crossing-over. It is possible, by suitable use of protein inhibitors, to prevent chiasma formation if cells are treated at the end of zygotene or in early pachytene. The picture which thus emerges is one in which late replication in certain chromosome regions is functionally associated with the pairing process but that crossing-over is a deliberate event which is normally set in motion after the completion of pairing.

LITERATURE CITED

Davies, D. R., and C. W. Lawrence, 1967 The mechanism of recombination in *Chlamydomonas reinhardii*. II. The influence of inhibitors of DNA synthesis on intergenic recombination. Mutation Res. **4:** 147–154.

Henderson, S. A., 1966 Time of chiasma formation in relation to the time of deoxyribonucleic acid synthesis. Nature **211:** 1043–1047.

Hotta, Y., M. Ito, and H. Stern, 1966 Synthesis of DNA during meiosis. Proc. Natl. Acad. Sci. U.S. **56:** 1184–1191.

Hotta, Y., L. G. Parchman, and H. Stern, 1968 Protein synthesis during meiosis. Proc. Natl. Acad. Sci. U.S. **60:** 575–582.

Ito, M., Y. Hotta, and H. Stern, 1967 Studies of meiosis *in vitro* II. Effcet of inhibiting DNA synthesis during meiotic prophase on chromosome structure and behavior. Develop. Biol. **16:** 54–77.

Lawrence, C. W., and D. R. Davies, 1967 The mechanism of recombination in *Chlamydomonas reinhardii*. I. The influence of inhibitors of protein synthesis on intergenic recombination. Mutation Res. **4:** 137–146.

Mather, K., 1938 Crossing-over. Biol. Rev. Cambridge Phil. Soc. **13:** 252–292.

Moses, M. J., and J. R. Coleman, 1964 Structural patterns and the functional organization of chromosomes. *The Role of Chromosomes in Development,* Edited by M. Locke, Academic Press, New York, pp. 11–50.

Parchman, L. G., and H. Stern, 1969 The inhibition of protein synthesis in meiotic cells and its effect on chromosome behavior. Chromosoma (in press).

Peacock, W. J., 1968 Chiasmata and crossing over. *Proceeding of Replication and Recombination of Genetic Material,* Edited by W. J. Peacock and J. R. D. Brock, Australian Academy of Science, Canberra.

Roth, T. F., and M. Ito, 1967 DNA dependent formation of the synaptinemal complex at meiotic prophase. J. Cell Biol. **35:** 247–255.

Sandler, L., D. L. Lindsley, B. Nicoletti, and G. Trippa, 1969 Mutants affecting meiosis in natural populations of Drosophila melanogaster. Genetics **60:** 525–558.

Stern, H., and Y. Hotta, 1967 Chromosome behavior during development of meiotic tissue. *The Control of Nuclear Activity,* Edited by L. Goldstein, Prentice Hall, Englewood Cliffs, N.J., pp. 47–76.

Analysis of DNA Synthesis during Meiotic Prophase in *Lilium*

YASUO HOTTA AND HERBERT STERN

1. Introduction

Two established facts about DNA metabolism in meiotic cells of *Lilium* are that most, if not all, of nuclear DNA replication occurs during the premeiotic S-phase, and that a small but significant amount of DNA synthesis occurs during two intervals of meiotic prophase, zygotene and pachytene (Hotta, Ito & Stern, 1966). Since chromosome pairing occurs during zygotene and since crossing over very probably occurs during pachytene, we have re-examined the nature of prophase DNA synthesis in order to determine whether such synthesis is functionally related to these two cytologically defined meiotic events.

The studies reported are based on the male meiotic cells (microsporocytes) of several pure species and hybrids of *Lilium*. The usefulness of these plants for a combined biochemical and cytological study of meiosis has been amply discussed elsewhere (Ito & Stern, 1967). The natural synchrony of the meiotic cells, the relatively long duration of the individual prophase stages (2 to 5 days), and the ability of the cells to progress through meiosis under *in vitro* conditions, have made it possible to analyze each interval of DNA synthesis separately. The principal results of these analyses point to the conclusion that a distinctive component of nuclear DNA does not replicate during the premeiotic S-phase but does so at the zygotene stage and that the synthesis of DNA during pachytene is of the repair replication type.

2. Materials and Methods

Lily bulbs were purchased from Oregon Bulb Farms (Gresham, Oregon) and grown as described previously (Ito & Stern, 1967). Unless indicated otherwise, the hybrid varieties

Bright Star and Cinnabar were used. Meiotic cells were explanted into culture media as described by Ito & Stern (1967). Those from Bright Star were cultured at 15°C and all others were cultured at 19°C. Isotopically labeled compounds were added to growth media after preculturing the cells for at least 12 hr. Although DNA metabolism appears to be unaffected by explantation, this procedure was nevertheless adopted as a precaution because of our observation that RNA metabolism is erratic during the first 6 to 12 hr following explantation.

(a) *Processing of DNA*

The methods used for the isolation of nuclei and the preparation of DNA were those described previously (Hotta, Bassell & Stern, 1965). Native DNA was denatured either by adjusting solutions to pH 12 with NaOH or by heating in a boiling water bath followed by quick cooling. The exposure time to denaturing conditions was 20 min. Lily DNA was totally denatured under either of these conditions. For centrifugation, denatured samples were placed directly into solutions of cesium chloride (pH 12). Samples for hybridization analysis were treated in the following way: 0·05 M-Tris buffer (pH 8) was added to the relatively concentrated alkaline DNA sample such that the buffer concentration would be 0·01 M on final dilution. The mixture of DNA solution and buffer were titrated with 1 N-HCl to pH 8·0 using a glass electrode, and then diluted to the desired volume with water.

A swinging bucket (Spinco no. SW65) or anglehead (Spinco no. 40) rotor was used for centrifugation of DNA in solutions of CsCl. With the swinging bucket, samples were centrifuged at 35,000 rev./min for 72 hr; with the anglehead, they were centrifuged at 40,000 rev./min for 60 hr. All centrifugations were carried out at 23°C. Fractions were collected dropwise from the bottom of each centrifuge tube. The refractive index of selected fractions was measured to determine the concentration of CsCl along the gradient. Buoyant densities assigned to DNA preparations have been derived from such refractometric measurements.

(b) *Hybridization analysis*

Alkaline denatured samples of DNA, dissolved in 4 × SSC (SSC is 0·15 M-NaCl, 0·015 M-sodium citrate) were passed through 27 mm nitrocellulose filters (Denhart, 1966; Gillespie & Spiegelman, 1965) in amounts up to 100 μg. All of the DNA was retained by the filters which were dried at room temperature for 2 hr followed by 6 hr at 80°C *in vacuo*. The filters were then immersed for 4·5 hr at 65°C in a solution of bovine serum albumin (Armour, fraction V) at 400 μg/ml., 0·01 M-Tris buffer (pH 7·3), 0·5 M-KCl, and 0·01 M-$MgCl_2$. Approximately 2 ml. of this solution were used/filter. Following the incubation period, the excess fluid was drained off and the filters dried in a vacuum oven at 60°C.

DNA samples to be tested against those fixed on the filters were dissolved in 1 × SSC or 0·1 M-NaCl, 0·001 M-EDTA, 0·3 M-Tris (pH 7·3) and fragmented either by sonication for 30 sec with a Bronson Sonifier cell disruptor (model 140D, setting no. 4) or by rapidly passing the solution 6 times through a 27 gauge syringe needle. The fragmented DNA was denatured by alkali or heat as described above. Portions of the denatured DNA solutions were diluted to the desired concentration of DNA. These diluted solutions contained 4 × SSC and were used as such in the hybridization test. Each filter was incubated with approximately 1·0 ml. of the solution at 65°C by gently shaking overnight or for other intervals of time as indicated in the text. Following incubation, the filters were washed quickly with a solution containing 0·01 M-Tris buffer, 4 × SSC, and 0·01 M-$MgCl_2$, adjusted to pH 9·0. The filters were then dried under an infrared lamp and placed in 5 ml. of a toluene–PPO–POPOP mixture (Spectrofluor, Amersham-Searle) for scintillation counting.

(c) *Dissociation of hybridized DNA*

To obtain melting profiles of the artificially formed hybrids, each of the washed filters was immersed in 1 ml. of 1 × SSC solution containing 0·005 M-Tris buffer (pH 7·4) and 0·001 M-$MgCl_2$. Filters thus immersed were heated to various temperatures for 20 min and then quickly chilled. The filters were transferred from the chilled solutions, washed with 2 ml. of SSC, dried and put into scintillation fluid for counting. The original suspension medium and the wash solution were combined, a small amount of carrier DNA added, and trichloroacetic acid added to a final concentration of 5% to precipitate the DNA. The precipitates were collected on glass filter papers which were placed in scintillation fluid after drying. In some cases the counts present in the original suspension medium were determined directly by adding 10 ml. of Bray's dioxan-type scintillation solution. No significant differences were found between the two methods of analysis.

(d) *Radioisotopes*

Carrier-free [^{32}P]orthophosphate, [^{3}H]thymidine (spec. act., 30 c/m-mole) were purchased from New England Nuclear. Unless otherwise indicated, the scintillation fluid used was Spectrofluor purchased from Amersham-Searle. Counting efficiency was approximately 40% for tritium and 100% for [^{32}P]phosphate. All data are reported in terms of counts obtained and are not corrected for the differences in efficiency.

3. Results

(a) *Analysis of zygotene DNA*

(i) *Composition*

In previous studies of meiotic cells from *Lilium longiflorum* and *Trillium erectum*, we reported that DNA synthesized during zygotene was intranuclear and had a base composition distinct from bulk DNA (Hotta *et al.*, 1966). We have now analyzed several other pure species and hybrids of *Lilium* and, as shown in Table 1, all of them display a similar pattern of DNA synthesis during the zygotene stage. The DNA synthesized can be resolved by isopycnic centrifugation into one to three density peaks (see Fig. 3(a)). The major peak is in the region of highest density.

Density profiles of the DNA were obtained by culturing zygotene cells in the presence of a radioactive precursor, centrifuging the extracted DNA to equilibrium in a CsCl gradient, and determining the number of counts in successive fractions obtained by collecting drops from the bottom of the centrifuge tube. The amount of label at each of the peaks and the number of peaks present varied somewhat from experiment to experiment, but the positions of the peaks were constant and characteristic for each species. All peaks were labeled if cells were exposed to isotope over the entire zygotene interval, a period of two to three days. If cells were exposed for only one day, preferential labeling of the major peak was usually observed. Since the different phases of the zygotene stage (early, mid or late) cannot be reliably distinguished by cytological techniques, and since the cells from even a single flower bud are most probably heterogeneous with respect to the substages of zygotene, no attempt was made to determine whether the DNA peaks were labeled in a temporal sequence. Regardless, however, of the duration of the labeling period, analysis of DNA digests showed that the DNA synthesized during zygotene in all varieties tested had G + C contents of 48 to 50% (Table 1). These values contrast with bulk nuclear DNA which has a G + C content of approximately 40%.

These results make it evident that a distinctive population of DNA molecules is replicated during the zygotene of meiosis and that such replication is a constant characteristic of all species thus far examined. Although careful measurements of the amount of DNA synthesized at zygotene have not been made for all of the different species analyzed, the similarity in specific activity of such DNA between species labeled under similar conditions suggests that the amounts made are similar, namely, about 0·3% of the genome (Hotta *et al.*, 1966). The high G + C content of zygotene DNA might suggest that it is the DNA that is homologous with ribosomal RNA especially in view of the extensive evidence that rDNA† synthesis occurs in various

† Abbreviation used: rDNA, DNA homologous to ribosomal RNA.

TABLE 1

Composition of bulk and zygotene DNA in liliaceous plants

Plant	Bulk DNA Density (g/ml.)	Zygotene DNA Density (g/ml.)	%G + C
T. erectum	1·705	1·713	—
L. longiflorum	1·702	1·712	49·8
L. speciosum	1·700	1·713	49·5
L. tigrinum	1·701	1·710	—
L. Cinnabar	1·701	1·712	48·9
L. Bright Star	1·701	1·711	47·8
L. Limelight	1·702	1·713	—

DNA was prepared from isolated microsporocyte nuclei after culturing meiotic cells in presence of [^{32}P]phosphate or [^{3}H]thymidine through the zygotene stage. The preparations were centrifuged to equilibrium in solutions of CsCl and fractions collected as described under Materials and Methods. Each of the densities listed under Bulk DNA represents the density of the CsCl solution in the fraction present at the peak position of the optical density profile. Identical values were obtained when somatic nuclei were used as a source of DNA. The buoyant densities of zygotene DNA were obtained from fractions at the peak position of the radioactivity profile. Usually, as indicated in Fig. 4, several peaks were evident in the zygotene-labeled material. The values listed represent in all cases those of the heaviest and major radioactivity peak. The column, % G + C, represents chemically determined values. Total DNA from ^{32}P-labeled zygotene cells was digested with DNase and snake venom phosphodiesterase and the individual deoxynucleotides were resolved by paper electrophoresis for radioactive counting (Bendich, 1957). The chemical analysis thus included the several fractions of zygotene DNA and the G + C values are therefore lower than what would be expected from the density position of the major peak alone.

oocytes during the pachytene interval (Gall, 1969; Brown & Dawid, 1968). This, however, is not the case. In connection with studies to be published separately, we found rDNA to have a higher G + C content (55%) and a greater buoyant density than zygotene DNA. No radioactivity was incorporated into rDNA during meiotic prophase.

(b) *The nature of zygotene DNA replication*

The DNA synthesized during the zygotene of meiosis is also present in somatic cells (Hotta *et al.*, 1966). Since the meiotic cells are derived directly from adult somatic cells, we sought to determine whether the zygotene synthesis is an additional synthesis in a fully replicated genome or a delayed synthesis of an incompletely replicated genome. In the first set of experiments, DNA–DNA hybridization techniques were used to determine the amount of DNA per genome at different stages of meiosis which was homologous with DNA synthesized during zygotene. If *extra* DNA is synthesized during zygotene then prezygotene cells should contain as much of the zygotene DNA as somatic cells but post-zygotene cells should contain a larger amount. If the extra DNA were unstable, the excess could disappear. This possibility has been excluded at least for the duration of the meiotic cycle by the demonstration that DNA labeled during zygotene retains the label after completion of meiosis (Hotta *et al.*, 1966). If, on the other hand, synthesis of zygotene DNA represents a *delayed* replication, then prezygotene cells should contain only half as much as somatic cells whereas post-zygotene cells should contain the same amount as somatic cells.

DNA synthesized exclusively during the premeiotic S-phase, zygotene or pachytene intervals, was labeled by conventional procedures and the total DNA purified from isolated nuclei. Labeled DNA was also prepared from somatic cells for use as a control. Specific activities were determined for each of the unfractionated DNA preparations. Each of these preparations was hybridized with fixed amounts of unlabeled DNA prepared from nuclei or somatic cells and from nuclei at different stages of the meiotic cycle. The amount of radioactivity in the hybrid was converted into μg of test DNA by calculating from the previously determined specific activity. The weight ratio of test to fixed DNA was expressed as % labeled:fixed DNA. The conversion used to calculate the weight of test DNA hybridized is of course valid only if the specific activity of the test DNA in the hybrid is the same as that of the original DNA. That this is probably so in most cases is indicated by the fact that under saturation conditions the ratio of test to fixed DNA approaches the theoretical value of 1.

The results of a series of hybridization experiments are shown in Figure 1. The one striking feature of these results are the values for hybrids between early meiotic DNA and DNA labeled during the zygotene stage. The ratios of test to fixed DNA in these hybrids is approximately 50%. Such low values are not found if the fixed DNA is prepared from post-zygotene meiotic cells or from somatic cells, nor are they found if the test sample is from DNA labeled at times other than zygotene. The simplest explanation of the data is that the fixed DNA prepared from prepachytene cells has half the number of zygotene sequences per genome as does DNA prepared from late meiotic stages or from somatic cells. Since cells exposed to isotope during the zygotene stage label these sequences exclusively, the calculated 50% ratio for prepachytene stages reflects only the amount of labeled zygotene DNA hybridized. Presumably, the specific activity of the test DNA in such hybrids is approximately 50% of that in the original zygotene DNA preparation.

Although not shown in Figure 1, data were also obtained using fixed somatic DNA.

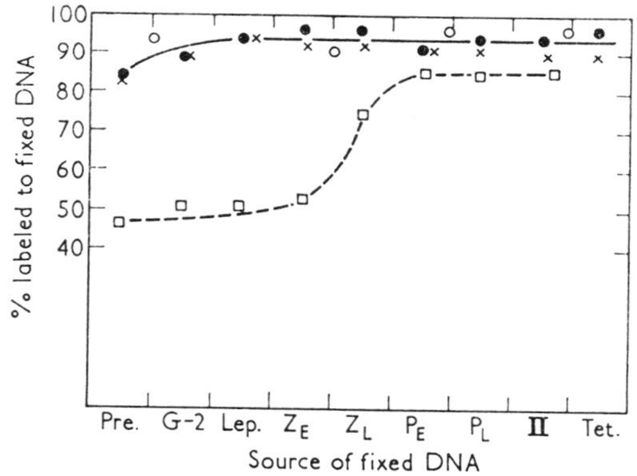

FIG. 1. Hybridization of ^{32}P-labeled DNA from somatic (—●—●—), premeiotic (—○—○—), zygotene (--□--□--), and pachytene (—×—×—) cells to DNA prepared from cells at successive stages of the meiotic cycle.

The stages are indicated on the abscissa. The earliest premeiotic stage (Pre.) used was late S-phase. Leptotene (Lep.), zygotene (Z), and pachytene (P) stages were identified cytologically. The subscripts E and L signify early and late, respectively. Cells at stages later than pachytene are grouped under II except for those which have completed meiosis and are designated as tetrads (Tet.). The conditions used for hybridization are described in the text. The values plotted are those obtained under saturation conditions which were determined for each of the stages shown on the abscissa.

Hybrids between somatic DNA and DNA labeled during the zygotene stage showed a calculated ratio of test to fixed DNA which was the same as that obtained with fixed DNA from post-zygotene meiotic cells. Thus, DNA sequences which are labeled during the zygotene stage appear to be present in approximately the same proportion in the genome of late meiotic cells as in the genome of somatic cells. It is therefore unlikely that the zygotene synthesis represents an extra round of DNA replication. On the contrary, since zygotene, DNA sequences are present to about half the extent in the genome of prezygotene cells, the synthesis observed at zygotene most probably represents the completion of a delayed replication.

Since the data on hybridization of zygotene DNA are of critical importance to interpretations of events in meiotic prophase, some of the experiments were repeated using purified zygotene DNA which was obtained by pooling fractions within the density range of 1·707 to 1·717 g/ml. from CsCl gradients. Since zygotene DNA sequences represent only about 0·3% of the total DNA, a sufficient amount had to be collected gradually from a large number of gradients. The pooled material was recycled three times through a CsCl gradient. The degree of purification achieved was undetermined but it was clearly sufficient to test for hybridization of zygotene-labeled DNA in the absence of large excesses of non-homologous DNA. Under the conditions of the test, 8 μg of enriched zygotene DNA in 0·5 ml. of medium was sufficient to saturate the zygotene sequences in 80 μg of fixed DNA. The results, shown in Table 2, confirm the

TABLE 2

Hybridization of purified zygotene DNA to DNA from different stages in the meiotic cycle

Source of fixed DNA	Radioactivity in hybrids (cts/min)	
	Expt 1	Expt 2
Premeiotic	105	111
Leptotene–early zygotene	126	121
Pachytene	252	287

Individual filters contained 80 μg of cold DNA prepared as described under Materials and Methods. Each filter was incubated with 8 μg of purified zygotene ^{3}H-labeled DNA (1000 cts/min/μg) in 0·5 ml. of 4 × SSC. The purification is described in the text. The values shown are for single experiments. One observation of incidental interest is the absolute amount of zygotene DNA bound to the DNA isolated from pachytene cells. Using an average value from the 2 experiments, the amount is 0·27 μg which is approximately 0·34% of the DNA on the filter. A similar value was obtained from calculations based on the incorporation of precursor during the zygotene stage (Hotta *et al.*, 1966).

data in Figure 1, namely, that DNA from post-zygotene cells contains about twice as many zygotene sequences as does DNA from prezygotene cells. An incidental result of this experiment is the calculation showing that 0·34% of fixed DNA on the filter is hybridized with zygotene DNA (see legend to Table 2). This value is very close to the 0·3% cited from an earlier study and thus indicates a reasonable degree of purification of the DNA used in this analysis (Hotta *et al.*, 1966).

Temperature melting profiles of renatured DNA were compared with those of native preparations in order to determine the extent to which the artificially produced hybrids were a measure of native complementarity (Church, Luther & McCarthy,

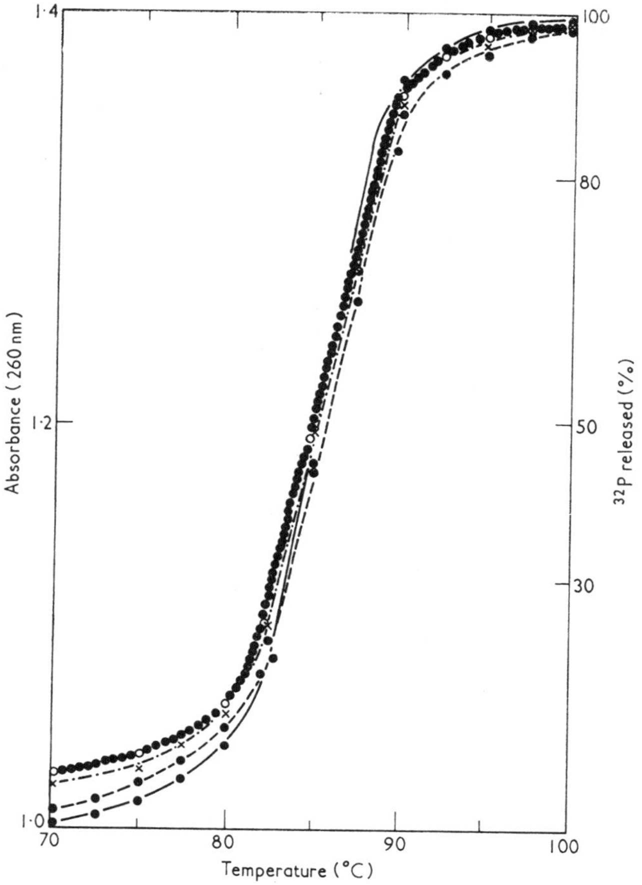

FIG. 2. Temperature denaturation curves of artificially formed DNA–DNA hybrids. These were prepared as described for the tests under Fig. 1. For comparison, the melting profile of a native DNA duplex is shown as a plot of temperature against absorbancy at 260 nm (—●—●—). The melting profiles of the artificially formed hybrids are plotted according to the percentage of ^{32}P radioactivity released. The procedure for dissociating the hybrids is described under Materials and Methods. ○●●●●○●●●○, S-phase; —·×—·—×·—, pachytene; --●--●--, zygotene.

1969). Radioactively labeled DNA was hybridized under saturation conditions to cold DNA fixed on filters, and the melting profiles of the hybrids were determined as described under Materials and Methods. The melting curves thus obtained are illustrated in Figure 2. The slopes of these curves clearly indicate hydrogen bonding in the artificially formed duplexes is not random. If compared with the melting profile of native DNA as determined by the hyperchromic effect, the complementarity in the artificial hybrids is slightly less but nevertheless approximates that present in the native molecule. A puzzling and still unexplained feature of the melting curves in Figure 2 is the small difference in the mean melting temperature (T_m) between total and zygotene-labeled DNA. Based on composition, a difference of 4 to 5 deg. C would

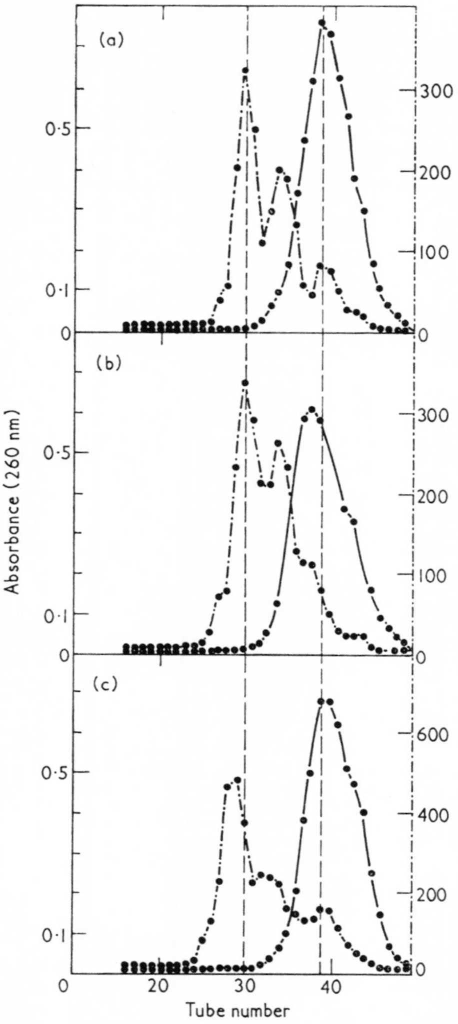

FIG. 3. Effect of BdUrd on buoyant density of DNA from cells exposed to density analog at either S-phase or zygotene stage.

The distribution profile of bulk DNA is indicated by absorbance at 260 nm and that of zygotene DNA by radioactivity. The vertical dashed lines represent density positions of non-BdUrd labeled DNA in the CsCl gradient. The line towards the right marks a density of 1·701 g/ml. which is the peak position of bulk DNA from the varieties Cinnabar and Bright Star (Table 1). The line towards the left marks a density of 1·712 g/ml. which is the peak position of zygotene DNA from Cinnabar, the variety used in these experiments. For density labeling, cells were exposed to 4×10^{-3}M-BdUrd. For radioactive labeling [^{3}H]deoxyadenosine (5 μc/ml.) was used. The BdUrd was chased by exposing cells to 10^{-2}M-thymidine. The reasons for the small density shift are discussed in the text.

(a) Cells explanted into culture medium at S-phase and 5 days later radioactive deoxyadenosine + cold thymidine were added to medium. Cells were harvested after 10 days of culture.

(b) Explanted cells exposed to BdUrd during first 3 days of culture to label DNA replicating during S-phase. Two days later the cells were exposed to tritiated deoxyadenosine and cold thymidine until they were harvested at the end of 10 days. Under these conditions, bulk DNA becomes density labeled.

(c) Explanted cells exposed to BdUrd + tritiated deoxyadenosine after being cultured for 5 days. Cells were harvested at the end of 10 days. Under these conditions, zygotene DNA becomes density labeled. —●—●—, O.D.; — · ●— · —●— ·, radioactivity.

be expected. Although we have consistently found in four tests that the T_m of zygotene-labeled artificial hybrids is 1 to 2 deg. C higher than the T_m of totally labeled DNA, we have in no case found a difference of the order of 5 deg. C. Despite this feature of the data, we have included the figure inasmuch as it does demonstrate that the artificial hybrids behave similarly to native ones at least with respect to melting properties.

The evidence for a delayed replication of zygotene DNA was further tested by

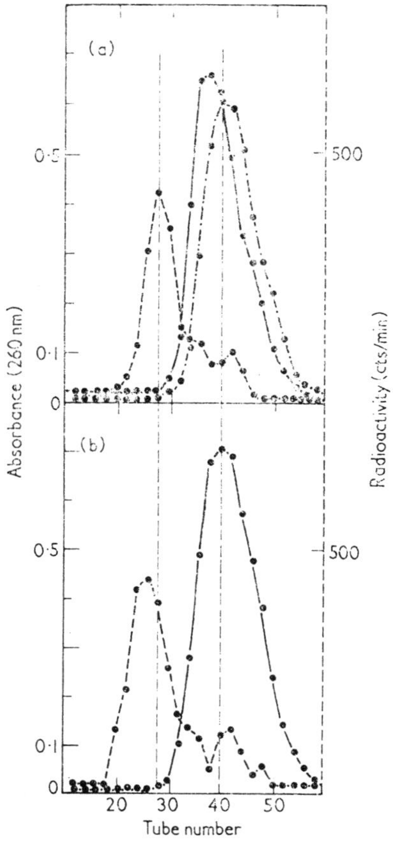

FIG. 4. Hybridization of ^{32}P-labeled zygotene DNA to successive fractions of DNA centrifuged to equilibrium on a CsCl gradient. Dashed vertical lines have same significance as in Fig. 3. Incubation procedure was the same as described under Fig. 3.

(a) Cells exposed to BdUrd during S-phase. The broken curve to the right is S-phase labeled DNA which was prepared separately and added as a marker. The optical density profile of the bulk DNA is shifted toward a higher density due to BdUrd incorporation. The hybridization profile (dashed curve to the left), however, peaks in the usual position indicating an absence of BdUrd incorporation into zygotene DNA during S-phase.

—●—●—, O.D.; -●--●-, hybridization profile; – · ●– · –●– ·, radioactivity.

(b) Cells exposed to BdUrd during zygotene stage. Hybridization profile shifted toward denser region due to BdUrd incorporation into zygotene DNA. Bulk DNA, as expected, is unaffected by BdUrd. The absence of a shift in the minor peak at tube no. 42 is unexplained.

—●—●—, O.D.; -●--●-, hybridization profile.

density labeling using BdUrd as a density analog. If zygotene DNA does not replicate during the premeiotic S-phase then exposure of cells at S-phase to BdUrd should cause a shift in the density of bulk but not of zygotene DNA. The respective densities of bulk and zygotene DNA were determined by isopycnic centrifugation. The position of zygotene DNA in the CsCl gradient was identified either by differential labeling or by hybridization with labeled zygotene DNA.

The usefulness of the technique for this particular set of experiments is however limited by the low degree of BdUrd substitution for thymidine in the DNA. The magnitude of the density shift is of the order of 0·002 g/ml. as evident from Figures 3 and 4. Most probably, the thymidine pool which is present during premeiotic S-phase and the zygotene–pachytene stages (Foster & Stern, 1958) accounts for the low level of BdUrd substitution, although other factors governing BdUrd metabolism cannot be excluded. Despite the low level of BdUrd incorporation, the density shifts observed are consistent and reproducible.

After exposing meiotic cells at the S-phase to BdUrd for one to three days, the precursor was removed and the cells cultured for an additional five to ten days in the presence of cold thymidine and [^{3}H]deoxyadenosine. The thymidine was used to swamp the BdUrd pool in the cells and the deoxyadenosine to label the zygotene DNA. The various gradient profiles of bulk and zygotene DNA shown in Figure 3 are drawn from cells which were either cultured without BdUrd, exposed to BdUrd only during the S-period, or exposed to BdUrd only during the zygotene period. The results confirm expectations. Bulk, but not zygotene, DNA shows a shift if BdUrd is present in the medium only during the S-phase whereas only zygotene DNA undergoes a shift if BdUrd is present during the zygotene interval.

Similar evidence was obtained by hybridizing successive fractions of BdUrd-DNA from a CsCl gradient with labeled zygotene DNA. The results of an experiment in which cells were exposed to BdUrd either during the premeiotic S-phase or during the zygotene stage are shown in Figure 4. A shift in density of bulk DNA occurs only in cells exposed to BdUrd during S-phase whereas the density of zygotene DNA is shifted only for cells exposed to BdUrd during the zygotene stage. The results of these and preceding experiments are consistent with the conclusion that synthesis of DNA during the zygotene stage of meiosis represents a delayed replication.

(c) *Relationship of zygotene DNA to chromosome reproduction*

The term "zygotene DNA" has been used in this paper as a matter of convenience. The DNA itself is not unique to meiotic cells. Its presence in all somatic tissues argues that it, like the bulk of the DNA, must be replicated during each cell cycle. Whether its synthesis in mitotic systems is also delayed is yet to be determined although some reports on late replicating DNA in mitotic cells are consistent with such possibility (Guttes & Guttes, 1969).

Evidence on this point was sought in premeiotic cells since under certain conditions of *in vitro* culture they may be induced to undergo a mitotic division (Stern & Hotta, 1969). The nature of these conditions is not fully understood but it is known that the course which premeiotic cells follow when explanted into standard culture media depends upon the developmental stage of the cells at the time of explantation. Cells at or close to the S-phase at the time of explantation usually undergo mitosis in culture. Those explanted during the premeiotic G-2 interval undergo meiosis but many of the cells are achiasmatic. The aim of this experiment was to determine whether the mitotic cells synthesized zygotene DNA at the conclusion of their S-phase.

In the absence of cytological criteria to distinguish between the different stages of premeiosis, length of flower bud is the most useful index of development (Erickson, 1948). There is considerable variation in the behavior of cultured meiotic cells which are explanted at particular premeiotic bud lengths. Cells from a single flower bud, however, usually behave with sufficient uniformity so that enough cultures can be obtained in which few if any meiotic cells are present. To conduct the experiment a relatively large number of flower buds was used. Premeiotic cells from individual flower buds were explanted into test tubes containing 1 ml. of culture medium.

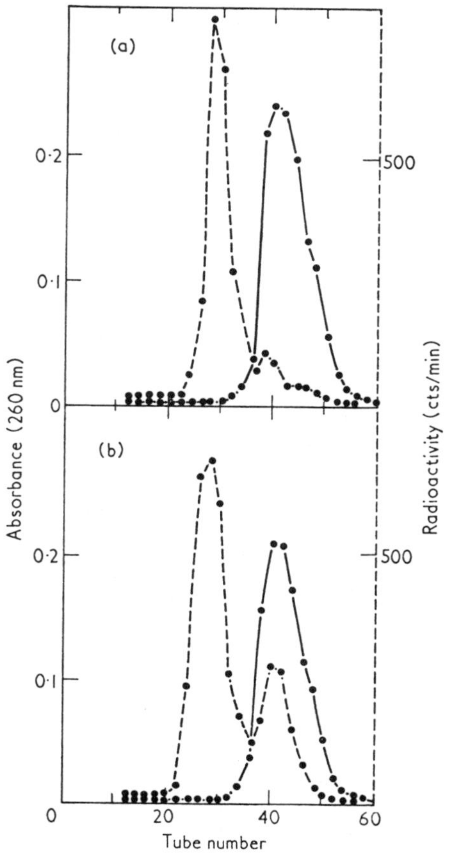

FIG. 5. Optical density and radioactivity profiles of DNA centrifuged to equilibrium in a CsCl gradient. The DNA was prepared from cells explanted into culture media to which 20 μc/ml. of [^{32}P]phosphate were added on the third day of culture.

(a) Cells which later entered an achiasmatic meiosis.

(b) Cells which reverted to a mitotic division.

Zygotene DNA is labeled in both cases as shown by the radioactivity profile which is typically that of zygotene DNA.

—●——●—, O.D.; --●--●--, radioactivity.

Isotope was added to the medium after three days of culture since S-phase-type synthesis rarely occurs in cultured premeiotic cells after this interval of time. Samples of cells were removed from the cultures one to three weeks later for cytological analysis. For this experiment only those cultures were retained in which the cells were uniformly meiotic or mitotic. These were pooled into separate groups and DNA prepared from each. The preparations were centrifuged in CsCl solutions and the sedimentation profiles thus obtained are illustrated in Figure 5. Zygotene DNA was synthesized in each of the cultures. The mitotic cells (Fig. 5(b)) also show some residual S-phase synthesis. Thus, premeiotic cells which are induced to revert to mitosis replicate zygotene DNA either in late S-phase or during the G-2 interval. That such synthesis is essential to the occurrence of mitosis has been indicated by experiments with similar groups of cells in which inhibitors of DNA synthesis (hydroxyurea, deoxyadenosine, 5-aminouracil) were added after three days of culture. In no case were divisions observed during the subsequent two weeks of culture.

In summary, these results indicate (1) that zygotene DNA synthesis is distinct from premeiotic S-phase synthesis in G + C content, and (2) that it represents completion of the normal semi-conservative replication of the genome. Meiotic cells thus differ from mitotic cells in that a particular fraction of DNA which may normally be "late labeling" is not replicated during the regular S-phase but remains unreplicated until the zygotene stage is reached.

(d) *Characteristics of synthesis during the pachytene stage*

Patterns of DNA synthesis during pachytene are variable both within and between species. Examples of the variations found are illustrated in Figure 7. In general, constant and distinctive peaks of radioactivity are not found. Although labeled DNA is usually present in the light region of the gradient, most of the DNA synthesized overlaps with bulk DNA. Pachytene synthesis when examined in a number of cell preparations appears more or less randomly distributed with respect to the density profiles of bulk DNA and thus contrasts with zygotene synthesis which occurs exclusively and constantly in the heavy region of the gradient. Since the pachytene interval extends over three to five days, it has been possible to compare early and late labeling patterns but no evidence has been obtained for any sequential pattern of synthesis. The specific activity of total DNA labeled during pachytene is usually considerably lower than that labeled during zygotene. Based on the amount of radioactivity incorporated at each stage, pachytene synthesis may be estimated as being one-third to one-fifth of the DNA synthesized at zygotene. The smaller isotope incorporation at pachytene cannot be explained by pool differences. Pachytene and zygotene cells have been labeled with either [^{32}P]phosphate, [^{3}H]thymidine or [^{3}H]BdUrd, but the relative amounts of incorporation are independent of the type of precursor used (Table 3). Differences in isotope uptake can also be excluded as a factor responsible for the lower level of incorporation since no significant differences in radioactivity of the acid-soluble pools were found between zygotene and pachytene cells (e.g. $1{\cdot}3 \times 10^{6}$ cts/min and $1{\cdot}25 \times 10^{6}$ cts/min of [^{3}H]thymidine at zygotene and pachytene, respectively).

The apparently dispersed distribution of DNA synthesis during pachytene within the genome was further examined using BdUrd as a precursor. If high molecular weight populations of DNA molecules were being replicated, incorporation of BdUrd should cause a shift of the replicated fragments toward the heavier end of the gradient.

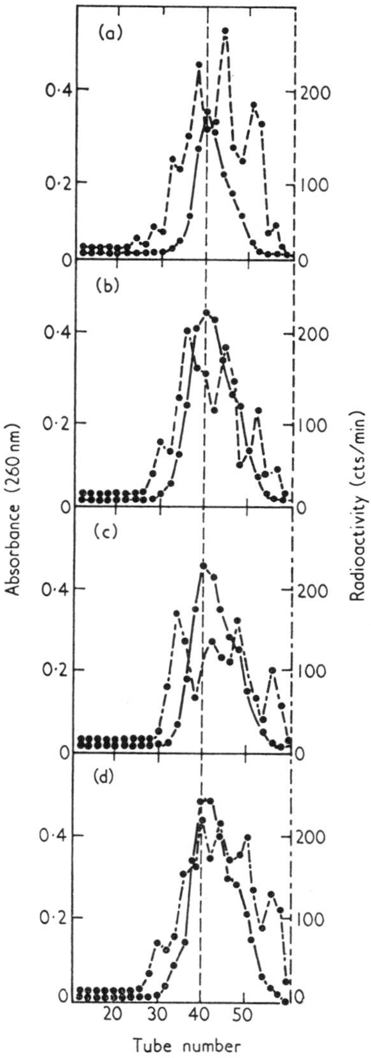

FIG. 6. Variations in pattern of DNA labeling in cells at the pachytene stage.

Extracted DNA was centrifuged to equilibrium on a CsCl gradient. Dashed vertical line marks density of 1·701 g/ml. Each set of curves was obtained from the cells of a single bud. The cells were cultured for 2 to 4 days from early through late pachytene. The isotopes used were: in (a) ^{32}P, 20 μc/ml.; in (b) ^{32}P, 10 μc/ml.; in (c) [^{3}H]thymidine, 10 μc/ml.; in (d) [^{3}H]thymidine, 10 μc/ml. —●——●—, O.D.; -●--●- and -·●-·-●·-, radioactivity in (a), (b) and (c), (d), respectively.

To test the efficiency of BdUrd incorporation, the meiotic cells were exposed to [^{3}H]BdUrd at zygotene and pachytene stages. The amounts incorporated per μg DNA are shown in Table 3. These data indicate that BdUrd substitutes for thymidine with the same efficiency at pachytene and zygotene. BdUrd-DNA prepared from pachytene cells, however, behaves differently from that of zygotene cells on CsCl

Table 3

Comparison of DNA precursor incorporation into cells at zygotene and pachytene

Stage	[^{3}H]BdUrd	[^{3}H]TDR/Thymidine	^{32}P
Zygotene	328	572	17×10^2
Pachytene	104	191	7×10^2
Zygotene/pachytene	3·1	3·0	2·4

Zygotene and pachytene cells were each harvested from 20 buds and divided into 3 groups. One group was exposed to 5 μc/ml. of [^{3}H]BdUrd; a second group was exposed to 5 μc/ml. of [^{3}H]thymidine; the third group was exposed to 5 μc/ml. of carrier-free [^{32}P]phosphate. The cells were exposed to label for 2 days. At the end of the incubation period, the activity in the acid-soluble pool was determined as was also the specific activity of the purified DNA. Values for the acid-soluble fraction are not shown in the Table as they were similar at zygotene and pachytene. The values for the DNA are expressed as cts/min/μg DNA. The ratios of zygotene to pachytene counts are shown in the bottom line.

gradients. This is illustrated in Figure 7. In contrast to zygotene BdUrd-DNA (Figs 3 and 4), pachytene BdUrd-DNA shows no density shift when centrifuged in a solution of CsCl. Attempts to detect a shift by shearing the pachytene DNA to an average molecular weight of 500,000 daltons were unsuccessful. Centrifugation of DNA samples in alkaline gradients showed no preferential shift of the pachytene label (Fig. 7) although we consistently observed a preferential broadening of the pachytene-labeled profile. A probable interpretation of these results is that the labeled pachytene regions are appreciably smaller in size than the fragments examined on the gradient. The preferential broadening of the pachytene profile under alkaline conditions might be due to a preponderance of single-stranded nicks in the labeled regions but we have no additional evidence to support this interpretation.

Another significant difference between zygotene and pachytene DNA synthesis is found in the hybridization experiments which are summarized in Figure 1. These experiments reveal no variation in the level of pachytene DNA during the meiotic cycle. Failure to detect a difference in levels such as found for zygotene DNA cannot be attributed to the sensitivity of the technique. Pachytene DNA was labeled exclusively during pachytene and could not have been contaminated with S-phase labeling which occurs about a week or more before pachytene. If contamination did occur from zygotene labeling, it would lead to an apparent variation in levels of pachytene DNA and this was not observed. The results therefore point to an absence of net DNA synthesis during the pachytene stage. If so, the segments of DNA synthesized during pachytene must have replaced pre-existing ones. Several attempts were made to measure the presence of excised DNA fragments during the pachytene stage but none of these proved successful.

Since hydroxyurea has been reported to inhibit semiconservative but not repair-type replication (Painter & Cleaver, 1967), the action of this inhibitor on DNA synthesis during S-phase, zygotene, pachytene and post-pachytene stages was examined. Post-pachytene cells were irradiated with ultraviolet light or X-rays before administration of isotope. Cells at late meiotic stages are particularly useful for such studies because DNA synthesis does not normally occur after the pachytene stage. The synthesis observed in these irradiated cells may therefore be regarded as

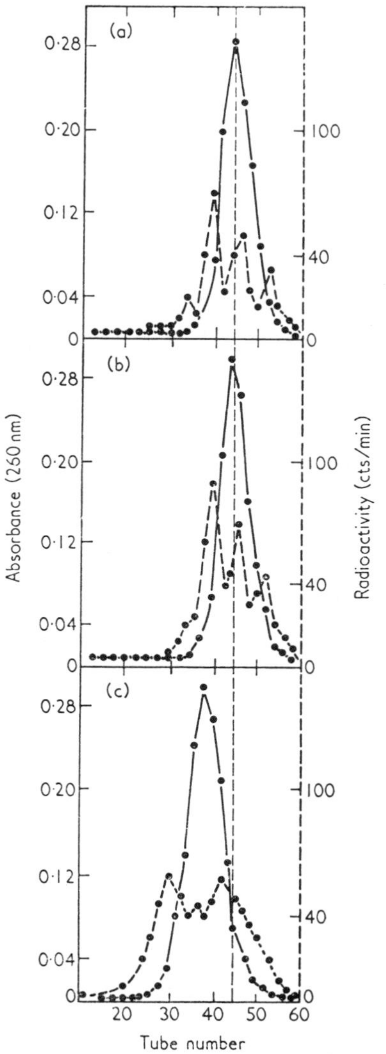

FIG. 7. The effect of BdUrd incorporation on the buoyant density of DNA prepared from pachytene cells. Incubation media contained 10 μc of [^{32}P]phosphate/ml. and, in (b) and (c), 4×10^{-3}M-BdUrd. The dashed vertical line marks density of 1·701 g/ml.

(a) Pachytene cells incubated for 3 days without BdUrd.

(b) Pachytene cells incubated in presence of BdUrd.

(c) Same as (b) but DNA denatured by exposure to alkaline conditions. —●——●—, O.D.; --●--●--, radioactivity.

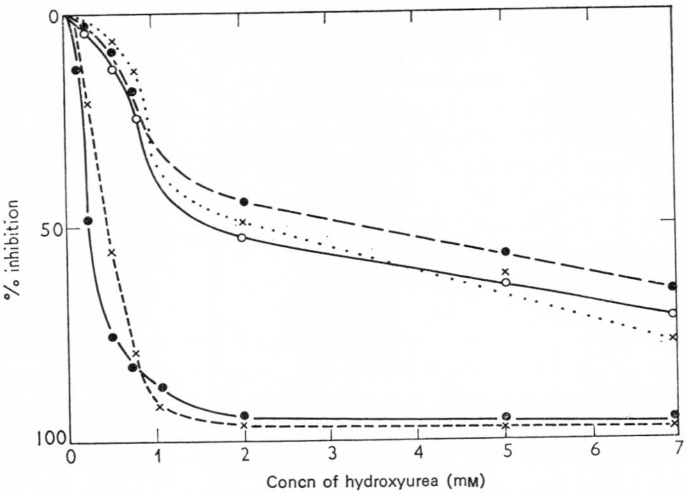

FIG. 8. Effect of hydroxyurea on DNA synthesis at different stages of meiosis and in response to radiation.

Cells were exposed to [^{32}P]phosphate for 2 days. For S-phase cells, the isotope concentration was 2 μc/ml.; for all other stages it was 20 μc/ml. The tetrad stage was used for irradiation since no DNA synthesis normally occurs at this stage. For u.v. irradiation a G15T8 (General Electric) 15 w lamp was used at a distance of 35 cm for an interval of 10 min. The cells were otherwise kept in the dark. DNA synthesis was measured immediately following irradiation. For each stage the levels of synthesis obtained in the absence of hydroxyurea are used to calculate percentage inhibition.

--×--×--, Zygotene; —●—●—, S-phase; ··×···×··, ultraviolet irradiation; —○—○—, pachytene; --●--●--, X-irradiation.

of the repair replication type. The results of this experiment are shown in Figure 8. Two types of behavior are evident. Premeiotic S-phase and zygotene cells respond strongly to hydroxyurea such that synthesis is almost completely inhibited at a concentration of 1×10^{-3}M. Pachytene and irradiated cells are much less responsive to the inhibitor. Even at a concentration of 7×10^{-3} M, only about 70% inhibition is achieved. Thus, to the extent that hydroxyurea discriminates between semiconservative and non-conservative repair replication, the results here obtained indicate that pachytene but not zygotene synthesis is of the repair replication type.

4. Discussion

The initial purpose in analyzing for DNA synthesis during meiotic prophase was to find evidence pertaining to the classical cytogenetic view that crossing over occurs during the pachytene stage when homologous chromosomes are paired (Rhoades, 1968). That a small but significant amount of DNA synthesis does in fact occur during meiotic prophase was reported in an earlier publication (Hotta *et al.*, 1966). Such evidence, together with a number of recent reports concerning the elimination of chiasmata by perturbing cells during the zygotene–pachytene stages (Henderson, 1966; Peacock, 1968), provide a satisfactory though not conclusive basis for the view that crossing over occurs during the pachytene stage. The experiments reported in this paper, however, indicate that the metabolism of DNA during meiotic prophase is more elaborate than would be expected if DNA synthesis during this interval were purely a reflection of chromosomal recombination. No DNA synthesis occurs during the first prophase stage, leptotene, but the subsequent zygotene and pachytene stages each have a distinctive and different pattern of DNA synthesis.

In *Lilium*, zygotene DNA synthesis occurs three to five days after completion of S-phase synthesis, the duration of the interval depending upon the particular species. In *Trillium erectum* the interval is much longer since the cells become dormant when they enter the leptotene stage and do not resume development until they have been exposed to a period of low temperature (Sparrow & Sparrow, 1949). It has been previously reported that the amount of DNA synthesized during the zygotene interval represents about 0·3% of the nuclear DNA. A most important fact revealed in this paper is that zygotene synthesis represents a delayed replication. Moreover, the distinctive buoyant density of zygotene DNA is characteristic of all species thus far tested. It would therefore appear that in meiotic cells, of liliaceous plants at least, a unique portion of the genome is delayed in its replication until the cells reach the zygotene stage. How this DNA relates to chromosome structure is still unknown. Although it has a high buoyant density, it is not rDNA. The autoradiographic data of Wimber & Prensky (1963) on meiotic prophase labeling and our own unpublished data indicate that it is probably not localized in a small region of the chromosome. Grain distribution over prophase-labeled nuclei appears to be random. The general fragmentation of chromosomes upon partial inhibition of zygotene DNA synthesis (Ito, Hotta & Stern, 1967) leads us to speculate that zygotene DNA sequences are distributed along the length of each of the chromosomes.

The best pointer we have to the function of zygotene DNA is that its synthesis coincides with the interval of chromosome pairing and that inhibition of its synthesis prevents the initiation and/or continuation of chromosome pairing (Stern & Hotta, 1969). The DNA synthesis at zygotene thus appears to be a necessary condition for chromosome pairing. That the synthesis itself, even when following the S-phase, is not a sufficient condition for pairing is indicated by the experiments with mitotic revertants described in this paper. Other factors which may arise as a consequence of development during the leptotene stage must operate during zygotene to effect chromosome pairing coincidentally with or subsequent to the initiation of DNA synthesis. None of the experiments thus far described, however, throws any light on the role which zygotene DNA synthesis might play in the pairing process itself. One attractive speculation is that the zygotene DNA regions serve as sites at which homologous chromosomes are first aligned. The kinetics of zygotene DNA hybridization are consistent with this possibility inasmuch as preliminary studies indicate an appreciable degree of molecular heterogeneity with respect to base sequence. Nevertheless, such a role, even if correct, cannot imply that the regions of zygotene DNA constitute the effective pairing regions of homologous chromosomes. As will be discussed below, to the extent that crossing over can be identified with repair replication during pachytene, the regions of zygotene DNA cannot be the exclusive sites of chromosomal recombination.

Recent experiments have indicated that chiasma formation may be eliminated by interfering with protein synthesis at the end of the zygotene stage (Parchman & Stern, 1969). Elimination of chiasmata by exposing meiotic cells to elevated temperatures after completion of zygotene has been clearly demonstrated by Peacock (1968). Thus, synthesis of DNA during the zygotene stage is not involved in the formation of chiasmata. Chiasma formation would appear to occur during pachytene and the characteristics of DNA synthesis at this stage are therefore particularly relevant to the process of chromosomal recombination. The experiments reported in this paper point to the conclusion that pachytene DNA synthesis is of the repair replication type. The DNA–DNA hybridization analysis indicates that DNA replication during the pachytene stage does not result in a net increase of DNA. The regions undergoing replication must therefore be replacing pre-existing ones. Moreover, unlike zygotene DNA, these regions vary in composition even within a single species and their average base composition approximates that of total nuclear DNA. Such characteristics are compatible with the more or less random occurrence of crossing over between homologous regions of paired chromosomes. Moreover, they are incompatible with a restriction of crossing over to the regions of zygotene DNA despite the attractiveness of the speculation that zygotene DNA may function in the alignment of homologous chromosomes. The failure of the bromodeoxyuridine substituted DNA synthesized during pachytene to show a "heavy shift" in buoyant density and the ineffectiveness of hydroxyurea in arresting pachytene DNA synthesis are characteristics which are commonly attributed to repair replication processes.

In general, the data reported in this paper reveal an unexpected complexity of DNA metabolism in the over-all process of crossing over. At least two distinct phases can be identified, one concerned with the pairing of homologs, the other with the formation of chiasmata. That the cytological differences between the zygotene and pachytene stages are matched by differences in DNA metabolism lends substance to the speculation that each of the two types of DNA synthesis has a unique functional role in the total crossing over process. The data, however, are insufficient to clarify the nature of these roles. In this respect, the experiments described in the accompanying paper (Howell & Stern, 1971) are particularly significant inasmuch as they provide corroborative evidence based on analyses of enzyme activities rather than on the characterization of the products of DNA synthesis.

This investigation was supported by National Science Foundation Grant GB5173X and supplemented by U.S. Public Health Service Grant HD03015 from the National Institute of Child Health and Human Development.

REFERENCES

Bendich, A. (1957). In *Methods in Enzymology*, ed. by S. P. Colowick & N. O. Kaplan, vol. 3, p. 715. New York: Academic Press.
Brown, D. D. & Dawid, I. B. (1968). *Science*, **160**, 272.
Church, R. B., Luther, S. W. & McCarthy, B. J. (1969). *Biochim. biophys. Acta*, **190**, 30.
Denhart, D. T. (1966). *Biochem. Biophys. Res. Comm.* **23**, 641.
Erickson, R. O. (1948). *Amer. J. Bot.* **35**, 729.
Foster, T. S. & Stern, H. (1958). *Science*, **128**, 653.
Gall, J. G. (1969). *Genetics*, **61**, 121.
Gillespie, D. & Spiegelman, S. (1965). *J. Mol. Biol.* **12**, 829.
Guttes, E. & Guttes, S. (1969). *J. Cell. Biol.* **43**, 229.
Henderson, S. A. (1966). *Nature*, **211**, 1043.
Hotta, Y., Bassell, A. & Stern, H. (1965). *J. Cell Biol.* **27**, 451.
Hotta, Y., Ito, M. & Stern, H. (1966). *Proc. Nat. Acad. Sci., Wash.* **56**, 1184.
Howell, S. H. & Stern, H. (1971). *J. Mol. Biol.* **55**, 357.
Ito, M., Hotta, Y. & Stern, H. (1967). *Devel. Biol.* **16**, 54.
Ito, M. & Stern, H. (1967). *Devel. Biol.* **16**, 36.
Painter, R. B. & Cleaver, J. E. (1967). *Nature*, **216**, 369.

Parchman, L. G. & Stern, H. (1969). *Chromosoma*, **26**, 298.
Peacock, W. J. (1968). In *Replication and Recombination of Genetic Material*, ed. by W. J. Peacock & R. D. Brock, p. 242. Canberra: Australian Academy of Science.
Rhoades, M. M. (1968). In *Replication and Recombination of Genetic Material*, ed. by W. J. Peacock & R. D. Brock, p. 229. Canberra: Australian Academy of Science.
Sparrow, A. H. & Sparrow, R. C. (1949). *Stain Technol.* **24**, 47.
Stern, H. & Hotta, Y. (1969). *Genetics Suppl.* **61**, 27.
Wimber, D. E. & Prensky, W. (1963). *Genetics*, **48**, 1731.

METHYLATION OF LILIUM DNA DURING THE MEIOTIC CYCLE

YASUO HOTTA AND NORMAN HECHT

INTRODUCTION

This study was undertaken to examine the possibility that DNA methylation might play a functional role in the metabolism of DNA during meiosis. Of particular interest was the now established pattern of DNA synthesis during the meiotic cycle in Lilium. Three principal intervals have been described[1]. The first of these is the premeiotic S-phase interval when the bulk of nuclear DNA is replicated. The second interval occurs during the zygotene stage when a high GC satellite component undergoes a replication delayed from the S-phase. Synthesis of this quantitatively minor component is coincident with and apparently essential to chromosome pairing. The third interval occurs during the pachytene stage and it has the characteristics of a repair replication[2].

The results obtained provide no indication of a functional role for methylation. They do nevertheless provide evidence that meiotic cells of higher plants behave no differently from somatic cells of mammals with respect to DNA methylation.

METHODS

Two varieties of hybrid lilies were used in these experiments, Cinnabar and Bright Star. Male meiotic cells were extruded from developing anthers and transferred to culture media as previously described[3]. All the experiments reported here were performed on cultured cells. No significant differences in behavior were found between the two varieties of lily used. The data are therefore generally reported without indicating variety. Cinnabar microsporocytes were cultured at 19° and Bright Star microsporocytes at 15°.

Biochemicals

5-Bromodeoxyuridine was purchased from "Calbiochem.". 1-[*Me*-^{3}H]Methionine (134 mC/mmole), [^{3}H]thymidine (10 C/mmole), and $^{32}P_i$ (25 mC/0.1 mg solid carrier-free) were purchased from New England Nuclear. Unless otherwise indicated premeiotic cells were cultured in radioisotope concentrations of 5 μC/ml, and zygotene or pachytene cells were cultured in concentrations of 10 μC/ml.

DNA purification

The DNA was isolated and purified as described in a previous publication[4]. Radioactivities were determined with a scintillation counter after samples were treated with 0.1 M NaOH at 70° for 20 min and the DNA then precipitated by acidification. This treatment served to remove contaminating RNA. Measurable radioactivity from contaminating protein in purified DNA was ruled out by preparing DNA from cells which had been cultured for 2 days in the presence of [^{3}H]leucine (purchased from Schwarz Chemicals, 50 C/mmole) at a concentration of 20 μC/ml. DNA prepared from these cells and purified by the procedure indicated had negligible radioactivity.

DNA fractionation

Nuclei isolated by the glycerin–sucrose procedure[5] were suspended in 0.05 M Tris–HCl (pH 7.2), 0.02 M ethylenediaminetetraacetate containing 100 μg/ml of ribonuclease, and incubated at 25° for 20 min[6]. 5% Sodium lauryl sulfate was then added to a concentration of 0.5 %. After incubation for 10 min at room temperature a half volume of water-saturated phenol was added and the mixture agitated with a "Junior Vortex Mixer" at a setting of 7 for 30 sec. 5 M NaCl was then added to a final concentration of 1.0 M (aqueous phase) and agitation repeated for an additional 30 sec. A volume of chloroform–amyl alcohol mixture (21:1, v/v) equal to that of the phenol was finally added and the whole agitated for an additional minute. The mixture was centrifuged and the aqueous and organic phases were removed separately leaving the interphase material in the centrifuge tube. Equal volumes of fresh aqueous and organic solutions were added to the interphase material and the suspension agitated for 1 min by the Vortex mixer. After centrifugation, the aqueous phases were combined and the organic phase discarded. The interphase material was collected in 70 % ethanol, washed with alcohol–ether to remove lipids and the DNA isolated by the pronase procedure[4]. DNA was also purified from the combined aqueous phases in the standard way. By this procedure, 10–25 % of the DNA from premeiotic S-phase cells and 2–5 % of the DNA from zygotene cells remained at the interphase.

Deoxyribonucleotide analysis

DNA samples previously treated with 0.1 M NaOH to remove RNA were digested in a solution of deoxyribonuclease (Worthington Biochemicals) containing 0.01 M Tris (pH 7.2), 0.003 M $MgCl_2$, and 100 μg/ml enzyme for 5 h at 37°. The pH was then adjusted to 9.0 with 0.02 M Tris–$MgCl_2$, 1 unit/ml of snake venom phosphodiesterase (Worthington) added and the mixture incubated for an additional 10–15 h. After deproteinization with chloroform–amyl alcohol and if necessary, lyophilization, the digest was placed on a 25 cm × 1 cm^2 column of Dowex 1-X_2 (acetate), and after washing the column with water to remove any nucleosides the deoxynucleotides were eluted according to the procedure of HURT *et al.*[7]. Using this procedure, less than 1 % of the DNA was converted to deoxynucleosides.

Base analysis

The 5-methyldeoxycytidylate and deoxycytidylate fractions were hydrolyzed with or without prior concentration by lyophilization in 88 % formic acid at 130° in a glass ampoule for 24 h. The hydrolysates were evaporated to dryness by lyophilization and the residues dissolved in 50–100 μl of water and spotted on Whatman No. 3 filter paper[8]. The chromatograms were developed with water-saturated *n*-butanol containing 0.01 vol. of 15 M NH_4OH by either ascending or descending procedure. The contents of each spot were determined by comparison with the position and absorption spectrum of known standards.

Precursor pools

Cells were disrupted by homogenization in cold 1.0 M HCl or 70 % ethanol and the soluble components collected by centrifugation. The clear supernatant fluids were either lyophilized or flash evaporated and the residues dissolved in 0.01 M Tris (pH 7.0) and deproteinized with a mixture of chloroform–amyl alcohol (21:1, v/v). A portion of the aqueous phase was placed on a Dowex 1-X_2 (acetate) column and cold deoxycytidylate and methyldeoxycytidylate added as carriers to identify the radioactive constituents of the extract. The water wash of the column which contained the free bases and nucleosides was concentrated and hydrolyzed in 88 % formic acid for resolution by paper chromatography[8]. The remainder of the material on the column was eluted in the standard manner for deoxynucleotides[7]. A second portion of the extract was placed on a DEAE-cellulose column previously equilibrated with a solution of 7.0 M urea–5 mM Tris–HCl (pH 7.5) to separate oligonucleotides[9]. A gradient developed from 0.0 to 0.5 M NaCl in 7.0 M urea, 5 mM Tris–HCl (pH 7.5) was used to elute the components. The eluates were then diluted 50 fold with water and passed through DEAE-cellulose paper to absorb the mono- or polydeoxynucleotides.

RESULTS

Site of DNA methylation

Meiotic cells cultured in the presence of [*Me*-^{3}H]methionine incorporate tritium label into the 5-methylcytosine component of DNA. A typical analytical result is shown in Fig. 1. In this experiment premeiotic S-phase cells were cultured for 2 days

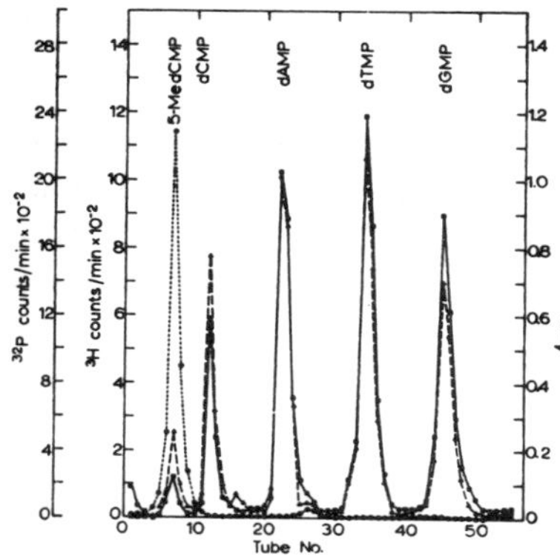

Fig. 1. Distribution of label among deoxyribonucleotides from DNA of cells exposed to $^{32}P_i$ and [*Me*-^{3}H]methionine. Premeiotic cells were exposed for two days to the isotopes during S-phase. The DNA prepared from the cells was digested enzymatically as described under METHODS. The position of the 5-methyldeoxycytidylate peak was identified by running a standard on another column. The other components were identified by their absorption spectra. Details of experimental procedure are provided under METHODS. The solid line, the broken line, and the dotted line indicate the absorbance, ^{32}P radioactivity and ^{3}H radioactivity, respectively.

in a medium containing $^{32}P_i$ and [*Me*-^{3}H]methionine. The DNA was then extracted and enzymatically degraded to monodeoxynucleotides which were fractionated by column chromatography. As would be expected for S-phase cells, the ^{32}P radioactivity tracks the curves for ultraviolet light absorbancy. The tritium label, on the other hand, is almost entirely confined to the 5-methyldeoxycytidylate fraction. A small amount of the tritium label, usually much less than 10 %, consistently elutes behind the deoxyadenylate fraction. The nature of this minor component is unknown. ^{3}H activity is below the limits of detection elsewhere in the elution profile .A similar profile of ^{3}H distribution was found for the DNA of cells exposed to labelled methionine at other stages of the meiotic cycle. Evidence for the exclusive presence of the tritium label in the pyrimidine moiety of 5-methyldeoxycytidylate was obtained by hydrolysis of the deoxynucleotides in 88 % formic acid and separation of the products by paper chromatography as described under METHODS. All of the tritium label was found in the 5-methylcytosine spot. An analysis of the acid and ethanol-soluble pool was carried out as outlined in METHODS in order to determine whether methylation might have occurred at the precursor level rather than in the polymer. In extracts from cells cultured for 2 days in the presence of [*Me*-^{3}H]methionine, no tritium label could be found in either 5-methylcytosine, 5-methyldeoxycytidine, or 5-methyldeoxycytidylate, all of which were added as carriers to the original extract. The absence of methylated cytosine or its derivatives in the DNA precursor pool is in line with other studies of DNA methylation in mammalian cells and is consistent with the view that methylation of DNA–cytosine residues occurs entirely at the level of the polymer.

Methylation activity in relation to the meiotic cycle

Cells at different stages of the meiotic cycle were cultured for 36 h in the presence of $^{32}P_i$ and [*Me*-^{3}H]methionine. The level of incorporation of each of the isotopes into DNA at the different stages of meiosis is shown in Fig. 2. Methylation,

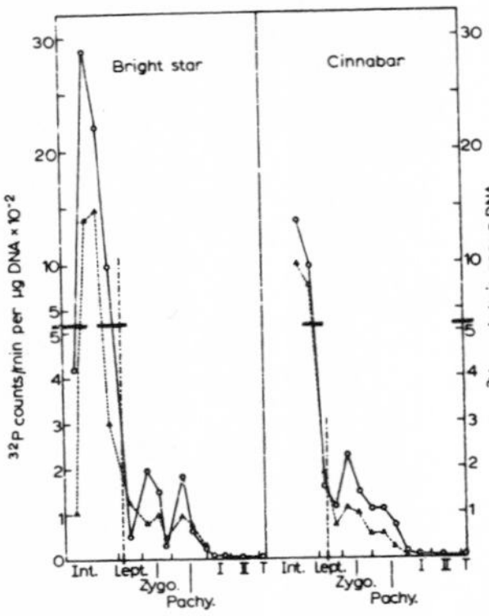

Fig. 2. Methylating activity in relation to DNA synthesis during the meiotic cycle. At each of the stages tested meiotic cells of Cinnabar or Bright Star lilies were exposed to isotope in culture media for 36 h. In the case of premeiotic S-phase cells 5 μC/ml each of $^{32}P_i$ and [*Me*-^{3}H]methionine were used. 10 μC/ml were used for all other stages. DNA synthesis is plotted as a function of ^{32}P radioactivity in the DNA (○–○). Methylation is plotted as a function of tritium incorporation (△- - -△). The stages indicated on the abscissa are as follows: interphase, leptotene, zygotene, pachytene, post-pachytene through 1st meiotic division (I), 2nd meiotic division (II) and tetrads (T).

as measured by tritium incorporation, parallels DNA synthesis as measured by ^{32}P incorporation. Methylation does not occur during intervals lacking DNA synthesis, and the amount of methylation at any particular interval is roughly proportional to the amount of DNA synthesis. Thus, although the three intervals of DNA synthesis during the meiotic cycle each have distinctive functions[2], no distinctive differences are detectable with respect to methylating activity.

The levels of methylation at the different phases of meiosis were compared by analyzing the distribution of ^{32}P radioactivity among the monodeoxynucleotides obtained by enzymatic digestion of DNA. At least four separate experiments were run for each of the meiotic stages analyzed. The results are summarized in Table I. The ratio of methyl[^{32}P]deoxycytidylate : [^{32}P]deoxycytidylate is a measure of the

TABLE I

DNA METHYLATION RELATIVE TO DNA SYNTHESIS DURING MEIOTIC CYCLE

Cells at the stages indicated were cultured in the presence of $^{32}P_i$ for 2–3 days. The isolated DNA was digested enzymatically as described under METHODS and the deoxynucleotides separated chromatographically. The number of separate tests performed for each stage is indicated in 2nd column. The radioactivity value for 5-methyldeoxycytidylate and deoxycytidylate are expressed as a percentage of the total radioactivity in the 5-deoxynucleotide peaks as shown in Fig. 1. The ratio of these percentages is given in the last column.

Stage	*Number of tests*	*^{32}P incorporation (%)* 5-*Methyldeoxycytidylate*	*Deoxycytidylate*	5-*Methyldeoxycytidylate* / *Deoxycytidylate*
Premeiotic S-phase	4	4.6	17.2	0.268
Zygotene	3	6.1	22.7	0.269
Pachytene	3	4.9	18.0	0.272

proportion of cytosine residues methylated during a particular interval. In general, the level of methylation during the meiotic cycle appears to be proportional to the number of cytosine residues in the DNA synthesized and is not significantly modified by the stage at which synthesis occurs.

Relationship of methylation to DNA synthesis

Methylation of DNA normally occurs only during intervals of DNA synthesis but, as observed by others, it is not immediately essential to DNA synthesis[10,11]. Inhibition of methylation in meiotic cells by high concentration of ethionine or norleucine has no corresponding effect on DNA synthesis. Results of such an experiment are shown in Fig. 3. Less than 10 % of DNA synthesis is inhibited in the presence of analog concentrations which inhibit 50 % of the methylation reaction. This differential response is the same for all stages of meiosis.

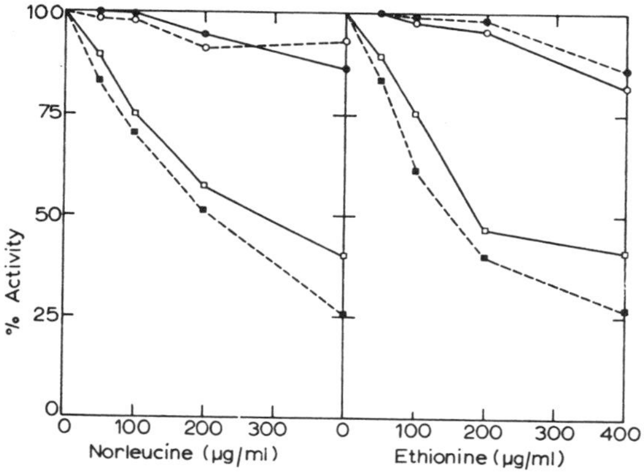

Fig. 3. Inhibition of DNA methylation by norleucine and ethionine. Cells were incubated with isotopic precursors as under Fig. 2 and with various concentrations of norleucine and ethionine. Broken lines are values for zygotene cells and solid lines are for S-phase cells. The circles are for ^{32}P uptake (DNA synthesis) and the squares are for ^{3}H uptake (methylation). All radioactivities are expressed as a percentage of the radioactivities in control cells incubated in the absence of inhibitors.

Methylation does not occur if DNA synthesis is inhibited. The effect of inhibitors of DNA synthesis on methylation is shown in Table II. For any particular concentration of inhibitor, the percentage inhibition of DNA methylation is approximately equal to the percentage inhibition of DNA synthesis. It would appear as though the level of methylation is determined by the number of cytosine residues made available as a consequence of new synthesis.

The capacity of meiotic cells to methylate DNA at most times in the meiotic cycle is demonstrated by experiments in which premeiotic or early meiotic cells were cultured in the presence of ethionine for periods up to 12 days. As shown in Fig. 3, ethionine inhibited methylation 50–90 % depending upon stage and duration of treatment. However, the same concentrations of ethionine had only a small retarding effect on the progress of the cells through the meiotic cycle. In order to determine whether cells had the capacity to methylate DNA at times other than the normal

TABLE II

EFFECT OF INHIBITION OF DNA SYNTHESIS ON METHYLATION

Meiotic cells were incubated in the presence of $^{32}P_i$ (10 μC/ml) and [*Me*-^{3}H]methionine (10 μC/ml) for 2 days. The DNA was purified as described under METHODS and the two radioactivities determined. About 100–200 μg DNA were used for each measurement. ^{32}P radioactivity is a measure of DNA synthesis and ^{3}H radioactivity is a measure of methylation. Bud lengths, rather than cytological examination, have been used to group the meiotic cells. Such lengths are fairly good indices of meiotic stage[3]. Deoxyadenosine and cyclohexamide were the inhibitors used. These were added together with the isotopic precursors at zero time.

Variety	*Bud length (mm)*	*Inhibitor*	*^{32}P (counts/min per μg DNA)*		*Inhibition* (%)	*^{3}H (counts/min per μg DNA)*		*Inhibition* (%)
			Control	*Test*		*Control*	*Test*	
Cinnabar	10–11	Deoxyadenosine (1 mM)	1050	730	30.5	21.6	16.9	26.3
	10–11	(2 mM)	1100	460	58.2	25.4	12.8	49.6
	10–11	(4 mM)	1090	300	72.5	25.2	8.7	65.5
	11–12	(4 mM)	740	24	67.5	19.8	7.4	66.7
	12–13	(4 mM)	11	4	63.6	4.2	1.5	64.3
	13–14	(4 mM)	6	2	66.6	2.1	1.1	47.7
Bright Star	13–14	(4 mM)	1290	460	64.3	9.2	2.6	71.7
	13–14	Cyclohexamide (1 μg/ml)	1290	923	67.2	9.2	2.8	69.5

synthetic intervals, groups of cells weie transferred from the ethionine medium to a methionine-containing one after they reached the desired meiotic stage. The results of this experiment are summarized in Table III. Regardless of the stage at which the

TABLE III

EFFECT OF THE PRESENCE OF ETHIONINE DURING DNA REPLICATION OR METHYLATION DURING SUBSEQUENT STAGES OF MEIOSIS

Microsporocytes at premeiotic interphase (Group I) and at interphase–leptotene (Group II) were exposed to 400 μg ethionine/ml for 24 h and then to 100 μg/ml for 7–12 days. At 7 days or later a batch of cells was removed, the stage of meiosis determined cytologically on a sample of these and the remainder transferred to an ethionine-free medium containing 20 μC/ml of [^{3}H]-methionine. The cells were cultured for 36 h in the presence of label and the DNA then isolated for determination of methylation. The cytological stages at the time of transfer to labelled medium are shown in the table. The abbreviations used are: Div. I, cells in stages from diakinesis through 1st anaphase; Div. II, cells in stages from binucleate form (diads) to tetranucleate form (tetrads). The grouping with the highest proportion of contracted chromosomes is diakinesis–1st anaphase. The "Control" value if from cells cultured through meiosis over a period of 12 days in the presence of 20 μC/ml of [^{3}H]methionine.

Group I		*Group II*	
Stage at transfer	*^{3}H (counts/min per μg DNA)*	*Stage at transfer*	*^{3}H (counts/min per μg DNA)*
Pachytene	6.68	Pachytene–1st metaphase	7.77
Diakinesis–1st anaphase	2.52	Div. I–Div. II	4.71
Diad–Div. II	6.16	—	
Tetrads	7.52	Tetrads	6.48
Control	15.25	Control	10.77

transfer was made, the cells showed an abnormally high methylating activity. Thus, all the post-pachytene stages which normally show no methylating activity have appreciable levels of activity. An interesting exception is the group of cells designated "diakinesis–anaphase I". This group contains a very high proportion of contracted chromosomes and the relatively low methylation for this group suggests that methylation may not occur in condensed metaphase–anaphase chromosomes. Clearly, however, methylation can occur at any other stage in the cycle if the methylation reaction is inhibited at a prior interval of DNA synthesis.

Timing of methylation in relation to DNA synthesis

Normally, methylation of DNA occurs only in those molecules undergoing replication. This fact is made evident in the absorbance and radioactivity profiles of DNA extracted from cells which had been exposed to label during zygotene and centrifuged to equilibrium in CsCl gradient. As previously reported, only a high GC satellite component is replicated during zygotene and, if labelled with $^{32}P_i$ the radioactive band of DNA is localized at the heavy end of the CsCl gradient. It may be seen from Fig. 4 that if zygotene cells are exposed to $^{32}P_i$ and [*Me*-^{3}H]methionine, methyl group radioactivity tracks that of ^{32}P. If bromodeoxyuridine is also added to the culture medium, both ^{32}P and ^{3}H radioactivities show a density shift. Methylation is thus exclusively associated in the nucleus with those molecules which are undergoing replication.

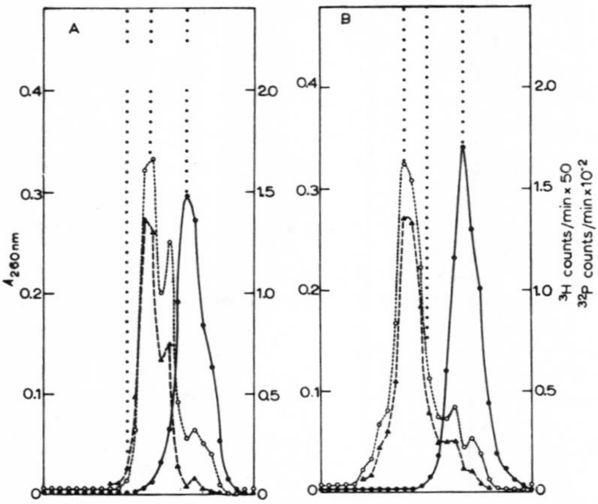

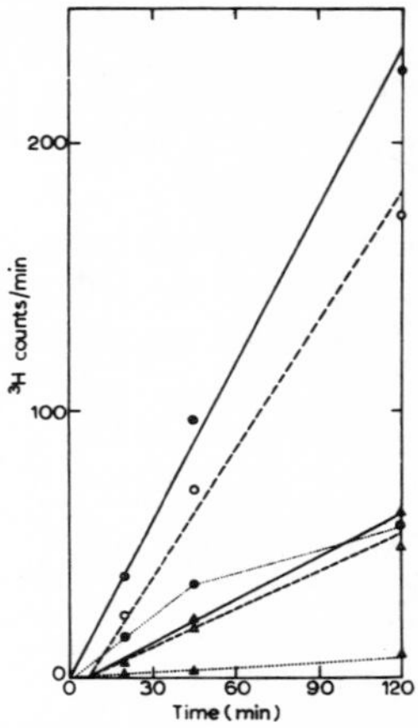

Fig. 4. Methylation of DNA at zygotene stage. Zygotene cells were cultured in the presence of label as described under Fig. 2. In B, the culture medium was supplemented with 2 mM of 5-bromodeoxyuridine to density label the DNA. The DNA purified from the cells was centrifuged to equilibrium in a CsCl gradient. The absorbance (●–●) and radioactivity (▲- - -▲, ^{3}H; ○- - -○, ^{32}P) profiles from cells cultured in the absence of and in the presence of 5-bromodeoxyuridine are plotted in A and B, respectively. The densities of the CsCl solution in g/ml are shown for peak value.

Fig. 5. Time course of DNA synthesis and methylation during premeiotic S-phase. Cells were incubated for the times indicated in the abscissa with 10 μC/ml of [^{3}H]deoxycytidine. Following incubation the DNA was purified and then digested enzymatically as described under METHODS. Deoxycytidylate and 5-methyldeoxycytidylate were separated chromatographically and their respective radioactivities determined. Prior to purification the DNA was separated into two fractions as described under METHODS. A set of three values is plotted at the different times for the radioactivities in deoxycytidylate (circles) and 5-methyldeoxycytidylate (triangles), respectively: Total radioactivity in DNA (solid lines); radioactivity of DNA in aqueous phase (broken lines); radioactivity of DNA in interphase (dotted lines). As discussed in the text, the interphase DNA is presumed to represent the nascent DNA.

A more precise determination of the temporal relationship between DNA synthesis and methylation is provided by short term labelling experiments with deoxycytidine. In this study the fractionation method developed by MUELLER and colleagues[6] for the separation of nascent DNA from already replicated DNA was used. The particulars of the procedure are discussed under METHODS. A general observation made in these experiments is that with labelled deoxycytidine as precursor, the lag between synthesis and methylation appears to be of the order of 2–10 min. By contrast, if ^{32}P is used as a marker of DNA synthesis, the lag is 10–60 min. Presumably, the differences in apparent lag time reflect differences in pool size and/or metabolism at the precursor level. However, the point of major interest in the experiments summarized in Fig. 5 is the virtual exclusion of methyl label from the nascent DNA which is the form trapped at the interphase between the organic and aqueous layers. Although the deoxycytidine label appears first in the interphase, the tritium label of the methyl group does not appear in the interphase to any significant extent but is identified almost entirely in the aqueous layer. Thus, although methylation occurs within 2 min or less following DNA replication, the freshly replicated DNA strand is not methylated until after it leaves the so-called nascent state.

DISCUSSION

The percentage (4.5) of methylated cytosine residues in the meiotic cell DNA of the hybrid lilies studied here is unexceptional, and is close to the value (4.0 %) reported for *Lilium usitatissimum*[12]. Indeed, the results of most of the analyses carried out on DNA methylation in meiotic cells of lilies are virtually identical with those reported for hepatomas in cell culture[11] for mouse fibroblasts[13]. The data are also consistent with the mechanisms of methylation shown for pea seedlings[14] and for rat spleen[15]. Methylation occurs at the polymer level in the pyrimidine moiety of the cytosine residue and, normally, such methylation takes place in association with DNA replication. The time course of methylation observed in lily meiotic cells is similar to that reported by LARK for *Escherichia coli*[10]. However, the analysis here performed would indicate that methylation does not occur during the earliest phase of DNA replication when the nascent DNA is bound in a tight complex with other cellular constituents. The nature of the complex will be described in detail elsewhere but is essentially the same as that described by MUELLER and coworkers[6].

The deeper purpose of this study has not been fulfilled. It was hoped that cells undergoing meiosis might furnish clues as to the function of DNA methylation. Suggestions of the type made by BOREK AND SRINIVASAN[16] that methylation might provide recognition sites in terms of nuclease action, although operative in bacteriophage systems[17], have not found concrete support in these studies. Although various kinds of DNA syntheses are separated in time during the meiotic cycle, and although one of the intervals is associated with scission-repair activity, no special relationship has been observed between the DNA synthesized during that interval (pachytene) and methylation. These analyses are of course by no means exhaustive, but if methylation does play some special role, that role requires techniques of detection other than those used here.

ACKNOWLEDGEMENTS

This work has been supported by grants from NSF GB-5173 × 3 and UC-CRCC.

REFERENCES

1 Y. HOTTA, M. ITO AND H. STERN, *Proc. Natl. Acad. Sci. U.S.*, 56 (1966) 1184.
2 Y. HOTTA AND H. STERN, *J. Mol. Biol.*, 55 (1971) 337.
3 M. ITO AND H. STERN, *Develop. Biol.*, 16 (1967) 36.
4 Y. HOTTA AND A. BASSEL, *Proc. Natl. Acad. Sci. U.S.*, 53 (1965) 356.
5 Y. HOTTA AND H. STERN, *Protoplasma*, 60 (1965) 218.
6 D. L. FRIEDMAN AND G. C. MUELLER, *Biochim. Biophys. Acta*, 174 (1969) 253.
7 R. O. HURT, A. M. MARKO AND G. C. BUTLER, *J. Biol. Chem.*, 204 (1953) 847.
8 G. R. WYATT, in B. CHARGAFF AND J. N. DAVIDSON, *The Nucleic Acids*, Vol. 1, Academic Press, New York, 1955, p. 243.
9 J. SEDET AND R. L. SINSHEIMER, *J. Mol. Biol.*, 9 (1964) 489.
10 C. LARK, *J. Mol. Biol.*, 31 (1968) 389.
11 T. W. SNEIDER AND V. R. POTTER, *J. Mol. Biol.*, 42 (1969) 271.
12 H. A. SOBER AND R. A. HARTE, *Handbook of Biochemistry selected data for Molecular Biology*, Chemical Rubber Co., Cleveland, 1968, p. H-39.
13 R. H. BURDON AND R. L. P. ADAMS, *Biochim. Biophys. Acta*, 174 (1969) 322.
14 F. KALOUSEK AND N. R. MORRIS, *Science*, 164 (1969) 721.
15 F. KALOUSEK AND N. R. MORRIS, *J. Biol. Chem.*, 244 (1969) 1157.
16 E. BOREK AND P. R. SRINIVASAN, *Ann. Rev. Biochem.*, 35 (1966) 275.
17 B. S. STRAUSS AND M. ROBBINS, *Biochim. Biophys. Acta*, 161 (1968) 68.

The Appearance of DNA Breakage and Repair Activities in the Synchronous Meiotic Cycle of *Lilium*

Stephen H. Howell and Herbert Stern

1. Introduction

One major problem which has commanded the attention of many investigators is the mechanism by which genetic recombination is effected during gametogenesis in higher organisms. Certain features of the mechanism have been outlined in classical cytogenetic studies of meiosis. The pairing of homologs, a precondition to crossing-over, was assigned to the zygotene stage of meiosis, and crossing-over itself to the pachytene stage (Rhoades, 1968). A variety of more recent studies has added substance to the classical outline. The synaptonemal complex, found in the zygotene and pachytene meiotic cells, has been identified as a unique structure associated with paired meiotic homologs (Moses, 1969). Autoradiographic studies of meiotic cells have supported the conclusion of classical cytogenetics that chiasmata are the physical counterparts of crossing-over and that chisasma formation occurs during the pachytene stage (Taylor, 1965; Peacock, 1968).

Several molecular models have been proposed to explain genetic recombination in higher organisms. Most models have been based to some extent on molecular processes identified in micro-organisms, and some have sought to rationalize the phenomenon of gene conversion as revealed by tetrad analysis (Whitehouse, 1963; Holliday, 1964; Stahl, 1969). All the models, however, suffer from a common fault, a lack of any evidence that the postulated mechanisms do in fact operate during the meiotic cycle.

The present study is addressed to an identification of those enzymic activities which occur in meiotic cells and which could account for the process of genetic recombination. The plan of study was based partly on the information accumulated concerning recombination in procaryotic organisms and partly on the evidence thus far obtained on DNA metabolism during meiosis in the eucaryotic lily (Hotta, Ito & Stern, 1966; Hotta & Stern, 1971).

We have assumed that the mechanism of genetic exchange in eucaryotes shows some elements in common with procaryotes, principally, that the mechanism for reciprocal recombination in eucaryotes parallels the breakage and reunion events found with parental DNA in bacteriophage (Meselson & Weigle, 1961). Of particular interest to our study have been the probable steps of recombination as dictated by the analysis of biparental recombinant molecules in phage (Anraku & Tomizawa, 1965): (1) *scission* of parental DNA strands in such a way as to allow for; (2) *molecular hybridization* by base pairing of complementary DNA strands and (3) *rejoining* by covalent bond formation between biparental strands. Nearly a full complement of enzymic activities necessary to effect such recombination steps has been demonstrated for *Escherichia coli* and T-even phage-infected *E. coli*: (see review Richardson, 1969). The activities which are generally included in such a complement and for which the meiotic cells have been analyzed are as follows: DNA endonuclease, polynucleotide ligase, phosphatase, polynucleotide kinase and phosphodiesterase.

Meiotic cells of the lily, microsporocytes, provide excellent material for examining these enzymic activities relative to the meiotic cycle. The cells develop synchronously and can be separated readily from the surrounding somatic tissue. Previous studies on the meiotic cycle in lily by Hotta *et al.* (1966) have indicated that unique DNA metabolic events occur during meiotic prophase at the time when genetic recombination is thought to occur. They found a small but discrete DNA synthesis period, in addition and subsequent to the normal S-phase DNA synthesis, in the zygotene–pachytene period of meiotic prophase. The meiotic prophase DNA synthesis was found necessary for the orderly progress of meiotic events and for maintaining the intactness of the meiotic chromosomes (Ito, Hotta & Stern, 1967). Several lines of evidence presented in the accompanying paper (Hotta & Stern, 1971) have led to the supposition that at least part of this meiotic prophase DNA synthesis is a DNA repair replication synthesis.

2. Materials

(a) *DNA*

Radioactive T7 DNA was prepared and isolated according to Thomas & Abelson (1966). Specific activities were 3 to 6×10^3 cts/min/μg for ^{3}H-labeled T7 DNA and 5 to 10×10^3 cts/min/μg for ^{32}P-labeled T7 DNA. For sedimentation or nitrocellulose filter analysis, DNA was used within 10 days of isotope incorporation. Uniformly ^{3}H-labeled *Bacillus subtilis* DNA, with a specific activity of 6×10^3 cts/min/μg was isolated according to Marmur (1961). Lily microsporocyte DNA was isolated according to Hotta, Bassel & Stern (1965). Specific activity of ^{3}H-labeled lily DNA, labeled in microsporocyte culture, was 2×10^3 cts/min/μg. Calf thymus DNA was purchased from Nutritional Biochemicals.

Precaution was taken to minimize mechanical breakage or enzymic degradation of DNA during isolation and handling.

(b) *Enzymes*

Alkaline phosphatase from *E. coli* was purchased from Worthington and purified to remove contaminating endonuclease according to Weiss, Live & Richardson (1968). Electrophoretically purified pancreatic deoxyribonuclease purchased from Worthington was used to produce the nicked T7 DNA substrate for ligase experiments. For quantitative nicking of DNA (Weiss, Live & Richardson, 1968) pancreatic deoxyribonuclease in standard vials was purchased from Worthington. Polynucleotide kinase from *E. coli* B/5 infected with either T4$^+$ or a temperature-sensitive ligase mutant T4 *tsA80* (kindly supplied by Dr R. S. Edgar) was purified according to Richardson (1965). Micrococcal nuclease was purchased from Worthington.

(c) *Lilies*

Lilies were grown under greenhouse conditions. Two horticultural varieties of *Lilium* hybrids, Cinnabar and Bright Star, were used according to seasonal availability. Buds, collected according to bud length, contained microsporocyte cells homogeneous with respect to cytological meiotic stage. Filaments of microsporocytes were extruded by squeezing individual buds under sterile conditions. Generally 50 to 150 buds/individual meiotic stage were used for each experiment.

(d) *Others*

[γ-^{32}P]ATP was prepared according to Weiss, Live & Richardson (1968) with high specific activities of 1 to 6×10^{10} cts/min/μmole for ligase assays and end-group analysis and lower specific activities of 2 to 8×10^{7} cts/min/μmole for kinase assays.

3. Methods

(a) *Preparation of microsporocyte extracts*

Microsporocyte filaments were cleaned of bud somatic tissue by repeated suspension in balanced salt solution (White, 1963) supplemented with 0·3 M-sucrose. Cleaned microsporocyte filaments were suspended in 1 to 6 ml. of 35% glycerol, 0·5 M-sucrose, 20 mM-Tris Cl (pH 7·0) and 20 mM-KCl for endonuclease, phosphodiesterase and phosphatase extraction or in 1 to 6 ml. of 10 mM-Tris Cl (pH 8·0), 0·5 mM-EDTA, 1·0 mM-dithiothreitol and 10 mM-KCl for polynucleotide kinase and ligase extraction. Microsporocytes were disrupted by sonication in a Sonifer cell disruptor (using standard or special microtip) at 60% maximum output. Sonication of the microsporocytes, monitored by the light microscope, was continued until all cells and their nuclei were disrupted. Centrifugation of the sonicated material at 40,000 ***g*** for 10 min separated a pellet of cell debris and a floating skin from an opalescent supernatant fluid which was collected with a pipet. Centrifugation and recovery of the supernatant fluid was repeated until the fluid was clarified. This fraction is referred to as the *microsporocyte extract*.

In the microsporocyte extract, the yield of protein/bud assayed according to Lowry, Rosebrough, Farr & Randall (1951) increased throughout meiotic prophase almost linearly with time and ranged from 0·25 mg protein/bud at leptotene stage to 1·0 mg protein at late pachytene.

Microsporocyte nuclei were isolated according to Hotta *et al.* (1965).

(b) *Partial purification of lily microsporocyte endonuclease*

Microsporocyte endonuclease was partially purified from microsporocyte extracts of the variety Bright Star at pachytene stage. The microsporocyte extracts were diluted with 6 vol. 10 mM-Tris Cl (pH 7·0) and 10 mM-KCl. The diluted extracts were fractioned by the addition of solid $(NH_4)_2SO_4$. The 50 to 80% $(NH_4)_2SO_4$ fraction dissolved in 10 mM-Tris Cl (pH 7·5) generally contained 80% of the total activity and the specific activity ranged from 9 to 12 units/mg. (Endonuclease units are defined in Methods section (d).) The 50 to 80% $(NH_4)_2SO_4$ fraction was further purified by DEAE-cellulose column chromatography. The sample was adsorbed onto a column (1 $cm^2 \times 12$ cm) of DEAE-cellulose equilibrated with 10 mM-Tris Cl (pH 7·5). The activity was eluted with a 300 ml. linear concentration gradient of 0·0 to 0·7 M-KCl in 10 mM-Tris Cl (pH 7·5). More than 50% of the microsporocyte endonuclease eluted in 10 ml. generally between 0·25 to 0·35 M-KCl. The specific activity of the peak fractions ranged from 75 to 95 units/mg. No phosphatase or phosphodiesterase activities could be detected in the peak fractions.

Another endonuclease activity found in Bright Star microsporocyte extracts was eluted from the DEAE column at 0·15 to 0·18 M-KCl. This activity was not found in Cinnabar microsporocyte extracts. The activity was identical to endonuclease activity extracted from either Bright Star or Cinnabar somatic tissue with respect to its pH optimum (pH 5·6) and co-chromatography on DEAE-cellulose. This activity is referred to as the *pH 5·6 endonuclease.*

(c) *Filter retention assay of endonuclease activity*

DNA endonuclease assays were carried out according to a procedure modified from Geiduschek & Daniels (1965). The standard assay in 0·25 ml. contained 1 μg of ^{3}H-labeled T7 DNA (3 to 6×10^3 cts/min/μg), 0·75 μmole $Mg(C_2H_3O_2)_2$, 25 μmoles $NaC_2H_3O_2$ (pH 5·2) and up to 3 units of endonuclease. The reactions were carried out at 37°C for 15 min. The reactions were ended by the addition of 0·1 ml. of 0·1 M-EDTA (pH 7·5) and the DNA was denatured at 20°C for 10 min by the addition of 0·2 ml. of 1·0 N-NaOH. Seven ml. of 0·15 M-KCl, 0·02 M-EDTA and 0·05 M-Tris Cl, pH 7·5, were added to the denatured mixture. The mixture was poured over a presoaked Type B6 membrane filter (Schleicher & Schuell, 27 mm diameter) in a Millipore microanalysis filter holder at a flow rate of about 0·5 ml./sec. The filter was washed with 21 ml. of the same KCl–buffer solution in 7-ml. portions. The filters were dried and counted in a liquid-scintillation spectrometer.

The cts/min retained on the filter for each of the 15-min reactions were expressed as a percentage of the cts/min retained for the zero time reaction (% cts/min retained). The units of endonuclease activity were calculated according to the relationship presented in section (d) below.

(d) *Standardization of the filter retention assay and calculation of* r_0 *values*

The amount of isotopically labeled T7 DNA retained on nitrocellulose filters in the endonuclease assay was standardized with respect to the average number of phosphodiester bonds hydrolyzed (r_0)/single strand of T7 DNA. Endonuclease assay procedures as described previously were carried out with ^{3}H-labeled T7 DNA which was digested with varying concentrations of pancreatic deoxyribonuclease according to Weiss, Live & Richardson (1968). The terminated reactions were denatured and separate samples of the reaction were adsorbed on to nitrocellulose filters (as described in section (c) of Methods) or subjected to zonal velocity sedimentation analysis on alkaline sucrose density gradients (see section (e) of Methods). The sedimentation analysis was used to determine the average number of bonds broken (r_0) in order to relate that value to the "% cts/min retained" values obtained in the filter assay. The average number of bonds broken (r_0) was calculated according to Montroll & Simha (1940) from their equation (15a)

$$\alpha = \frac{(p + 2) + (p + 2)^2 - 2(1 + p)\,(M_w/M_0 + 1)}{(1 + p)\,(M_w/M_0 + 1)}$$

where $r_0 = \alpha p$ and where $p + 1$ is the number of monomeric units in a polymer. The value of $p + 1$ for single-stranded T7 DNA = 41,000 nucleotides if M_w, the molecular weight of the monomeric unit or the average nucleotide molecular weight, is considered to be 320 daltons for T7 DNA (Volkin, Astrachan & Countryman, 1958) and M_0, the initial molecular weight, for single-stranded T7 DNA is considered as $1{\cdot}32 \times 10^7$ daltons (Studier, 1965). M_w was determined from the sedimentation coefficient ($S^0_{20,w}$) of sedimented single-strand T7 DNA fragments. All distributions of T7 DNA fragments were unimodal and fairly symmetrical. The above relationships were applied when $r_0 > 2$ for T7 DNA, but when $r_0 < 2$ then equation (15b) of Montroll & Simha (1940) was similarly applied.

A plot of % cts/min retained *versus* log r_0 is presented in Fig. 1. The shape of the resulting curve varies with salt concentration of the solution in which the sample is applied to the filter, so that, except at extremes of r_0 ($r_0 < 2$ or $r_0 > 2 \times 10^3$), the % cts/min retained is higher with increased salt concentrations. At the salt concentration used in the standard assay (0·15 M-KCl) the curve approximates a straight line. The equation of the line from a linear regression analysis of the experimental points is % cts/min $= -27{\cdot}4 \log r_0 + 98{\cdot}4$. (95% confidence limits for the slope and the y intercept are $\pm 1{\cdot}2$ and $\pm 2{\cdot}7$, respectively.)

The plot (Fig. 1) indicates an increased sensitivity of the endonuclease assay for the initial stages of digestion (when r_0 is small), but it also reveals that the assay cannot distinguish any value of r_0 between 0 and 1. So, for these experiments when % cts/min retained > 97%, then r_0 was given the value of 0.

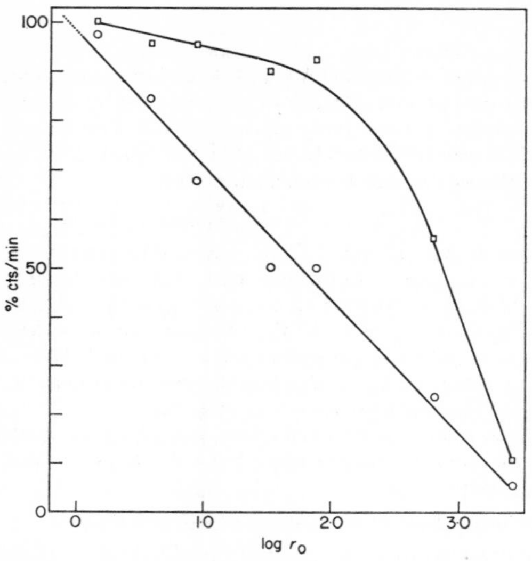

FIG. 1. Plot of % cts/min retained of limitedly digested ^{3}H-labeled T7 DNA *versus* log r_0 (r_0 = average number of breaks/single-strand T7 DNA molecule). Denatured T7 DNA fragments were adsorbed on to nitrocellulose filters in buffered 0·15 M-KCl solution (—○—○—), or in buffered 0·50 M-KCl solution (—□—□—). Linear regression analysis of values using buffered 0·15 M-KCl solution give % cts/min = −27·4 log r_0 + 98·4. (· · · · · ·) Extrapolation of regression line through 100% cts/min retained which is log $r_0 \cong 0$.

One unit of endonuclease activity in the standard assay is defined as that activity which will increase r_0 by 1/min i.e. that activity which will hydrolyze on the average one phosphodiester bond/single-strand T7 DNA molecule/min.

(e) *Sedimentation analysis of endonuclease digested DNA*

Zonal sedimentation analysis of endonuclease-digested T7 DNA was carried out on alkaline and neutral sucrose gradients. For alkaline gradients 0·1 ml. endonuclease-digested T7 DNA was denatured by addition of 25 μl. of 1 M-NaOH. The mixture was layered onto 5 to 20% (w/v) linear sucrose gradients in 0·9 M-NaCl, 0·1 M-NaOH and 0·001 M-EDTA. Alkaline gradients were centrifuged for 2·5 hr at 50,000 rev./min (5°C) in an SW50.1 rotor in a Beckman L2-65 ultracentrifuge. For neutral gradients, 0·1 ml. was layered onto 5 ml. 5 to 20% (w/v) linear sucrose gradients in 1·0 M-NaCl, 0·01 M-Tris Cl (pH 7·5) and 0·001 M-EDTA. Neutral gradients were centrifuged for 3·0 hr as described above.

Zonal sedimentation analysis of lily endonuclease-digested T7 DNA was carried out using the partially purified microsporocyte or somatic tissue activities described in section (b) of Methods. The reaction mixtures in 0·25 ml. contained 5 μg ^{3}H-labeled T7 DNA, 0·75 μmole $Mg(C_2H_3O_2)_2$, 25 μmoles $NaC_2H_3O_2$ (pH 5·2 or 5·6) and 7 units of lily endonuclease. In a time-course of digestion at 37°C, reactions were terminated at 0, 3 and 10 min by chilling and adding 20 μl. of 0·1 M-EDTA. The reaction mixtures were dialyzed in microdialysis chambers against 10 mM-Tris Cl (pH 7·5) and 10 mM-EDTA.

(f) *End-group analysis of microsporocyte endonuclease-digested DNA*

For end-group analysis of endonuclease-digested DNA, the partially purified microsporocyte endonuclease from variety Bright Star was used. The reaction mixture (1·0 ml.) contained 40 μg T7 DNA, 5 μmoles $Mg(C_2H_3O_2)_2$, 100 μmoles $NaC_2H_3O_2$ (pH 5·2) and approx. 20 units of microsporocyte endonuclease. The digestion was terminated after

5 min by addition of 50 μl. of 0·1 M-EDTA (pH 7·5) and by gentle shaking with 10 mM-Tris Cl (pH 7·5) buffer-saturated phenol. Phases were separated by centrifugation, the aqueous phase removed and dialyzed exhaustively against 50 mM-Tris Cl (pH 7·5).

The digested T7 DNA was subjected to zonal sedimentation analysis and to end-group analysis according to Weiss, Live & Richardson (1968).

(g) *Polynucleotide ligase and kinase*

Polynucleotide ligase assays modified from Weiss, Jacquemin-Sablon, Live, Fareed & Richardson (1968) for optimum conditions with the microspororocyte activity. The reaction mixtures in 0·25 ml. contained 12·5 μmoles Tris Cl (pH 7·5), 1·25 μmoles $MgCl_2$, 1·25 μmoles dithiothreitol, 25 nmoles ATP, 12·5 nmoles 5′-^{32}P-labeled phosphoryl-T7 DNA ($10{\cdot}2\times10^{-2}$ pmoles internal phosphomonoesters) and 17·8 μg protein in microsporocyte extracts. The standard ligase assay measures the quantity in pmoles of internal [^{32}P]phosphomonoesters rendered phosphatase resistant.

Polynucleotide kinase assays were performed according to Richardson (1965) under conditions optimum for microsporocyte activity. 1·0 mM-potassium phosphate (pH 7·5) was added to inhibit phosphatase.

(h) *Repair of endonuclease digested DNA*

T7 DNA, limitedly digested with the partially purified microsporocyte endonuclease as in section (e) of Methods or with pancreatic deoxyribonuclease, was post-treated with microsporocyte extracts (17 μg protein) from the zygotene–pachytene stage. The post-treatment was carried out under the conditions of the standard ligase assay except that the temperature was reduced to 21°C. The reactions were terminated and phenol extracted as in section (f) of Methods. The T7 DNA reaction products were subjected to sedimentation analysis according to section (e) of Methods.

(i) *Phosphatase and phosphodiesterase assays*

Phosphatase assays from Garen & Levinthal (1960) and phosphodiesterase assay from Koerner & Sinsheimer (1957) were modified for optimum conditions with lily activities. Reactions were modified by using 50 mM-$NaC_2H_3O_2$ (pH 5·5) for phosphatase assays and using 75 mM-$NaC_2H_3O_2$ (pH 4·2) for phosphodiesterase assays.

4. Results

The results obtained from our study are presented in the following order: (a) demonstration of an endonuclease activity in meiotic cells; (b) the temporal pattern of this activity in relation to meiotic stage; (c) the production of single strand or "repairable" breaks by the endonuclease activity; (d) the production of 5′-hydroxyl termini by the action of the endonuclease activity; (e) demonstration of a polynucleotide ligase activity in meiotic cells and its temporal pattern during the meiotic cycle; (f) demonstration of a polynucleotide kinase activity in meiotic cells and its temporal pattern during the meiotic cycle and (g) the repair of endonuclease-produced breaks by the activities of polynucleotide ligase and kinase.

(a) *Properties of the pH 5·2 microsporocyte endonuclease activity*

Crude or partially purified extracts from Cinnabar lily tissues contain either one of two types of endonuclease depending upon the tissue of origin (Fig. 2). Meiotic cells (microsporocytes) yield an activity that has a sharp pH optimum at 5·2. Repeated trials under various conditions to find other endonuclease activities in meiotic tissues over a range extending from pH 4·0 to 9·5 were unsuccessful.

The microsporocyte endonuclease preferentially hydrolyzed double-strand DNA substrates. The activity of the extracts with denatured DNA substrates (>0·15 unit/mg protein) was approximately 10% the activity with native DNA (1·8 units/mg protein). DNA from other sources challenged with these extracts was digested with comparable efficiency. The activity of the enzyme under standard assay conditions with

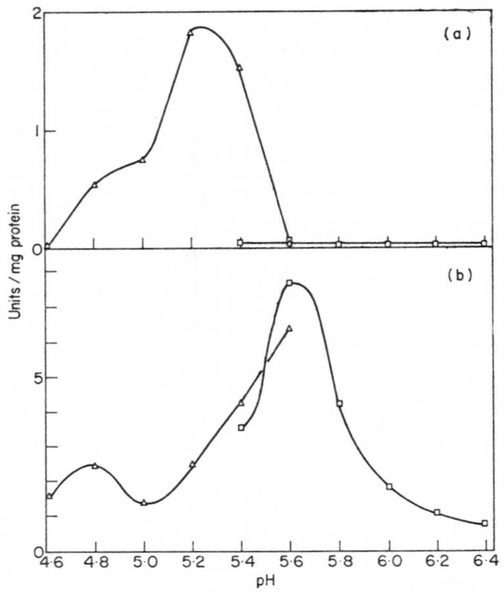

FIG. 2. pH optima profiles for endonuclease activities from microsporocyte (pachytene stage) extract in (a) and from somatic tissue (perianth) extract in (b). Substrate for assay was native ^{3}H-labeled T7 DNA. 0·1 M-Sodium acetate buffer from pH 4·6 to 5·6, —△—△—; 0·075 M-Tris–maleate buffer from pH 5·4 to 6·4, —□—□—.

T7 phage DNA was 1·8 units/mg protein, with *B. subtilis* DNA 1·7 units/mg protein and with *Lilium* DNA 1·5 units/mg protein. Filter retention assays performed using *B. subtilis* and *Lilium* DNA gave only semi-quantitative results, since the initial population of substrate DNA was not as homogeneous in molecular size as the population of T7 DNA molecules used in the standard assay. The microsporocyte endonuclease attack on these substrates was judged to be endo- and not exonucleolytic since the products of limited digestion (approx. 60% cts/min retention by filter assay) showed no mononucleotides or small oligonucleotides. All radioactively labeled digestion products remained at the origin when analyzed by descending paper chromatography using 1-propanol–ammonia–water solvent of Hanes & Isherwood (1949).

The pH 5·2 microsporocyte endonuclease demonstrates properties shared by other "acid" endonucleases (see review by Bernardi, 1968). The activity possesses a non-stringent requirement for added divalent cations. Optimum activity (1·8 units/mg protein) was found at 3 mM-Mg^{2+} concentration, while with no added Mg^{2+} the activity was 85% of optimum. High Mg^{2+} concentrations were inhibitory: at 75 mM-Mg^{2+} the activity fell to 30% of that at 3 mM. The activity is inhibited by bihelical RNA. 50% inhibition in the standard assay was found at 20 μg/ml. of *E. coli* sRNA (obtained from Dr T. C. Pinkerton) and the minimum concentration which gave minimum detectable inhibition (>10%) was 4 μg/ml. Concentrations of this RNA which gave effective inhibition of the endonuclease were far above the estimated concentrations of soluble RNA found in crude extracts. Ribonuclease treatment of microsporocyte extracts failed to enhance endonuclease activity. The microsporocyte extract endonuclease did not appear restricted to cytoplasmic components because 40 to 50% of the total cell endonuclease activity was found in nuclear fractions.

Somatic tissues of variety Cinnabar were tested for the presence of the microsporocyte endonuclease activity. Extracts from such tissues generally showed a higher level of endonuclease activity, e.g. perianth tissue 7·6 units/mg, leaf tissue 6·2 units/mg, compared to microsporocyte tissue 1·8 units/mg. However, activities from such tissues were optimal at pH 5·6 rather than at the meiotic tissue optimum of pH 5·2 (Fig. 2(b)).

That these differences in pH optima reflected real differences in properties between the somatic and meiotic ativities rather than some trivial difference in the nature of the extracts was indicated by several observations. Partial purification by ammonium sulfate precipitation and DEAE-cellulose column chromatography as described in Methods was not accompanied by a shift in pH optima of the respective activities. Furthermore, the activities which elute at different KCl concentrations (0·25 to 0·35 M for 85% recovery of the pH 5·2 endonuclease and 0·14 to 0·19 M for 90% recovery of the perianth pH 5·6 endonuclease) could be recovered separately from DEAE after previous mixing of somatic and meiotic extracts. Even where the two activities coexist in the same tissue, as in the case of the meiotic cells of Bright Star, they can be separated with the same elution pattern as described for the mixed extracts.

(b) *Meiotic stage variations in microsporocyte endonuclease activity*

Extracts from lily Cinnabar, microscoporocytes taken at different meiotic prophase stages were tested at limiting enzyme concentrations for the pH 5·2 microsporocyte endonuclease activity. It was found that the microsporocyte specific endonuclease activity varied with meiotic prophase stage. In a representative experiment illustrated in Figure 3, it can be seen that the endonuclease activity was not detectable in microsporocyte extracts in premeiotic interphase or in early prophage stages. Endonuclease

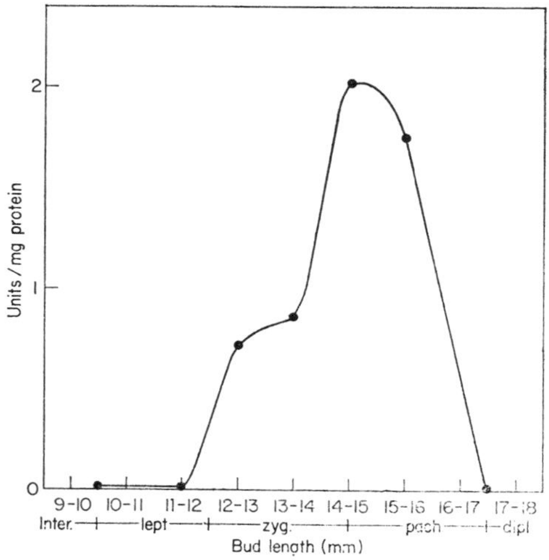

FIG. 3. Endonuclease activity in microsporocyte extracts at premeiotic interphase (inter.) and various meiotic prophase stages (Lept., leptotene; zyg., zygotene; pach., pachytene; dipl., diplotene.) Bud length is correlated with meiotic stage of microsporocytes.

activity was detectable in early zygotene stages and reached maximum levels in extracts from zygotene–pachytene stage microsporocytes. The endonuclease specific activity declined through the pachytene stage and reached nearly undetectable levels at diplotene. It should be pointed out that, in six repeated experiments performed over two months during the lily growing season, the profile of specific endonuclease activity *versus* meiotic stage shown in Figure 3 and the amount of maximum activity was generally reproduced, but that the amount of endonuclease activity at any given stage of midprophase could vary by as much as 50% between different series of experiments.

No factor, inhibitory or stimulatory to the microsporocyte endonuclease activity, was found by mixing extracts prepared at different meiotic prophase stages. Mixtures of 9 to 10-mm bud (late interphase) extracts (0·02 unit, 131 μg protein) with 14- to 15-mm (zygotene–pachytene) extracts (0·28 unit, 120 μg protein) gave 0·27 unit of activity in the standard assay.

As a comparison, the perianth tissue (somatic tissue which surrounds the microsporocytes in the bud) was tested for variations in the pH 5·6 endonuclease activity relative to the meiotic stage of the microsporocytes. The specified activity of perianth extracts was similar in 9- to 10-mm buds with interphase microsporocytes (6·8 units/mg protein) and in 14- to 15-mm buds with pachytene stage microsporocytes (7·6 units/mg protein).

(c) *Single-strand breaks by the microsporocyte endonuclease*

The finding that the microsporocyte endonuclease was correlated in time with the zygotene–pachytene stage warranted further examination of the properties of the endonuclease activity. A question particularly relevant to DNA breakage and rejoining activity is the nature of the endonuclease-produced break.

A comparison of velocity sedimentation properties of endonuclease-digested DNA before and after denaturation provides a means for determining whether the microsporocyte endonuclease produces single- or double-strand breaks in bihelical DNA molecules. Figure 4 shows alkaline sucrose density-gradient profiles of denatured DNA and neutral gradient profiles of undenatured DNA for each of the following treatments of the original native T7 DNA: (a) not digested, (b) limitedly digested with pH 5·2 endonuclease, (c) limitedly digested with pH 5·6 endonuclease.

The denatured form of T7 DNA previously digested with the pH 5·2 microsporocyte endonuclease sediments on alkaline gradients at increasingly slower rates with time of digestion. From the estimation of the sedimentation coefficient of the resulting T7 DNA fragments, the average number of breaks (r_0) per T7 DNA single strand was calculated to be 3·5 ($S^0_{20,\mathrm{w}} = 26$ s) for three minutes digestion and to be 8·0 ($S^0_{20,\mathrm{w}} =$ 20 s) for ten minutes digestion. Undenatured DNA from the same digests, sediments on neutral gradients at nearly the same rate as the undigested control ($S^0_{20,\mathrm{w}} = 37$ s) (Studier, 1965), with the exception of some trailing from the main peak. The neutral gradient profiles of the microsporocyte endonuclease-digested DNA exclude the possibility that the microsporocyte acts principally to break single strands, for double-strand breaks would reduce neutral gradient sedimentation rates of the digested DNA in accordance with the values of r_0 obtained from alkaline gradient profiles.

Unlike the action shown by the microsporocyte endonuclease, both the denatured and native forms of T7 DNA, previously digested with the pH 5·6 endonuclease, show decreasing sedimentation rates with increasing time of digestion. From alkaline

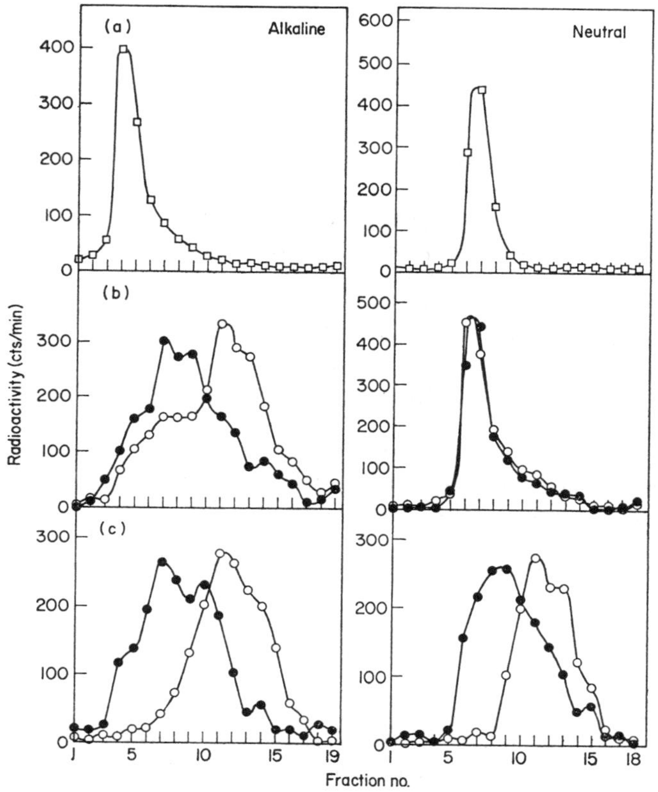

FIG. 4. Zonal velocity sedimentation analysis on alkaline and neutral sucrose density gradients of ^{3}H-labeled T7 DNA limitedly digested with lily endonucleases. In (a) undigested T7 DNA (—□—□—), in (b) T7 DNA digested with the pH 5·2 microsporocyte endonuclease and in (c) T7 DNA digested with the pH 5·6 somatic tissue endonuclease. Digested for 3 min (—●—●—) or for 10 min (—○—○—). Recovery of radioactivity from gradient profiles was virtually 100% of total input cts/min.

gradients, the r_0 value for single-strand T7 DNA for three minutes digestion was 3·2 ($S^0_{20,w} = 27$ s) and for ten minutes digestion was 12 ($S^0_{20,w} = 18$ s). On neutral gradients for three minutes digestion, r_0 for double-strand DNA = 3·0 ($S^0_{20,w} = 25$ s), and for ten minutes digestion $r_0 = 11$ ($S^0_{20,w} = 18$ s). Since an equivalent number of breaks per DNA molecule is found for a single sample by both alkaline and neutral gradient analysis, then the pH 5·6 somatic tissue endonuclease must produce predominantly double-strand DNA breaks.

The finding that the microsporocyte endonuclease produces initially single-strand breaks contrasts with the action of the pH 5·6 endonuclease from Lily somatic tissue and the other "acid" endonucleases which have been reported to hydrolyze DNA by double-strand breaks (Bernardi, 1968). The action of other acid endonucleases is thus to produce unrepairable DNA breaks whereas the breaks produced by the microsporocyte endonuclease do not exclude the possibility of repair.

TABLE 1

End-group analysis of endonuclease-digested T7 DNA

	T7 DNA	pH 5·2 Microsporocyte endonuclease	Phosphatase	Polynucleotide kinase	Radioactivity (cts/min)	No. 5′- OH termini‡ / intact single-strand T7 DNA molecule	
(1)	+	−	−	−	0	—	
(2)	+	−	−	+	927	0·08	Δ1·25
(3)	+	+†	−	+	14,383	1·33	
(4)	+	−	+	+	9854	0·91	Δ1·06
(5)	+	+†	+	+	22,368	2·07	

† for these samples $1 < r_0 < 2$.

‡ $$\frac{\text{No. 5}'\text{- OH termini}}{\text{intact single-strand T7 DNA molecule}} = \frac{\text{cts/min }^{32}\text{P incorporated}}{\text{spec. act. of }[\gamma\text{-}^{32}\text{P}]\text{ATP } (1{\cdot}89\times10^{7}\text{ cts/min/nmole})} \times \frac{4{\cdot}0\times10^{4}\text{ nucleotides}}{\text{n-moles T7 DNA } (23{\cdot}0\text{ nmoles})}$$

(d) *End-group analysis of microsporocyte endonuclease-produced DNA termini*

End-group analysis of partially digested T7 DNA can be used to determine whether the phosphodiester bond hydrolyzed by the microsporocyte endonuclease is adjacent to the 3′ or 5′ carbon of the deoxyribose moiety. For such analysis, polynucleotide kinase from phage T4 infected *E. coli* is used to catalyze the quantitative transfer of [^{32}P]orthophosphate from [γ-^{32}P]ATP to DNA termini (Weiss, Live & Richardson, 1968). Polynucleotide kinase exhibits a selectivity for 5′ termini but does not catalyze the transfer of a phosphate group to a 5′ terminus that is already phosphorylated (Richardson, 1966). Polynucleotide kinase, therefore, quantitatively marks the appearance of 5′-hydroxyl termini, or in conjunction with phosphatase, the appearance of any 5′ termini produced by microsporocyte endonuclease hydrolysis of T7 DNA.

The results of the end-group analysis are shown in Table 1. The substrate in this analysis was limitedly digested to a similar molar concentration of DNA breaks as obtained in the standard assay using the radioactively labeled substrate. Line 3 indicates that 1·25 5′-hydroxyl end groups are produced on each single strand of T7 DNA by limited digestion with the partially purified microsporocyte endonuclease. That such 5′-hydroxyl end-groups are absent before endonuclease digestion is shown in line 2. Line 5 indicates that the endonuclease-digested DNA treated with phosphatase can be additionally phosphorylated to nearly the same extent (1·06 end groups/T7 DNA single-strand molecule) as without phosphatase treatment. Since a difference in nearly one 5′ terminus per strand is found between phosphatase-treated and untreated endonuclease digests, it may be concluded that the partially purified endonuclease is not expressing a sufficient phosphatase activity to obscure the analysis.

Because new 5′ termini are phosphorylated to the same extent before or after phosphatase treatment, the microsporocyte endonuclease must yield 5′-hydroxyl and 3′-phosphoryl groups by hydrolyzing the phosphodiester bond adjacent to 5′-deoxyribose.

(e) *Properties and meiotic stage variations of microsporocyte polynucleotide ligase*

Extracts from meiotic prophase microsporocytes possess a demonstrable polynucleotide ligase activity. Ligase activity was assayed on T7 DNA fragments produced by pancreatic DNase as a function of the increase in either the sedimentation rate or in phosphatase-resistant radioactivity from internal 5′[^{32}P]phosphomonoesters. (Table 2 and Fig. 5) Both assays measure the capacity of the microsporocyte extracts to join

TABLE 2

Polynucleotide ligase assay reaction requirements

Components	pmoles/hr /mg protein
Complete	7·04
minus ATP	0·32
minus ATP, +25 nmoles NAD	0·91
minus Dithiothreitol	1·23
minus Mg^{2+}	0·24
minus Microspore extract	0·10
minus Microspore extract, + 29 μg protein in bud tissue extract	$<$0·10

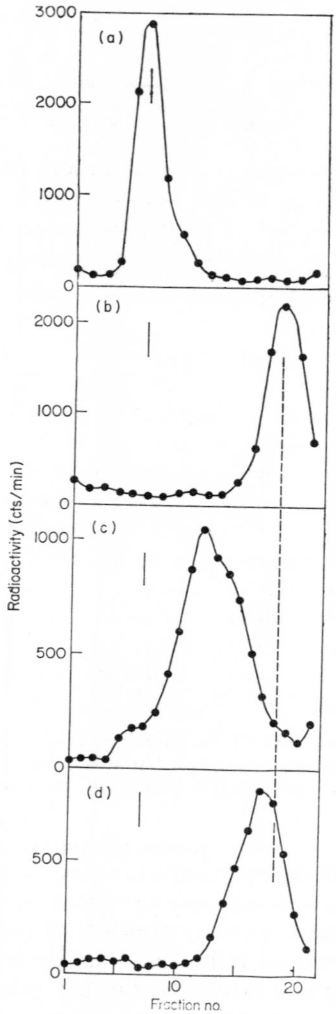

FIG. 5. Determination of polynucleotide ligase in microsporocyte extracts by alkaline sucrose gradient sedimentation analysis of T7 DNA fragments.

Joining activity is observed as the increase in sedimentation of DNA fragments. (a) Intact, denatured T7 DNA; (b) T7 DNA fragments from limited digest with pancreatic deoxyribonuclease; (c) post-treatment of DNA fragments with microsporocyte extracts and complete ligase reaction mix (complete); (d) post-treatment of DNA fragments with extracts and reaction mix minus ATP (-ATP control). Line is position of ^{32}P-labeled T7 DNA marker.

juxtapositioned 3′-hydroxyl and 5′-phosphoryl groups at DNA single-strand breaks.

The polynucleotide ligase specific activity in microsporocyte extracts (Table 2) is 500-fold less than that reported for crude bacterial extracts following phage infection (Weiss, Jaquemin-Sablon, Live, Fareed & Richardson, 1968), but twofold greater than that reported for mammalian mitotic cells (Lindahl & Edelman, 1968). The microsporocyte ligase activity requires ATP as a co-factor (Table 2 and Fig. 5) similar to the co-factor requirement for the phage-induced ligase (Weiss & Richardson, 1967) and the mammalian cell ligase (Lindahl & Edelman, 1968). The microsporocyte ligase is dependent on a sulfhydryl reagent and divalent cations (Mg^{2+}) in the reaction mix, and shows a pH optimum at pH 7·4.

Extracts prepared from microsporocytes at various meiotic stages and assayed at limiting enzyme concentrations showed that the specific ligase activity varied with meiotic stage of microsporocytes (Fig. 6). The premeiotic interphase cell extracts contained high specific ligase activity. Ligase activity decreased during leptotene,

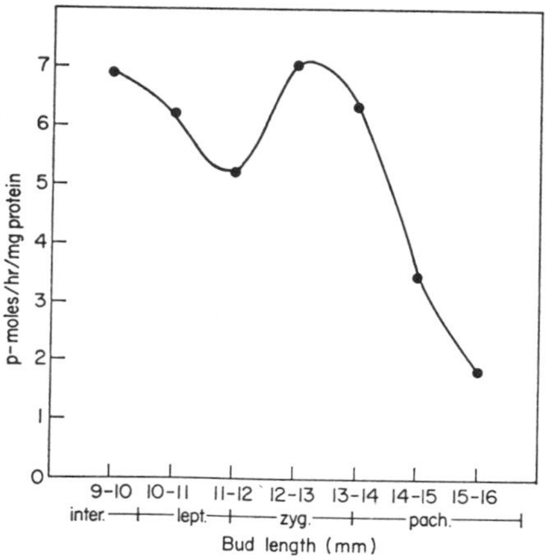

FIG. 6. Polynucleotide ligase activity in microsporocyte extracts at premeiotic interphase (inter.) and various meiotic prophase stages (lept., leptotene; zyg., zygotene; pach., pachytene). The standard assay measures the quantity in pmoles of internal phosphomonoesters rendered phosphatase resistant.

returned to maximum levels at zygotene–pachytene and then declined steadily throughout pachytene. This same pattern was reproduced in several experiments. The activity was not detectable by these assay procedures in microsporocytes after pachytene or at any time in somatic lily tissue (Table 2). If the specific activity of ligase in microsporocytes is expressed per cell rather than per mg protein, then the specific activity of zygotene–pachytene extracts is twofold greater than that of interphase extracts.

(f) *Properties and meiotic stage variations of microsporocyte polynucleotide kinase*

Optimum conditions for polynucleotide kinase activity in microsporocyte extracts were similar to those described by Richardson (1965). The assay measured the capacity of extracts to transfer [^{32}P]orthophosphate from [γ-^{32}P]ATP to 5′-hydroxyl terminated DNA. To determine the amount of polynucleotide kinase activity at various meiotic prophase stages, microsporocyte extracts were prepared as in the ligase assays and tested at limiting enzyme concentrations (Fig. 7). Polynucleotide kinase activity was present in premeiotic interphase extracts and showed slightly increasing specific activities until mid-pachytene stage when the activity sharply declines. Phosphatase assays were carried out on the same extracts to determine if the variations in kinase activity were actually a reflection of variations in phosphatase activities in the same extracts. Figure 7 indicates that the phosphatase-specific activities are high, but remain fairly constant in these extracts and, therefore, cannot account for the variations in assayable kinase activity.

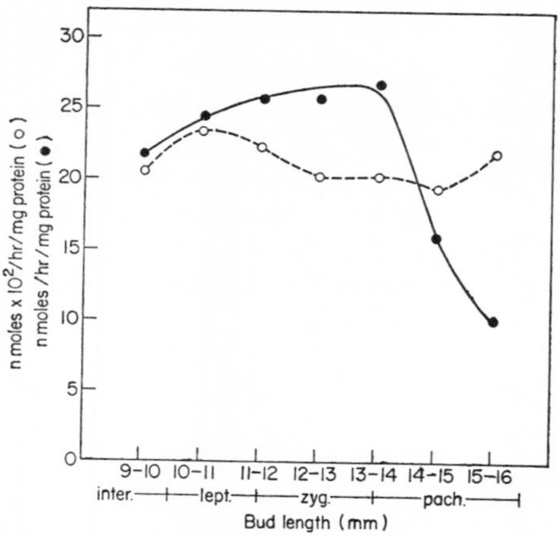

FIG. 7. Polynucleotide kinase (—●—●—) and acid (pH 5·6) phosphatase (--○--○--) activities in microsporocyte extracts at premeiotic interphase (inter.) and various meiotic prophase stages (lept., leptotene; zyg., zygotene; pach., pachytene). The standard kinase assay, measures the quantity in nmoles of [^{32}P]orthophosphate from [γ-^{32}P]ATP rendered acid-insoluble by transfer to 5′-hydroxyl terminated DNA fragments. The standard phosphatase assay measures the quantity in nmoles of *p*-nitrophenol liberated from *p*-nitrophenyl phosphate.

A per cell instead of a per mg protein comparison of specific activities indicates that the kinase activity at zygotene–pachytene peak is twofold greater than the activity at premeiotic interphase. The specific kinase activity of lily somatic tissue (e.g. perianth tissue) is 20% of the maximum activity found in meiotic tissue.

The specific activity of microsporocyte polynucleotide kinase in crude extracts is comparable to that found in extracts from phage-infected *E. coli* (Richardson, 1965) and fourfold greater than that reported for rat liver nuclei extracts (Novogradsky, Moshe, Traub & Hurwitz, 1966).

Microsporocyte polynucleotide kinase appears to have the same DNA substrate requirements as the phage-induced kinase (Richardson, 1966). Although analysis of the kinase reaction products was not carried out, reaction requirements indicate that the enzyme will only phosphorylate 5′-hydroxyl termini on DNA. The microsporocyte activity efficiently phosphorylates 5′-hydroxyl-terminated DNA (micrococcal nuclease DNA digests) at a rate of 25 nmoles/hr/mg protein, whereas it phosphorylates 5′-phosphoryl terminated DNA (pancreatic endonuclease DNA digests) at a rate <2 nmoles/hr/mg protein. Without added DNA substrate, rates were <1 nmole/hr/mg protein.

(g) *Repair of microsporocyte endonuclease digested DNA*

The question of whether the enzyme activities present in the crude extracts of meiotic cells could in fact both nick and repair DNA is of particular interest to the problem of meiotic recombination. This possibility was tested by first limitedly digesting ^{3}H-labeled T7 DNA with microsporocyte endonuclease and then incubating the digestion product with microsporocyte extracts under conditions favoring both polynucleotide kinase and ligase activity. Figure 8(a) is a velocity sedimentation profile of denatured T7 DNA which has been limitedly digested, before denaturation, with the partially purified pH 5·2 microsporocyte endonuclease. The profile is neither fully unimodal nor symmetrical, but, as an approximation, the average number of

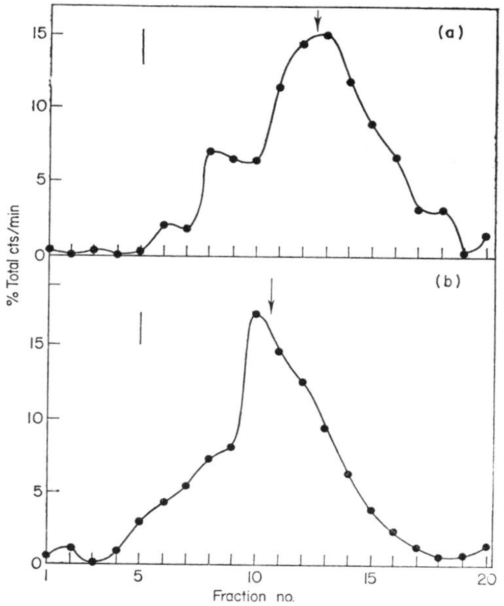

FIG. 8. (a) Alkaline sucrose gradient sedimentation profile of T7 DNA treated with microsporocyte endonuclease.

(b) Sedimentation profile of the endonuclease-treated DNA post-treated with microsporocyte extracts containing polynucleotide kinase and ligase activities. Line is position of sedimented intact T7 DNA marker. Arrow is least squares determined mean of the T7 DNA fragment distribution. Sedimentation coefficient is calculated from position of arrow.

breaks (r_0) per single strand T7 DNA molecule is ten. Figure 8(b) shows the distribution of endonuclease-digested T7 DNA post-treated with microsporocyte extracts containing, among other activities, polynucleotide kinase and ligase activities. An increase in the sedimentation rate for the population of single-strand DNA fragments was observed. The value of r_0 for the population of repaired molecules was estimated as 5. Thus, the difference in r_0 for the DNA before and after incubation with the microsporocyte extracts shows that at least 50% of the breaks produced by the microsporocyte endonuclease are repairable by microsporocyte extracts. To date, no complete repair of nicked T7 DNA to populations of whole molecules has been achieved with microsporocyte crude extracts using either pancreatic endonuclease or microsporocyte endonuclease-digested T7 DNA as substrate. Two reasons for difficulty in achieving complete repair appear probable. (1) Under conditions favorable to polynucleotide kinase and ligase activities, microsporocyte extracts exhibit some DNA hydrolytic activities including endonuclease. (2) DNA digested in extended reactions with the partially purified microsporocyte endonuclease begins to show some unrepairable breaks.

5. Discussion

In these studies, a set of complementary DNA breakage and repair activitities was found in extracts of synchronously developing microsporocytes at the late zygotene–

pachytene stage of meiosis. The co-ordinate appearance of endonuclease, polynucleotide kinase and ligase activity points to the conclusion that these activities, acting in concert, sponsor chromosomal DNA breakage and rejoining events during the zygotene–pachytene stage.

In the accompanying paper (Hotta & Stern, 1971), it was demonstrated that the synthesis of a small amount of DNA, called pachytene DNA, occurs during the same interval in which the breakage and repair activities, found in this study, appear. Further, the DNA synthesized in this interval appears to be the product of breakage and repair events, in that pachytene DNA shows properties of repair replication DNA synthesis.

The properties described by Hotta & Stern (1971) which characterize pachytene DNA as the product of repair replication DNA synthesis are as follows. First, the amount of pachytene DNA synthesized is small and variable. Second, pachytene DNA is not satellite DNA, but it has a base composition and buoyant density which reflects that of the bulk DNA. Third, pachytene DNA appears to be a turn-over synthesis of DNA and not a late replication of some portion of the genome. Last, pachytene DNA appears to be non-conservatively replicated.

Stimulation of a repair replication DNA synthesis at a defined time in meiotic prophase would require an internal mechanism for the breakage of DNA strands in that interval. The transient microsporocyte endonuclease activity found in the present study appears as the agent for effecting such breakage and for stimulating DNA repair replication at the pachytene stage. Alternative explanations to account for the stimulation of DNA synthesis at pachytene, such as the transient appearance of DNA polymerase or an abrupt expansion of DNA precursor pools, lack experimental support. The level of DNA polymerase activity and the size of deoxyriboside or deoxyribotide precursor pools are relatively invariant during meiotic prophase (Hotta, unpublished observations).

The stimulation of DNA synthesis by endonucleolytic attack suggested by these studies, is well documented, on the other hand, by *in vitro* DNA synthesis experiments using purified *E. coli* DNA polymerase. The explanation offered for stimulated synthesis is that endonucleolytic attack increases polymerase binding sites and provides new template for primed DNA synthesis (Englund, Kelly & Kornberg, 1969; Kelly, Cozzarelli, Deutscher, Lehman & Kornberg, 1970). In the case of repair replication DNA synthesis in the lily microsporocyte, it is undetermined if the endonuclease action alone would suffice in opening new template for primed synthesis.

The transient appearance of this endonuclease activity during meiotic prophase would seem to be a unique characteristic of the microsporocytes and not a reflection of metabolic changes in the surrounding bud tissue, because the pH 5·6 endonuclease activity in the bud tissue shows no correlated variation. Neither does it appear that the microsporocyte endonuclease activity is simply one in a co-ordinate expression of many DNA hydrolytic activities at a particular period in the meiotic cycle. The microsporocyte endonuclease activity varies against a relatively constant background of other DNA hydrolytic activities, such as acid phosphatase and phosphodiesterase. The new 3′ terminus produced by the endonuclease, whether it is found in the original "nick" or in a subsequent "gap" (e.g. produced by exonucleolytic attack on the 5′ terminus at the nick) may not serve, without modification, to prime repair replication synthesis. The microsporocyte endonuclease was found in the present study to give 3′-phosphoryl nicks. Therefore, subsequent modification of the 3′ terminus by phosphatase or 3′ → 5′exonuclease action may be required for stimulation of primed repair replication DNA synthesis in the microsporocyte.

The appearance of the microsporocyte endonuclease activity during the zygotene–pachytene stages of meiotic prophase was accompanied by an increase in the level of an important repair activity, polynucleotide ligase. Maximum levels of ligase activity were found at times in the meiotic cycle when it might be expected that DNA metabolic processes would require such activity. Thus, at late premeiotic interphase just following S-phase DNA synthesis, the polynucleotide ligase activity is relatively high. If the normal S-phase replication of DNA in lily microsporocytes is discontinuous, as it is in other reported organisms (Okazaki, Okazaki, Sakabe, Sugimoto & Sugino, 1968; Schandl & Taylor, 1969), then the presence of polynucleotide ligase would be needed to join newly replicated fragments. Unfortunately, to date no measurement of ligase activity during premeiotic S-phase has been made due to the technical difficulty in obtaining clean microsporocyte preparations at the early premeiotic intervals. Nevertheless, what is of interest is that the highest ligase activities transiently appear in microsporocytes at the time of prophase repair replication synthesis, when a comparably small amount of DNA is synthesized. However, the cell might have similar requirements for rejoining DNA at the zygotene–pachytene period as it does immediately following the normal S-phase DNA synthesis period.

If the microsporocyte ligase and endonuclease activities are to function together in chromosomal breakage and rejoining during pachytene, then one aspect of their mutual incompatibility must be resolved. As mentioned above, the microsporocyte endonuclease produces 3′-phosphoryl and 5′-hydroxyl end-groups at nicks, but the microsporocyte ligase, like other reported polynucleotide ligases, requires juxtapositioned 3′-hydroxyl and 5′-phosphoryl end groups for joining (Zimmerman, Little, Oshinsky & Gellert, 1967). However, this difficulty appears to be overcome as indicated by the ability of the microsporocyte extracts to rejoin endonuclease-produced breaks in DNA, albeit incompletely.

The capacity of microsporocyte extracts to rejoin these DNA breaks can be accounted for by the presence of two enzymic activities which can modify endonuclease-produced DNA termini. One of these activities is the non-specific phosphatase, present in these extracts at any meiotic prophase stage, which could dephosphorylate 3′ termini. The other activity is polynucleotide kinase which could readily account for phosphorylation of 5′ termini.

The kinase activity, like the ligase activity, showed maximum activity in extracts from the zygotene–pachytene stage. Thus, it appeared possible that microsporocyte polynucleotide kinase and ligase co-operate in the joining of microsporocyte endonuclease-produced DNA breaks. Both activities could be expressed in the same microsporocyte extract under joining conditions because both activities require ATP. However, confirmation of such a multistep mechanism for joining of 3′-phosphoryl–5′-hydroxyl breaks will only be obtained by purification of individual activities.

Based on the observations of DNA breakage and rejoining activities in this study, it is proposed that these activities function in the process of genetic recombination. Such a function is inferred from the temporal correlation of these activities with both the repair replication synthesis of DNA (Hotta & Stern, 1971) and the cytologically determined period of crossing-over (Peacock, 1968). All these phenomena occur during the late zygotene–pachytene stage of meiotic prophase.

A speculative scheme which describes the breakage and rejoining of chromosomal

DNA for genetic recombination has been constructed within the context of the observed DNA metabolic events in pachytene (Fig. 9). The scheme is based on the enzymic activities identified in the meiotic cells of lily and based partially on certain

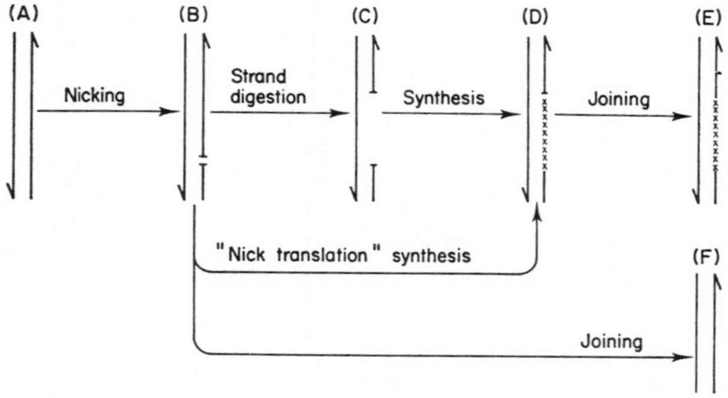

Fig. 9. Proposed scheme of DNA metabolic events during the zygotene–pachytene stage of meiosis in lily microsporocytes. Individual diagrams represent a short region of duplex DNA from a single chromatid. See text for explanation.

activities characterized in microbial systems (Richardson, 1969). The scheme begins with the nicking of duplex DNA in the meiotic chromatids (B). This step is considered as a controlling step, because all subsequent DNA metabolic activities in this scheme depend upon an initial nicking step. A deliberate nicking of chromosomal DNA at late zygotene to early pachytene is hypothesized in lieu of nicks resulting from incomplete repair of discontinuous DNA synthesis at S-phase since high levels of ligase activity are found in that premeiotic interval. The nicked DNA may be subject to one of several metabolic fates. A nick could be rejoined without incorporation of nucleotides (F). A nick might be repositioned by the process of "nick translation" described for *E. coli* DNA polymerase (Kelly *et al.*, 1970) and then joined (B-D-E) or a nick might be enlarged to a gap by strand digestion (C) and the gap closed by synthesis (D) followed by rejoining. Either one of the latter processes would result in repair replication synthesis without a net synthesis of DNA.

The possible relationship of these DNA metabolic events to a scheme of chromosomal recombination is shown in Figure 10. Two DNA duplexes (X) from homologous non-sister chromatids are multiply nicked during zygotene–pachytene when chromosomes are paired. Most frequently the duplexes will be repaired and rejoined without a reciprocal recombination event (Z). However, infrequently the non-covalently joined duplexes will successfully recombine, presumably by base pairing interaction between homologous strands of the DNA duplexes as described by Whitehouse (1963) and Thomas (1966). The product of such an event will be DNA duplexes (Y) which are reciprocally recombined (biparental recombinant molecules). The success of a recombination event may be governed by the nicking of DNA and by the metabolic fate of DNA following nicking. Strand digestion to yield partially single-stranded DNA may be required to provide a flexible DNA structure for the juxtapositioning of homologous DNA strands. Likewise, repositioning of a nick may be required to

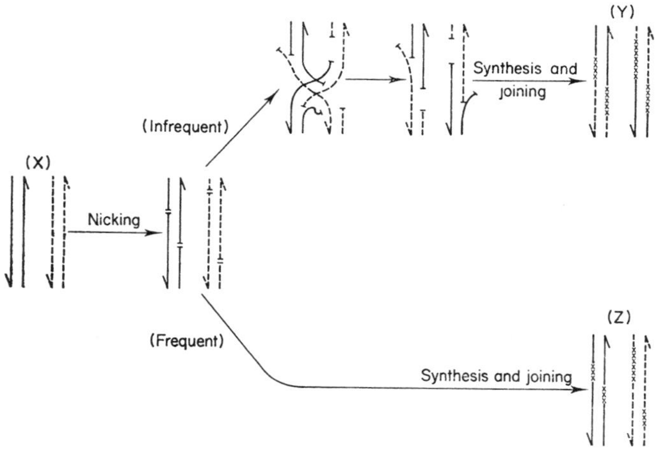

FIG. 10. Proposed scheme of recombination resulting from DNA metabolic events during zygotene–pachytene. Individual diagrams represent short regions of duplex DNA from homologous non-sister chromatids. The crossover event in the first diagram of the upper line is transposed in the following one for clarity. See text for explanation.

bring nicks into proximity with a single duplex or to alter the geometry between nicks on homologous duplexes.

The metabolic fate of DNA duplexes which are already reciprocally recombined may be significant to genetic information transfer. If such biparental recombinant molecules (Y) were subject to repair replication DNA synthesis, then this event may be recognizable in genetic analysis as a conversion or non-reciprocal recombination event. It should be pointed out, however, that completion of a reciprocal recombination event in this scheme is not contingent on repair replication synthesis.

One postulate of this scheme, that successful recombination arises infrequently from many breakage and rejoining events in meiotic prophase, was drawn from the following estimation of the frequency of repair replication events: The amount of DNA synthesized during repair replication DNA synthesis is about 0·1% of the genome. The amount of DNA per genome or per nucleus in lily Bright Star microsporocytes is $1{\cdot}1 \times 10^{-10}$ g (Coleman & Stern unpublished observation). Therefore, $1{\cdot}1 \times 10^{-13}$ g DNA or $6{\cdot}6 \times 10^{10}$ daltons of DNA per nucleus are synthesized at this time. DNA synthesized during repair replication gives no "heavy shift" on CsCl density gradients even when the bulk DNA is sheared to 5×10^{5} daltons (Hotta & Stern, 1971). Since the accuracy in the determination of the density shift in the gradient analysis is limited to about 20% of the density alteration of the fully substituted hybrid, then the minimum number of repair replicated regions can be estimated within this error according to the equations of Bonhoeffer & Gierer (1963). For every meiotic nucleus the minimum number of repair replication regions is $6{\cdot}6 \times 10^{5}$. If one assumes that each DNA repair replication region is the product of one single-strand break, then there must be, at minimum, $6{\cdot}6 \times 10^{5}$ single-strand breaks per nucleus. This is a minimum estimate, because in the proposed scheme (Fig. 10) some breakage and reunion events are expected to occur without DNA synthesis. An estimate of the number of recombination events per nucleus in lily can only be reached by cytologically observed chiasmata frequencies. In general, an average of three chiasmata per chromosomal pair or 36 chiasmata per nucleus are found in lily (Sen, 1969). Each crossover event must require no less than four single-strand breakage events. Then, the estimated number of single-strand breaks involved in successful recombination per nucleus is about $1{\cdot}4\times10^2$. Therefore the upper limit of probability that a single breakage and repair event will be involved in a successful recombination act is $1{\cdot}4\times10^2/6{\cdot}6\times10^5=2\times10^{-4}$.

The elementary events of genetic recombination proposed in this scheme for a eucaryotic organism are much the same as those proposed for procaryotes. Nonetheless, all similarities between procaryotes and eucaryotes in their elementary recombination events should not obscure the differences which are known to exist in the over-all process. Eucaryotic organisms must be able to handle vastly more genomic material than procaryotes in the process of orderly recombination. While in a procaryote it seems possible that both random breakage events and random encounters between homologous DNA strands might lead to successful recombination, it is difficult to conceive of the same sort of randomness in eucaryotic recombination. This difficulty is illustrated by the low frequency of breakage and rejoining events in a eucaryotic meiotic genome such as that calculated above for lily. $6{\cdot}6\times10^5$ nicks per genome is only one break per $3{\cdot}3\times10^5$ phosphodiester bonds. Obviously, random single-strand breakage events throughout the entire genome at this frequency would not give any four breaks in proximity for a crossover event. More ordered recombinational events in eucaryotes may be mediated by the elaborate pairing structure, the synaptonemal complex, or by undetermined specificities of the DNA breakage and repair enzymes.

The observations on the appearance of the DNA breakage and repair activities support the conclusion that recombination occurs principally during the late zygotene–pachytene stages of meiotic prophase and not during the premeiotic S-phase (Grell & Chandley, 1965). However these experiments do not exclude the possibility of lower frequency recombination events outside of meiotic prophase, such as during premeiotic S-phase or the mitotic cell cycle. What emerges from these studies of lily meiotic prophase is that an efficient mechanism for genetic recombination is afforded by the co-ordination of chromosome pairing and the breakage and repair activities during meiosis.

This work was supported by National Science Foundation Grant GB 5173X and supplemented by U.S. Public Health Service Grant HD 03015 from the National Institute of Child Health and Human Development. One of us (S.H.H.) was supported by a National Institutes of Health postdoctoral fellowship U.S. Public Health Service no. 5 FO 2 GM 33202. We are grateful to Drs Donald R. Helinski and W. James Peacock for criticism of the manuscript and to Drs Yasuo Hotta and Norman B. Hecht for helpful discussions during the course of this study.

REFERENCES

Anraku, N. & Tomizawa, J. (1965). *J. Mol. Biol.* **12**, 805.

Bernardi, G. (1968). In *Advances in Enzymology*, ed. by F. F. Nord, vol. 31, p. 1. New York: Interscience.

Bonhoeffer, F. & Gierer, A. (1963). *J. Mol. Biol.* **7**, 527.

Englund, P. T., Kelly, R. B. & Kornberg, A. (1969). *J. Biol. Chem.* **244**, 3045.

Garen, A. & Levinthal, C. (1960). *Biochim. biophys. Acta*, **38**, 470.

Geiduschek, E. P. & Daniels, A. (1965). *Analyt. Biochem.* **11**, 133.

Grell, R. F. & Chandley, A. C. (1965). *Proc. Nat. Acad. Sci., Wash.* **53**, 1340.

Hanes, C. S. & Isherwood, F. A. (1949). *Nature*, **164**, 1107.

Holliday, R. (1964). *Genet. Res.* **5**, 282.
Hotta, Y., Bassel, A. & Stern, H. (1965). *J. Cell. Biol.* **27**, 451.
Hotta, Y., Ito, M. & Stern, H. (1966). *Proc. Nat. Acad. Sci., Wash.* **56**, 1184.
Hotta, Y. & Stern, H. (1971). *J. Mol. Biol.* **55**, 337.
Ito, M., Hotta, Y. & Stern, H. (1967). *Devel. Biol.* **6**, 54.
Kelly, R. B., Cozzarelli, N. R., Deutscher, M. P., Lehman, I. R. & Kornberg, A. (1970). *J. Biol. Chem.* **245**, 39.
Koerner, J. F. & Sinsheimer, R. L. (1957). *J. Biol. Chem.* **228**, 1049.
Lindahl, T. & Edelman, G. E. (1968). *Proc. Nat. Acad. Sci., Wash.* **61**, 680.
Lowry, O. H., Rosebrough, N. J., Farr, A. L. & Randall, R. J. (1951). *J. Biol. Chem.* **193**, 265.
Marmur, J. (1961). *J. Mol. Biol.* **3**, 208.
Meselson, M. & Weigle, J. J. (1961). *Proc. Nat. Acad. Sci., Wash.* **47**, 857.
Montroll, E. W. & Simha, R. (1940). *J. Chem. Phys.* **8**, 721.
Moses, M. J. (1969). *Genetics*, (*Suppl.*) **61**, 41.
Novogrodsky, A., Moshe, T., Traub, A. & Hurwitz, J. (1966). *J. Biol. Chem.* **241**, 2933.
Okazaki, R., Okazaki, T., Sakabe, K., Sugimoto, K. & Sugino, A. (1968). *Proc. Nat. Acad. Sci., Wash.* **59**, 598.
Peacock, W. J. (1968). In *Replication and Recombination of Genetic Material*, ed. by W. J. Peacock & J. R. D. Brock, p. 242. Canberra: Australian Academy of Science.
Rhoades, M. M. (1968). In *Replication and Recombination of Genetic Material*, ed. by W. J. Peacock & J. R. D. Brock, p. 229. Canberra: Australian Academy of Science.
Richardson, C. C. (1965). *Proc. Nat. Acad. Sci., Wash.* **54**, 158.
Richardson, C. C. (1966). *J. Mol. Biol.* **15**, 49.
Richardson, C. C. (1969). *Ann. Rev. Biochem.* **38**, 795.
Schandl, E. K. & Taylor, J. H. (1969). *Biochem. Biophys. Res. Comm.* **34**, 291.
Sen, S. K. (1969). *Nature*, **224**, 178.
Stahl, F. W. (1969). *Genetics*, (*Suppl.*) **61**, 1.
Studier, F. W. (1965). *J. Mol. Biol.* **11**, 373.
Taylor, J. H. (1965). *J. Cell Biol.* **25**, 57.
Thomas, C. A. (1966). In *Progress in Nucleic Acid Research and Molecular Biology*, ed. by J. N. Davidson & W. E. Cohn, p. 315. New York: Academic Press.
Thomas, C. A. & Abelson, J. (1966). In *Procedures in Nucleic Acid Research*, ed. by G. L. Cantoni & D. R. Davies, p. 553. New York: Academic Press.
Volkin, E., Astrachan, L. & Countryman, J. L. (1958). *Virology*, **6**, 545.
Weiss, B., Jacquemin-Sablon, A., Live, T. R., Fareed, G. C. & Richardson, C. C. (1968). *J. Biol. Chem.* **243**, 4543.
Weiss, B., Live, T. R. & Richardson, C. C. (1968) *J. Biol. Chem.* **243**, 4530.
Weiss, B. & Richardson, C. C. (1967). *Proc. Nat. Acad. Sci., Wash.* **57**, 1021.
White, P. R. (1963). In *The Cultivation of Animal and Plant Cells*, p. 60. New York: Ronald Press.
Whitehouse, H. L. K. (1963). *Nature*, **199**, 1034.
Zimmerman, S. B., Little, J. W., Oshinsky, C. K. & Gellert, M. (1967). *Proc. Nat. Acad. Sci., Wash.* **57**, 1841.

Physiology and Biochemistry of Meiotic Events

DEVELOPMENTAL CONTROL OF HETEROCHROMATIZATION IN COCCIDS[1]

SPENCER W. BROWN

THE coccids are a relatively small group of highly specialized homopteran insects. They are parasitic on plants and quite sedentary in behavior. Coccids are well known among cytogeneticists for their diverse and complicated genetic systems. Of these, the systems involving heterochromatization are probably of greatest general interest because of their bearing on the gene action problem.

The heterochromatin in the coccids is unlike typical heterochromatin in several respects. It is restricted to the male sex; it is formed anew in each generation during early embryogeny and discarded at the completion of meiosis.

The first definitive analysis of a coccid chromosome system of this sort was made by SCHRADER (1929) who observed that one entire chromosome set became heterochromatic during the early development of the male and remained so throughout his life. Meiosis was modified so that the heterochromatic set was segregated from the euchromatic set during the second meiotic mitosis. Only the euchromatic derivative proceeded to form sperm; the heterochromatic derivative slowly degenerated. (See also NUR, 1967a, on the chromosome system of this coccid, *Gossyparia spuria* (Modeer).)

Other coccids had already been shown to have genetic systems in which the males came from unfertilized eggs just as in the bees and wasps and other Hymenoptera. SCHRADER (1929) and shortly thereafter SCHRADER and HUGHES-SCHRADER (1931) suggested that the heterochromatin system could be homologized with true male haploidy. If the heterochromatic set were of paternal origin and if it were genetically inert, then the male would be genetically equivalent to a true haploid male arising from an unfertilized egg (see especially HUGHES-SCHRADER's review, 1948).

The common mealybug, *Planococcus citri* (Risso), has proved amenable to laboratory culture and has been used to test the SCHRADERS' hypotheses. In order to avoid confusion with certain other coccid systems involving heterochromatization, that found in the mealybug will be called simply the mealybug system; technically it is referred to as the lecanoid system.

Parental origin of the heterochromatic set was readily demonstrated by X-ray analysis. The chromosomes of the coccids, like those of apparently all other hemipterans (HUGHES-SCHRADER and RIS, 1941; HUGHES-SCHRADER and SCHRADER, 1961) have diffuse centromeres so that fragment chromosomes are potentially capable of survival, and reciprocal translocations can be formed in

[1] Research in coccid genetics at Berkeley has been supported by continuing grants from the National Science Foundation, currently GB7131.

all possible ways because acentrics and dicentrics are not produced. Parental contributions can thus be marked cytologically. From experiments involving over 800 embryos and at X-ray doses yielding 30 to 90 per cent of embryos with altered chromosomes, it became obvious that the euchromatic set was of maternal origin and the heterochromatic set was of paternal origin. Except for a small percentage of spontaneous changes, maternal irradiation induced aberrations in the euchromatic set of the sons and paternal irradiation induced them in the heterochromatic set (BROWN and NELSON-REES, 1961). It is also interesting to note that the X-ray treatment itself did not detectably influence heterochromatization. Thus the first of the SCHRADERS' hypotheses was verified; the heterochromatic set is of paternal origin.

The meiotic sequence.—Meiosis in the mealybug male is highly modified. The maternal, euchromatic set and the paternal, heterochromatic set are strikingly different during the first prophase, and no pairing occurs. Both sets divide equationally at first anaphase. During the second meiotic division, the euchromatic and heterochromatic sets are segregated to opposite poles. Only the euchromatic derivative forms sperm; the heterochromatic derivative slowly degenerates (HUGHES-SCHRADER, 1935, 1948; BROWN, 1959; BROWN and NUR, 1964; NELSON-REES, 1963).

According to the second of the Schraders' hypotheses, the heterochromatic set is genetically inert. Genes transmitted from the father to the son would thus not be expressed. If the cytological observations are correct, the paternal, heterochromatic set is discarded at the end of the meiotic sequence. The son would thus not transmit the genes received from the father. Or, conversely, the mealybug male would be like the true haploid male of the Hymenoptera in expressing and transmitting only maternal genes.

The gene for salmon eye color results in a light eye color quite distinct from the very dark red eye color of the wild-type mealybug. In crosses of stocks of wild-type and salmon, salmon behaved as a typical recessive in the daughters: the salmon phenotype appeared only when both parents were salmon (Table 1). In the sons, however, the maternal gene was always the one expressed. In the critical case (Table 1, line 3) the wild-type allele transmitted by the father was not expressed and the phenotype of the sons was the same as that of their salmon mothers. Thus, the second of the Schraders' hypotheses can be substantiated; the

TABLE 1

Crosses of wild-type (S) and salmon eye color (s) stocks of mealybugs

	Mother	Father[1]	Phenotype of Daughters	Phenotype of Sons	Expected genotype of sons[1]
1.	SS	S(S)	wild type	wild type	S(S)
2.	SS	s(s)	wild type	wild type	S(s)
3.	ss	S(S)	wild type	salmon	s(S)
4.	ss	s(s)	salmon	salmon	s(s)

[1] The gene from the paternal parent is given in parentheses.

TABLE 2

Test crosses of the two types of male heterozygous for salmon eye color

	Mother	Father[1]	Phenotypes of Daughters	Phenotypes of Sons
1.	ss	S(s)	wild type	salmon
2.	ss	s(S)	salmon	salmon

[1] The gene from the paternal parent is given in parentheses.

paternal, heterochromatic set is genetically inactive in the male mealybug (BROWN and WIEGMANN, unpubl.; BROWN and NUR, 1964).

The two types of heterozygous males (Table 1, lines 2 and 3) can be further tested by backcrossing them to females from the salmon stock (Table 2). Again, the sons expressed only the maternal contribution. The daughters, however, showed that the heterozygous males transmitted only the gene which they received from their mothers. Thus the inference from the post-meiotic elimination of the heterochromatic set has been verified; the mealybug male transmits only the genes received from the mother.

Similar results have been obtained with two other marker genes affecting wing morphology, shattered and banjo; in the latter, each wing is shaped like a banjo because of the collapse of the proximal part of the wing. No female coccid has wings, so these genes cannot be tested for dominance under standard diploid conditions in the female. However, the results with the males are exactly comparable to those outlined above for salmon.

To summarize, the genetic tests show that the mealybug male expresses and transmits only maternal genes. The mealybug male is thus genetically analogous to the true haploid male which begins life from an unfertilized egg and has never received a paternal endowment. The mealybug genetic system is based on the heterochromatization of the paternal complement in the sons and its post-meiotic elimination.

X-ray and hybridization analyses.—Conclusions identical to those reached from the studies of marker genes had previously been reached from earlier studies with X rays (BROWN and NELSON-REES, 1961). When fathers were treated with increasing doses of X rays, the number of their daughters declined sharply in a typical fashion. Such dominant lethality has been well understood for a long time and seems simply attributable to the induction of duplications and especially deficiencies which cannot be tolerated by the zygote. After paternal treatment, however, the number of sons did not decrease until doses in excess of 30,000 r or rep had been reached and some sons appeared even after doses of 90,000 r. Thus, it was not possible to induce dominant lethality in the sons in an expected manner.

The above facts indicated again that the paternal, heterochromatized set was genetically inert, but certain complications soon became apparent. NELSON-REES (1962) analyzed sons surviving after high dosage paternal irradiation and discovered two results of interest. The surviving sons contained an approximately normal amount of heterochromatin, even though the material was grossly rearranged in extra long and extra small chromosomes, and the surviving sons were increasingly sterile following increasing paternal doses. The sterilizing effect was apparent even in whole mounts of male mealybugs in which the densely staining sperm bundles were visible in the abdomens of the control males. There were very few bundles following paternal treatment of 45,000 r or rep and none after the same at 90,000 r. Since meiosis was normal in these survivors, the infertility was probably attributable to breakdown in bundle formation.

In addition to these results of Nelson-Rees, Nur and Chandra (1963) observed that crosses between different species of mealybugs invariably failed completely. If the paternal set were completely inert in the sons, it might be expected that the heterochromatic chromosomes of one species could be replaced by those of another and that the males would survive. But this was not the case. Hybrid embryos were produced but the males did not develop further than the females.

Reactivation of the heterochromatic set.—The problem of the functioning of the paternal set was finally resolved by Nur in a series of elegant experiments which seem to put the finishing touch on the relationship of heterochromatization to gene action. It had been known for some time that some tissues of the male mealybug lacked heterochromatin. Nur suggested that in these tissues the heterochromatic set had reverted to the euchromatic state. After high dosage paternal irradiation, these tissues showed a dominant-lethal effect; they were grossly abnormal while the neighboring tissues, with heterochromatin, were unaffected. The Malpighian tubules provided an especially clear example. Normally these are long slender Y-shaped structures. After high dosage paternal treatment the tubules were much reduced in length and also thickened because of the abnormally large size of their cells (Nur, 1967b).

Heterochromatin is also lacking from the cells of the cyst walls of the testes. Here there is no characteristic morphology to reveal the damage, but these cells normally function in sperm bundle formation (Nur, 1962). It thus seems quite likely that the sterility of the sons after high dosage paternal treatment is due to a dominant lethal type of effect in the cells of the cyst wall.

Nur also examined development in the sons from species crosses and found abnormalities in the same tissues as in the X-ray series. He estimated that the damage due to hybridization was about equal to that produced by 60,000 r. There were, however, no surviving males in the species crosses. The hybridization effect was therefore uniform and unlike the random effect of radiation.

The results with the Malpighian tubules do demonstrate that the genetic activity of the paternal set is regained on deheterochromatization. There is thus a complete parallel between heterochromatization and gene action. The genetic activity of the paternal chromosomes is turned off when they are heterochromatized and turned back on again in those certain tissues in which the heterochromatization is reversed (Nur, 1967b).

Heterochromatin and sex determination.—Deheterochromatization is obviously under developmental control because it occurs in certain tissues and not in others. What about the original heterochromatization that occurs in the male embryos? This is a problem in sex determination and has proved a very difficult problem indeed.

It was early recognized by the Schraders that the chromosomes were apparently identical in both sexes of mealybugs (HUGHES-SCHRADER, 1948). The basic, sex determining mechanism of the more primitive coccids is XX for the female and XO for the male, and no XY types have been found. Very obviously something had happened to the X chromosome when the heterochromatin systems were evolved.

The sex ratios of the mealybug are highly variable among the progenies of different mothers, and the individual progenies are large enough, two to four hundred, to give obviously significant results. The sex ratios can be modified by temperature and humidity and also by aging the female before she is mated. By keeping the mealybug female virginal for 5 to 6 weeks after she is first ready to mate, she can be induced to produce almost all sons. Such results as these indicated that sex determination was at least in part not genetic.

The mystery of sex determination in the mealybug system was considerably clarified by the work of NUR (1963) on a parthenogenetic species of coccid, *Pulvinaria hydrangeae*. In this particular case, meiosis was normal and a haploid egg was produced. The egg divided once and its two products fused to form the zygote substitute. Barring mutation, the zygotes were always one hundered per cent homogygous. Because the mother herself came from a completely homozygous zygote, all the embryos in one mother should have been genetically identical, again barring mutation.

In a small percentage of embryos, typical male-type heterochromatization occurred. The chromosome number was sufficiently large to rule out random effects. The diploid number of the species was sixteen. In the heterochromatic embryos, precisely eight chromosomes were heterochromatic and eight were euchromatic.

The formation of the zygote substitute can now be reconsidered. The egg divided and its two daughter nuclei fused to make a diploid nucleus. Obviously something had happened to one of these nuclei, prior to the fusion, which predisposed its chromosomes to later heterochromatization. Thus, the heterochromatization which differentiates the males and the females in the mealybug system is under control of some sort of factor or influence in the egg itself. Because it seems highly unlikely that a special method of inducing heterochromatization would have been evolved for a parthenogenetic species producing only a few male embryos, a similar system must be functioning in the sexual species. According to this concept, the chromosomes of the sperm would be predisposed for heterochromatization if the egg which they entered had sufficient of the responsible factors; otherwise, heterochromatization would not occur and the embryos would develop as females.

Because of the nature of the parthenogenetic sequence, the females must have been completely or very nearly completely homozygous. Therefore, segregation of genetic factors could not have accounted for the male embryos and developmentally determined differences among the eggs must have been solely responsible.

Endomitotic failure of heterochromatin.—In addition to the simple change of deheterochromatization, Nur (1966) found that the heterochromatic set may fail to reproduce during the endomitotic sequences responsible for formation of the large nuclei of some of the tissues in male mealybugs. A single set of heterochromatic chromosomes may be maintained (and be readily countable), while the originally haploid euchromatic set has achieved the diploid, tetraploid, or octoploid states by endomitosis. At the change to 16-ploidy the heterochromatic set may also divide endomitotically to the diploid state. Such differential reproduction for hetero- and euchromatin had been reported earlier by Rudkin (1963, 1965) for the polytene chromosomes of Drosophila and may prove to be a fairly common type of behavior of heterochromatin.

Endomitotic sequences are truncated mitotic sequences, and perhaps this is an explanation for the failure of the heterochromatin to reproduce in the endomitotic sequences. Baer (1965) and Sabour (pers. comm.) have found a differential labelling of the euchromatin and heterochromatin of the mealybug. Pulse labelling experiments have indicated that the heterochromatin synthesizes its DNA after the euchromatin has largely finished with its synthetic activity. As shown by Lima-de-Faria (1959), late labelling seems to be a general feature of heterochromatin. It may be suggested that the S period is also truncated during the endomitotic sequence and perhaps this is the reason for the failure of the heterochromatic set to reproduce during the endomitotic sequence.

Developmental aspects of dosage compensation.—During the first and most of the second instars, male and female mealybugs are morphologically indistinguishable and approximately the same size. Inertness of the paternal heterochromatic set leaves the male with only one active set of genes per nucleus while the female has two. The male could compensate either by increasing the number of cells per unit tissue volume or by regular endomitosis to give two active sets per nucleus. Recent work by Berlowitz, Loewus, and Pallotta (1968) indicates that the males compensate by increase in cell number. Male embryos (290 μ) had 1.45 times as many cells as female embryos. This ratio was apparently a forecast of subsequent development during which the heterochromatin of the different tissues would vary in behavior as previously described by Nur (see above). The accuracy of this developmental forecast was substantiated by DNA determination at late second instar, when the sexes can be distinguished because of the pink color of the male. The males had 1.47 times as much DNA as the females. Not only is heterochromatization developmentally determined but the developmental sequence itself is accurately geared to compensate for the genetic inertness of the heterochromatic set. Ratios of cell numbers, however, varied at second instar from 0.99 for Malpighian tubules to 4.46 for brains. As previously noted, heterochromatization is reversed in the Malpighian tubules and a 1.0 ratio would be expected; the very high cell number in brain tissue of males may also reflect their much greater motility and much better sensory apparatus.

The schedule for heterochromatization in the mealybug may now be summarized. Heterochromatization of the paternal chromosome sets occurs in the male mealybug during early embryogeny and is accompanied by or results in genetic inertness. The determination of sex, and concomitantly of heterochromatization, resides in the egg. Eggs predetermined to give male embryos give embryos with heterochromatization; eggs predetermined to give female embryos do not. This, then, is a type of differentiation recognizable in the two classes of eggs produced.

Within the developing male the heterochromatic set has a wide potential. It may replicate in step with the euchromatic set, as it always seems to do in the germ line. In endomitosis, it may reduplicate along with the euchromatic set, or reduplicate at a much slower rate, or not at all. Finally, it may revert to the euchromatic state and regain its genetic function.

The entire program, from beginning to end, seems to be under developmental control. To date we have no evidence whatsoever of any gene or genes responsible for any phase of heterochromatization. The heterochromatic set can be grossly fragmented and rearranged by high dosage paternal irradiation, yet each chromosome, large or small, retains its heterochromatic nature and behavior in mitosis and meiosis. It is, therefore, unlikely that there are individual controlling sites on each chromosome responsible for regulating its state. It is unlikely, also, that there are one or a few directly active controlling sites for the entire paternal set, because the heterochromatic set occurs in the same nucleus as the euchromatic and both sets would be exposed to the same influence.

Genetic determination of the various features of the mealybug heterochromatin system must work, therefore, by way of control of developmental processes. We have no direct evidence as yet of any such genetic factors, but if we turn to comparative cytology we find evidence of profound changes. In one group of of coccids, the Aclerdidae, heterochromatization is confined to the germ line (Brown, unpubl.). In several different groups, the behavior of the heterochromatic chromosomes during meiosis is profoundly altered as well as the meiotic sequences itself (Brown and Nur, 1964). In more distantly related types, the armored scale insects, the paternal chromosomes are eliminated altogether at about the same stage at which they are heterochromatized in the mealybug. Occasionally, the elimination process fails, and the paternal chromosomes appear as heterochromatic.

Two important questions remain unanswered. Why did the mealybugs evolve a system like male haploidy in which sex was determined by factors operating in the mother? The answer to this question will undoubtedly come from ecological genetics. Secondly, why did the mealybug evolve a system in which a haploid-like condition was achieved by heterochromatization? The answers here will probably have to come from several different fields. The problem is too complex to warrant even a suggestion at this juncture.

LITERATURE CITED

BAER, D., 1965 Asychronous replication of DNA in a heterochromatic set of chromosomes in *Pseudococcus obscurus*. Genetics **52**: 275–285.

BERLOWITZ, L., M. W. LOEWUS, and D. PALLOTTA, 1968 Heterochromatin and genetic activity in mealybugs I. Compensation for inactive chromatin by increase in cell number. Genetics **60**: 93–99.

BROWN, S. W., 1959 Lecanoid chromosome behavior in three more families of the Coccoidea (Homoptera). Chromosoma **10**: 278–300.

BROWN, S. W. and W. A. NELSON-REES, 1961 Radiation analysis of a lecanoid genetic system. Genetics **46**: 983–1007.

BROWN, S. W., and U. NUR, 1964 Heterochromatic chromosomes in the coccids. Science **145**: 130–136.

HUGHES-SCHRADER, S., 1935 The chromosome cycle of Phenacoccus (Coccidae). Biol. Bull. **69**: 462–468. ——— 1948. Cytology of coccids (Coccoidea-Homoptera). Advan. Genet. **2**: 127–203.

HUGHES-SCHRADER, S., and H. RIS, 1941 The diffuse spindle attachment of coccids, verified by the mitotic behavior of induced chromosome fragments. J. Exp. Zool. **87**: 429–456.

HUGHES-SCHRADER, S., and F. SCHRADER, 1961. The kinetochore of the Hemiptera. Chromosoma **12**: 327–350.

LIMA-DE-FARIA, A., 1959 Differential uptake of tritiated thymidine into hetero- and euchromatin in *Melanoplus* and *Secale*. J. Biophys. Biochem. Cytology **6**: 457–466.

NELSON-REES, W. A., 1962. The effects of radiation damaged heterochromatic chromosomes on male fertility in the mealybug, *Planococcus citri* (RISSO). Genetics **47**: 661–683. ——— 1963. New observations on lecanoid spermatogenesis in the mealybug, *Planococcus citri*. Chromosoma **14**: 1–17.

NUR, U., 1962 Sperms, sperm bundles and fertilization in a mealybug, *Pseudococcus obscurus* Essig (Homoptera: Coccoidea). J. Morphol. **111**: 173–199. ——— Meiotic parthenogenesis and heterochromatization in a soft scale, *Pulvinaria hydrangeae* (Coccoidea: Homoptera). Chromosoma **14**: 123–139. ——— 1966 Non-replication of heterochromatic chromosomes in a mealybug, *Planococcus citri* (Coccoidea: Homoptera). Chromosoma **19**: 439–448. ——— 1967a Chromosome systems in the Eriococcidae (Coccoidea: Homoptera). II. *Gossyparia spuria* and *Eriococcus araucariae*. Chromosoma **22**: 151–163. ——— 1967b Reversal of heterochromatization and the activity of the paternal chromosome set in the male mealy bug. Genetics **56**: 375–389.

NUR, U., and H. S. CHANDRA, 1963 Interspecific hybridization and gynogenesis in mealybugs. Am. Naturalist **97**: 197–202.

RUDKIN, G. T., 1963 The structure and function of heterochromatin. Genetics Today. (Proc. 11th Intl. Cong. Genet.) **2**: 359–374. ——— 1965 Non-replicating DNA in giant chromosomes. Genetics **52**: 470.

SCHRADER, F., 1929 Experimental and cytological investigations of the life-cycle of *Gossyparia spuria* (Coccidae) and their bearing on the problem of haploidy in males. Zeit. Wiss. Zool. Abt. A. **134**: 149–179.

SCHRADER, F., and S. HUGHES-SCHRADER, 1931 Haploidy in metazoa. Quart. Rev. Biol. **6**: 411–438.

STUDIES ON OÖGENESIS IN THE POLYCHAETE ANNELID *NEREIS GRUBEI* KINBERG. I. SOME ASPECTS OF RNA SYNTHESIS

MEREDITH C. GOULD AND PAUL C. SCHROEDER

Oögenesis in the marine polychaete *Nereis* appears to include at least two distinct growth phases. Oöcyte growth is slow for much of the period available for oögenesis, and more rapid during the period of somatic metamorphosis which precedes spawning (Clark and Ruston, 1963). Both somatic metamorphosis and rapid oöcyte growth can be initiated by decapitation, which is thought to deprive the animal of a prostomial source of an inhibitory hormone (Durchon, 1952; Hauenschild, 1956). Inasmuch as the suggestion has been made that an alteration in RNA metabolism is associated with the increased growth rate of the oöcytes (Durchon and Boilly, 1964), we have examined the types of RNA normally formed during the rapid growth period, which in *Nereis grubei* commences when the oöcytes are 90–100 μ in diameter. Oöcyte morphology changes considerably through this period (Spek, 1930); in *N. grubei* the oöcyte is blue-green and opaque for most of the period of rapid growth. When it reaches 170–180 μ in diameter, the color changes to a brilliant yellow-green. Still later, the lipid yolk coalesces into large translucent droplets around the nucleus, while granules of jelly precursor material condense at the cortex of the oöcyte, which is about 200 μ in diameter at maturity. Mature oöcytes thus contain a yellow-green and strikingly layered cytoplasm, and are distinct in appearance from the opaque, homogeneous, blue-green 160 μ oöcytes. RNA synthesis during the rapid growth period has been examined in cells in each of these morphological stages.

Materials and Methods

Animals

Female specimens of *N. grubei* of appropriate age were collected at several points in central California (Moss Beach, San Mateo County; Pacific Grove, Monterey County). They were maintained in the laboratory at 11–15° C in sea water containing 100 μg/ml streptomycin base and 100 units/ml penicillin.

Incubation of oöcytes with tritiated uridine

Animals anesthetized in 7.3% $MgCl_2$ were passed through a series of five changes of sterile sea water, with vigorous agitation between changes. The animal was then split open; the oöcytes were removed in as sterile a manner as possible, placed upon sterile Nitex nylon bolting cloth (44 μ or 64 μ mesh) and rinsed repeatedly with sterile sea water until phase microscopic examination indicated the absence of all coelomic cells. The oöcytes were then placed in 5 ml sterile sea water containing 35 μc of H^3-uridine (New England Nuclear. Specific activities are given in the figure legends).

Those species of *Nereis* which breed in the heteronereid state appear to swarm soon after the oöcytes have reached the mature condition. Females deprived of males during this period will swim actively for a period of hours, then sink to the bottom. Soon thereafter, the oöcytes and the whole coelomic mass become gelatinous and the oöcytes begin to degenerate. The "fresh mature" oöcytes were taken from an animal actively swimming at the time of oöcyte removal. The "old mature" oöcytes were from an animal which had been swimming the previous day. These cells maintained excellent morphological integrity, but could be expected to disintegrate within 24–48 hours.

Upon completion of the incubation period, an aliquot of cells was withdrawn and fixed for one hour in acetic acid-ethanol (1:3) for autoradiography. The balance was frozen in a dry-ice-acetone bath for subsequent RNA extraction.

Autoradiography

After fixation the oöcytes were stained lightly with eosin to enhance visibility, imbedded in paraffin and sectioned at 5 μ. After paraffin removal, the sections were dipped in ice-cold 5% trichloroacetic acid for 15 minutes (not longer), rinsed, and coated with Ilford K5 liquid nuclear emulsion. Exposure was from 10 days to 3 weeks. They were then developed for 6 minutes in Kodak D19 (18° C) and stained in Mayer's hemalum and celestine blue (Doniach and Pelc, 1950). Successful RNase controls required the use of 1 mg/ml RNase (Worthington) for 17 hours (*i.e.* overnight) at 37° C. [Tweedell (1966) also found somewhat extreme conditions to be necessary for this reaction in *Pectinaria* oöcytes.]

RNA extraction

RNA was extracted by the cold-phenol method of Brown and Littna (1964), including DNase digestion (Worthington, 1x crystallized). The concentration of sodium dodecacyl sulfate for the initial phenol extraction was increased to 2%. With two exceptions (the preparations in Figures 8b and 8c), the samples were also digested with 50–100 μg Pronase (Calbiochem; previously autodigested) for an additional 30 minutes at 18–20° C. The samples were then brought to 0.5% SDS and 0.1 *M* NaCl and re-extracted with an equal volume of ice-cold phenol. The aqueous layer was re-extracted with chloroform at least twice more, and the RNA again precipitated with the addition of two volumes of absolute ethanol.

Analysis of extracted RNA

For sucrose gradient analysis, aliquots (1 ml or less) of the RNA solutions in .01 *M* sodium acetate, pH 5 were layered on top of 25 ml linear 5–20% sucrose gradients, made up in 0.01 *M* sodium acetate, 0.10 *M* NaCl and 10^{-4} *M* EDTA.

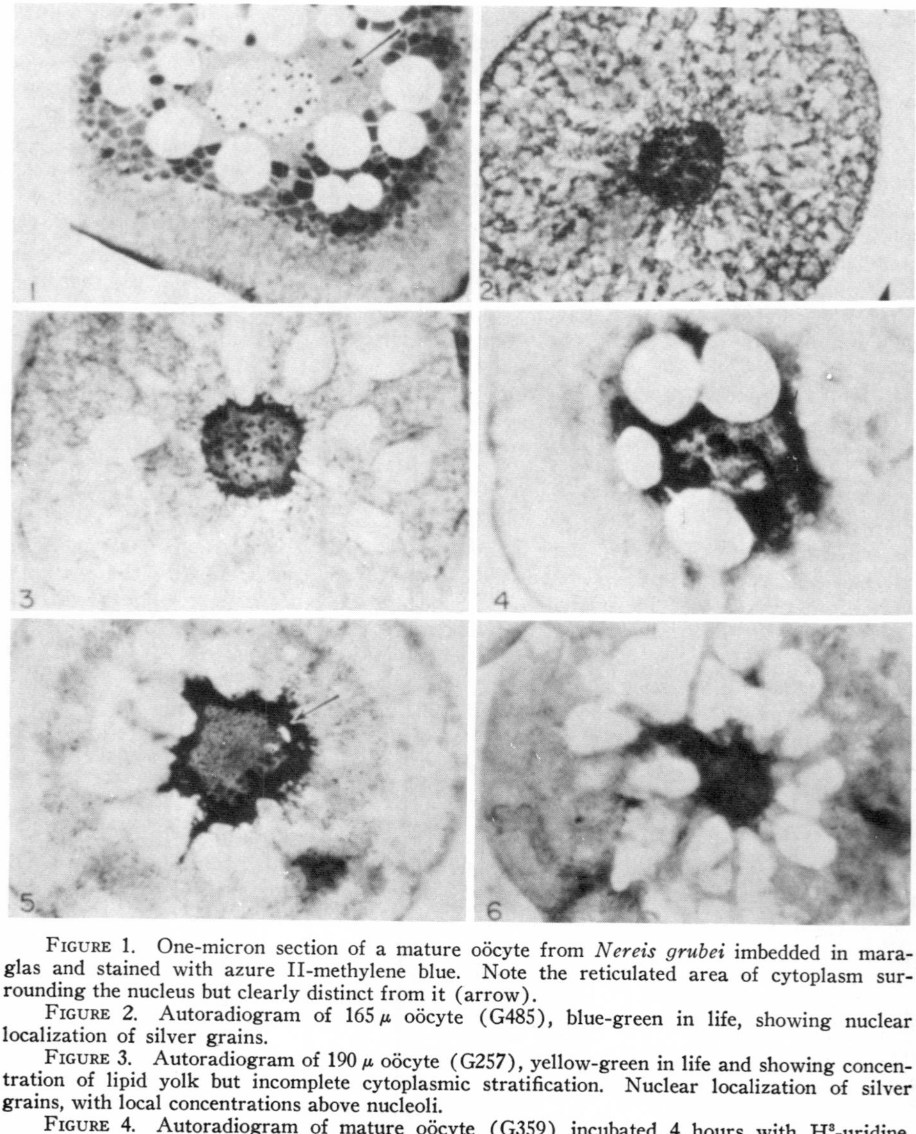

FIGURE 1. One-micron section of a mature oöcyte from *Nereis grubei* imbedded in maraglas and stained with azure II-methylene blue. Note the reticulated area of cytoplasm surrounding the nucleus but clearly distinct from it (arrow).

FIGURE 2. Autoradiogram of 165 μ oöcyte (G485), blue-green in life, showing nuclear localization of silver grains.

FIGURE 3. Autoradiogram of 190 μ oöcyte (G257), yellow-green in life and showing concentration of lipid yolk but incomplete cytoplasmic stratification. Nuclear localization of silver grains, with local concentrations above nucleoli.

FIGURE 4. Autoradiogram of mature oöcyte (G359) incubated 4 hours with H^3-uridine. Nuclear concentration of label, with relatively little over basophilic perinuclear cytoplasm.

FIGURE 5. Autoradiogram of older mature oöcyte (G548) with both nuclear and cytoplasmic label; the cytoplasmic label is concentrated over the basophilic perinuclear cytoplasm.

FIGURE 6. Autoradiogram of oöcyte from same sample as Figure 5, treated 17 hours with RNAase. Slight residual nuclear label, disappearance of nucleoli, perinuclear cytoplasmic basophilia and most silver grains.

Aliquots for RNase digestion were adjusted to pH 8 with 0.1 *M* Tris-HCl and incubated 10 minutes or longer at 18–20° C with 10–20 μg of boiled RNase (Worthington). The extracted RNA showed none of the resistance to the enzyme shown by that in the autoradiograms. The gradients were centrifuged 9–10 hours at 2° C and 25,000 RPM in the SB-110 rotor of an International B-60 ultracentrifuge. Fractions were diluted to 3 ml with 0.01 *M* sodium acetate, pH 5, for determination of the optical density at 260 mμ in a Beckman DU spectrophotometer, precipitated in the cold with 0.1 mg bovine albumen carrier and 0.3 ml 50% trichloroacetic acid, collected on glass fiber filters (Reeve-Angel) and counted in a Nuclear-Chicago liquid scintillation spectrometer with 4 ml toluene scintillator (4 g PPO and 0.2 g POPOP per liter of toluene).

DNA base composition

Two gamete-bearing male *Nereis* were homogenized in 5 ml SET (0.1 *M* NaCl, 0.1 *M* EDTA, 0.01 *M* Tris-HCl, pH 8) and warmed immediately to 60° C. Twelve mg Pronase (Calbiochem; previously autodigested) and sodium dodecyl sulfate to 0.5% were added, and incubated for 2½ hours. The solution was then cooled to room temperature and extracted by gentle rocking with an equal volume of water-saturated phenol. Following centrifugation at 12,000 g the aqueous layer was removed, re-extracted with an equal volume of chloroform-isoamyl alcohol (24:1) and centrifuged as before. Two volumes of absolute ethanol were added to the aqueous extract; the precipitate was collected and redissolved in 4 ml SET diluted 1:1 with water. The solution was then incubated with 200 μg RNase (Worthington; previously boiled 10 minutes) for 2½ hours at 37° C. Following addition of 500 μg Pronase, the incubation was continued for an additional hour. The DNA was then precipitated with ethanol, redissolved in SET, and extracted twice with phenol and once with chloroform-isoamyl alcohol as before. The resulting ethanol precipitate was redissolved in SET for CsCl density gradient analysis. We are indebted to Dr. Philip Hanawalt for carrying out this analysis. The guanidine-cytidine content of the DNA was calculated from its buoyant density in CsCl by the method of Schildkraut, Marmur and Doty (1962).

Results

Buoyant density of Nereis *DNA*

The buoyant density in CsCl of a sample of *Nereis* DNA extracted as described above was found to be about 1.700 g cc^{-3}, which corresponds to a guanidine-cytidine content of 40–41%. Although this information could not be put to the use originally intended in this analysis, we feel that it may be helpful to record this figure for the benefit of future investigations of nucleic acid metabolism in *Nereis*.

Autoradiography

The autoradiograms indicate the presence of labeled uridine in the nucleoli of cells in each size class. Label is also present over non-nucleolar regions of the germinal vesicle in all samples (Figs. 2–5). However, prominent cytoplasmic label

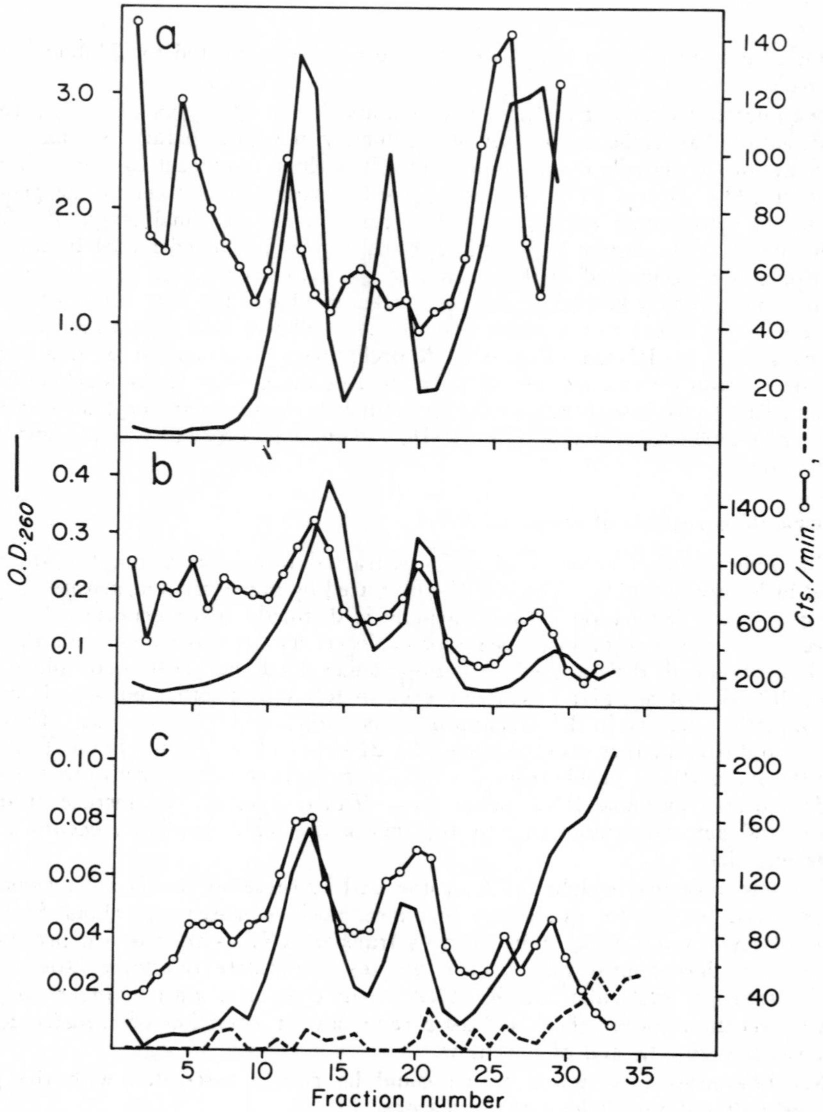

FIGURE 7. Sucrose gradient analysis of RNA from oöcytes of immature *Nereis grubei*. Open circles: radioactivity; solid line: optical density at 260 mμ; broken line: radioactivity after treatment with RNase. Specific activity of H^3-uridine applied in each case is given in parentheses following the incubation time. (a) 100 μ oöcytes, 4 hours incubation (24.5 C/m*M*). *Urechis caupo* carrier RNA added. (b) 165 μ oöcytes, 4 hours incubation (3.6 C/m*M*). No carrier RNA added. (c) 190 μ oöcytes, 4 hours incubation (24.5 C/m*M*). No carrier RNA added.

was found only in the two samples of mature oöcytes incubated for 21 hours with the precursor.

One-micron sections of mature oöcytes imbedded in Maraglas (Fig. 1, arrow) reveal that the basophilic region seen at the nuclear margin in paraffin sections (*e.g.* Fig. 5, arrow) is actually cytoplasmic. This basophilia is present to only a limited extent in 190 μ oöcytes (Fig. 3) and appears to represent a facet of the general layering of cytoplasmic elements which occurs during the final stage of oöcyte development. Cytoplasmic labeling in mature oöcytes incubated for 21 hours with H^3-uridine is concentrated in this region of the cytoplasm (Fig. 5) although the region is only slightly labeled in mature oöcytes incubated for four hours (Fig. 4). Both the incorporated radioactivity and the basophilia of this region are removed by treatment with RNase (Fig. 6). A preliminary examination of this region with the electron microscope indicates that it does not consist of the stacked annulate lamellae (also basophilic) found in artificially accelerated oöcytes of *Nereis diversicolor* by Durchon and Boilly (1964); (Durchon, 1967, p. 130 and our own observation).

Sedimentation analysis of extracted RNA

The sedimentation patterns of RNA extracted from *Nereis* oöcytes are presented in Figures 7 and 8. The RNA synthesized by immature and mature oöcytes during a 4-hour incubation with H^3-uridine is distinctly heterogeneous (Figs. 7a and 8a). However, prominent peaks of radioactivity are associated with the 28s and 18s ribosomal RNAs (sedimentation values have not been determined for *Nereis* RNAs, but are given with reference to RNA from other eukaryotic cells) seen by optical density in the preparations from 165 μ and 190 μ oöcytes (Figs. 7b and 7c) and from mature oöcytes labeled for 21 hours (Figs. 8b and 8c). Furthermore the radioactivity profile from the 100 μ oöcytes is consistent with the presence of 45s and 30s ribosomal RNA precursors. Taken together, the sucrose gradient profiles and autoradiograms suggest that ribosomal RNA synthesis occurs at all stages examined.

The low molecular weight RNA synthesized by these oöcytes is not adequately characterized by sucrose gradients; therefore, positive statements about 5s ribosomal (Brown and Littna, 1966) and 4s transfer RNA synthesis are precluded. However, it does appear that mature oöcytes accumulate relatively little radioactivity into low molecular weight RNA when compared with younger oöcytes. This observation could reflect a drastic reduction or cessation of transfer RNA synthesis towards the end of oögenesis.

No differences were found which could be clearly associated with the progressively altered morphology of the oöcytes.

In evaluating these results, the fact that the oöcytes were incubated in sea water rather than coelomic fluid should be kept in mind. We have observed that eleocytes adhering to growing coelomic oöcytes readily separate from them upon transfer to sea water, indicating some change at the oöcyte surface. Furthermore, Gonse (1957a, b) has reported differences in respiratory levels in oöcytes from the sipunculid *Phascolosoma vulgare* which he attributes to an effect of sea water upon cell permeability to sugars.

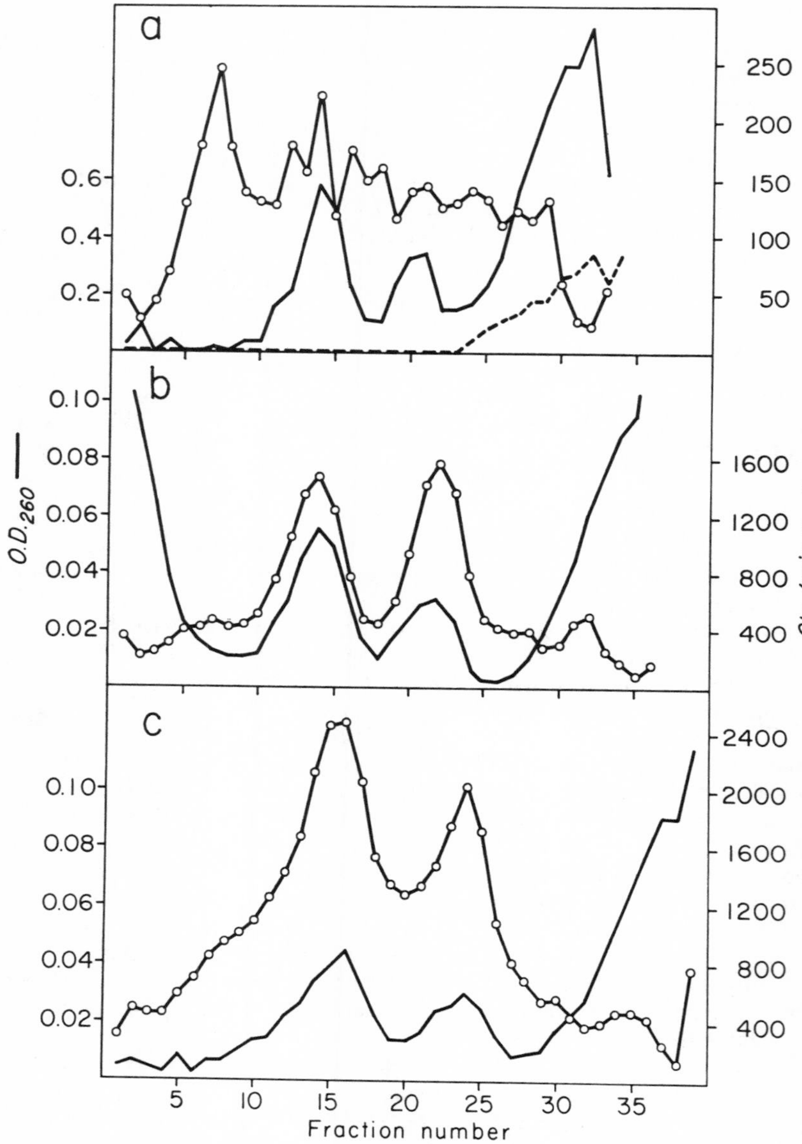

FIGURE 8. Sucrose gradient analysis of RNA from mature oöcytes. Legend as for Fig. 7. (a) Mature oöcytes 4 hours incubation (24.5 C/m*M*). No carrier RNA added. (b) "Fresh" mature oöcytes, 21 hours incubation (8 C/m*M*). No carrier RNA added. (c) "Older" mature oöcytes, 21 hours incubation (8 C/m*M*). No carrier RNA added.

Discussion

These studies clearly demonstrate that mature *Nereis* oöcytes, as well as immature oöcytes at the stages examined, are able to synthesize RNA in sea water in the absence of any accessory cells. An active synthesis of RNA has also been observed in the immature oöcytes from a variety of other marine invertebrates. As demonstrated autoradiographically these include sea urchins (Ficq, 1964; Piatigorsky and Tyler, 1967), starfish (Ficq, 1955b), amphibians (Ficq, 1955a), the polychaete worms *Pectinaria* (Tweedell, 1966), *Autolytus* (Allen, 1967) and *Ophryotrocha* (Ruthmann, 1964), and the echiuroid worm *Urechis* (Gould, unpublished data). Sucrose gradient analysis of the RNA synthesized by immature sea urchin (Gross, Malkin and Hubbard, 1965; Piatigorsky and Tyler, 1967) and *Urechis* (Gould, unpublished data) oöcytes has indicated considerable ribosomal RNA synthesis.

Reports of RNA synthesis by mature oöcytes, however, are fewer. Ribosomal RNA, transfer RNA, and RNA with a more DNA-like base composition are synthesized by mature *Urechis* oöcytes (Gould, unpublished data). Small amounts of heterogeneously sedimenting radioactive RNA have been detected in mature sea urchin oöcytes (Siekevitz, Maggio and Catalano, 1966), but studies with mature sea urchin oöcytes have generally been frustrated by the poor penetration of RNA precursors into these eggs (Piatigorsky, Ozaki and Tyler, 1967). The finding that mature *Nereis* oöcytes incorporate considerable amounts of precursor into RNA further discredits the idea that all mature oöcytes are in a state of metabolic inhibition (see, for example, Monroy, 1965, p. 77).

It should be noted, however, that our results contrast somewhat with the autoradiographic observations of Dhainaut (1965) on the oöcytes of the non-metamorphosing *Nereis diversicolor*. Mature oöcytes (200 μ) of this species were found to contain little cytoplasmic label and relatively little nuclear label after 12 hours in the presence of injected H^3-uridine. This author concludes that RNA synthesis virtually ceases in the mature oöcytes of this species, although we have found notable incorporation in three runs with mature oöcytes from *N. grubei,* which metamorphoses to spawn. It would be interesting to know whether this difference is related to the different modes of reproduction in the two species, or to differences of technique.

Previous investigators have found it difficult to study RNA metabolism in immature oöcytes of a given stage in marine invertebrates. Thus Piatigorsky *et al.* (1967) reported results obtained with mixtures of mature oöcytes and immature oöcytes of variable age, obtained from the sea urchin *Lytechinus pictus*. These authors have also examined the RNA accumulated in mature oöcytes gathered from animals injected with the precursor weeks or months previously (Piatigorsky and Tyler, 1967). However, the study of synthetic processes during echinoid oögenesis is hampered by the nature of the sea urchin ovary, in which oöcytes develop asynchronously while imbedded in a matrix of several other cell types (Holland and Giese, 1965). Oöcytes in a given stage of development cannot be conveniently isolated for chemical analysis. The ease with which relatively homogeneous populations of oöcytes in a given stage of oögenesis may be obtained makes *Nereis* an excellent organism in which to investigate the stage specificity of the synthesis of ribosomes and other materials (*e.g.* yolk components) which accumulate during oögenesis. The fact that the growth rate of these cells is influenced by an inhibitory hormone also makes them attractive biological objects as a hormone target tissue consisting of a single, isolable cell type.

This research was supported by NSF Postdoctoral Fellowship 46015 (PCS), NIH Predoctoral Fellowship 5-F1-GM-21,696 (MCG), NIH Grant GM-10,060 (to N. K. Wessels) and NSF Grant GB-6424X (to H. A. Bern).

LITERATURE CITED

ALLEN, M. JEAN, 1967. Nucleic acid and protein synthesis in the developing oocytes of the budding form of the syllid *Autolytus edwardsi* (Class Polychaeta). *Biol. Bull.*, **133**: 287–302.

BROWN, D., AND E. LITTNA, 1964. RNA synthesis during the development of *Xenopus laevis,* the South African clawed toad. *J. Mol. Biol.,* **8**: 669–687.

BROWN, D. B., AND E. LITTNA, 1966. Synthesis and accumulation of low molecular weight RNA during embryogenesis of *Xenopus laevis. J. Mol. Biol.,* **20**: 95–112.

CLARK, R. B., AND R. J. G. RUSTON, 1963. The influence of brain extirpation on oogenesis in the Polychaete *Nereis diversicolor. Gen. Comp. Endocrinol.,* **3**: 529–542.

DHAINAUT, A., 1965. Contribution à l'étude du métabolisme de l'A.R.N., par incorporation de ^{3}H-uracile, au cours de l'ovogénèse chez *Nereis diversicolor* O. F. Müller (Annélide Polychète). *Bull. Soc. Zool. France,* **89**: 408–413.

DONIACH, I., AND S. R. PELC, 1950. Autoradiographic technique. *Brit. J. Radiol.,* **23**: 184–192.

DURCHON, M., 1952. Recherches expérimentales sur deux aspects de la réproduction chez les annélides Polychètes: l'épitoquie et la stolonisation. *Ann. Sci. Natur.* (*Zool.*), Ser. XI, **14**: 119–206.

DURCHON, M., 1967. *L'endocrinologie des vers et des mollusques.* Masson and Cie, Paris, 241 pp.

DURCHON, M., AND B. BOILLY, 1964. Étude ultrastructurale de l'influence de l'hormone cérébrale des Neréidiens sur le développement des ovocytes de *Nereis diversicolor,* O. F. Müller (Annélide Polychète) en culture organotypique. *C. R. Acad. Sci., Paris,* **259**: 1245–1247.

Ficq, A., 1955a. Étude autoradiographique du métabolisme des protéines et des acides nucléiques au cours de l'oogénèse chez les batraciens. *Exp. Cell Res.*, **9**: 286–293.
Ficq, A., 1955b. Étude autoradiographique du métabolisme de l'oocyte d'*Asterias rubens* au cours de la croissance. *Arch. Biol.*, **66**: 509–524.
Ficq, A., 1964. Effets de l'actomonycine D et de la puromycine sur le métabolisme de l'oocyte en croissance. *Exp. Cell Res.*, **34**: 581–594.
Gonse, P., 1957a. L'ovogénèse chez *Phascolosoma vulgare* III. Réspiration exogène et endogène de l'ovocyte. Effet de l'eau de mer. *Biochim. Biophys. Acta*, **24**: 267–278.
Gonse, P. H., 1957b. L'ovogénèse chez *Phascolosoma vulgare* IV. Étude chromatographique des sucres du plasma. Action de différents substrats et du malonate sur la réspiration de l'ovocyte. *Biochim. Biophys. Acta*, **24**: 520–531.
Gross, P. R., L. I. Malkin and M. Hubbard, 1965. Synthesis of RNA during oogenesis in the sea urchin. *J. Mol. Biol.*, **13**: 463–481.
Hauenschild, C., 1956. Hormonale Hemmung der Geschlechtsreife und Metamorphose bei dem Polychaeten *Platyneris dumerilii*. *Z. Naturforsch.*, **11b**: 125–132.
Holland, N. D., and A. C. Giese, 1965. An autoradiographic investigation of the gonads of the purple sea urchin (*Strongylocentrotus purpuratus*). *Biol. Bull.*, **128**: 241–258.
Monroy, A., 1965. *The Chemistry and Physiology of Fertilization.* Holt, Rinehart and Winston, New York, 150 pp.
Piatigorsky, J., and A. Tyler, 1967. Radioactive labeling of sea urchin eggs during oogenesis. *Biol. Bull.*, **133**: 229–244.
Piatigorsky, J., H. Ozaki and A. Tyler, 1967. RNA and protein-synthesizing capacity of isolated oocytes of the sea urchin *Lytechinus pictus*. *Dev. Biol.*, **15**: 1–22.
Ruthmann, A., 1964. Zellwachstum und RNS-Synthese im Ei-Nährzellverband von *Ophryotrocha puerilis*. *Z. Zellforsch. Mikrosk. Anat.*, **63**: 816–829.
Schildkraut, C. J., J. Marmur and P. Doty, 1962. Determination of the base composition of deoxyribonucleic acid from its bouyant density in CsCl. *J. Mol. Biol.*, **4**: 430–443.
Siekevitz, P., R. Maggio and C. Catalano, 1966. Some properties of a rapidly labeled ribonucleic acid species in *Sphaerechinus granularis*. *Biochim. Biophys. Acta*, **129**: 145–156.
Spek, J., 1930. Zustandsänderungen der Plasmakolloide bei Befruchtung und Entwicklung des *Nereis*—Eies. *Protoplasma*, **9**: 370–427.
Tweedell, K. S., 1966. Oocyte development and incorporation of H^3-thymidine and H^3-uridine in *Pectinaria* (*Cistenides*) *gouldii*. *Biol. Bull.*, **131**: 516–538.

MEIOSIS IN *ORNITHOGALUM VIRENS* (LILIACEAE)

II. *Univalent Production by Preprophase Cold Treatment*

KATHLEEN CHURCH and D. E. WIMBER

Many physical and chemical alterations in the environment of the meiocyte can modify the frequency of chiasmata and hence, genetic crossing-over. Chiasma formation in *Ornithogalum virens*, a liliaceous plant, can be disrupted by cold treatment. The following report is a temporal analysis of the effect of cold temperature on chiasma formation. The results suggest that there is an event(s) occurring during the premeiotic DNA synthetic period or the premeiotic G2 period that is important for chiasma formation. Modification of this event by cold temperature leads to an achiasmate condition at diplotene of meiosis.

MATERIALS AND METHODS

Young racemes of *O. virens* on which some of the buds had progressed to some stage in meiosis were given ^{3}H-thymidine (20 μCi/ml, spec. act. 6.4 Ci/mM; New England Nuclear Corp.) by a wick feeding technique previously described [3]. Labelling was performed at room temperature for a period of 12 h. Immediately following the administration of the isotope, the plants bearing the labelled inflorescences were placed in a growth chamber at 9°C under continuous illumination. Beginning on day 3 of the cold treatment, inflorescences were removed from the plants at 24 h intervals up to and including day 15. The racemes were fixed in 3:1 (absolute ethanol/glacial acetic acid) for 24 h and stored in 70% ethanol. In any particular bud the meiotic cells are fairly closely synchronized, i.e. only 1 or 2 stages of short duration may be found. Squashes of anthers from several buds per inflorescence were prepared from all of the timed samples. Chiasma counts were made on cells at diplotene. The frequency of univalents at metaphase was also observed. The slides were then dipped in liquid photographic emulsion (Eastman Kodak NTB) and exposed in light tight boxes for a period of 6 months. They were stained with Delafield hematoxylin.

A second experiment was performed which was identical to that just described except that sampling was begun on day 10 and samples were collected every 12 h until day 14½.

RESULTS

Cells at diplotene, in this plant, contain three acrocentric bivalents that are very similar in shape and size. The nucleolar organizing

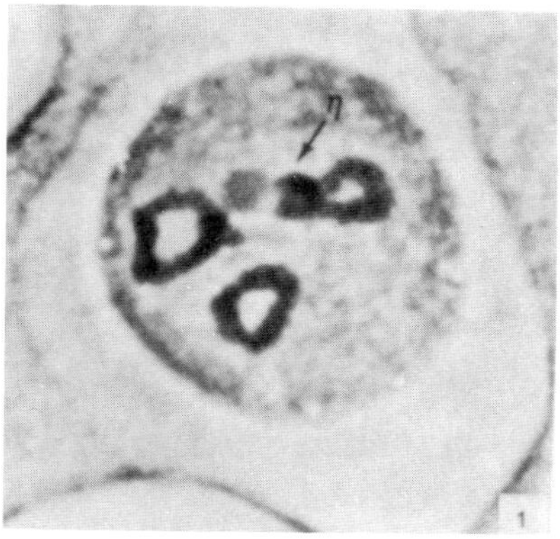

Fig. 1. Untreated cell at late diplotene showing 6 chiasmata. *N*, Nucleolar organizing bivalent.

bivalent is the only one that is easily recognized (fig. 1). Each bivalent usually contains two chiasmata. Bivalents with one chiasma are observed less frequently, and bivalents with three chiasmata are rarely seen. Thus, a cell usually contains a total of 5 or 6 chiasmata. The mean chiasma frequency per cell remained constant for a period of 12 days at 9°C (figs 2, 3). Univalents were practically never observed in diplotene cells for the first 12 days of the experiment. On day 13 in the first experiment (fig. 2) and day 12½ in the second experiment (fig. 3) a drastic drop in the chiasma frequency per cell was observed. Many cells contained 6 univalents and all cells contained at least two univalents (figs 4, 5). This low chiasma frequency was maintained throughout the rest of the experiment.

Occasional univalents at metaphase and anaphase began to appear on day 7 (figs 6, 7). The appearance of these univalents was not preceded by a reduction of chiasma frequency at diplotene. At the cold temperature the chromosomes become contracted and it appears that these univalents which occur at low frequency beginning at day 7 are caused by enhanced terminalization of chiasmata between diplotene and metaphase I.

Prepachytene states in *O. virens* are difficult to recognize. Pachytene, diplotene, and the later stages of meiosis can be readily identified. Observations of the autoradio-

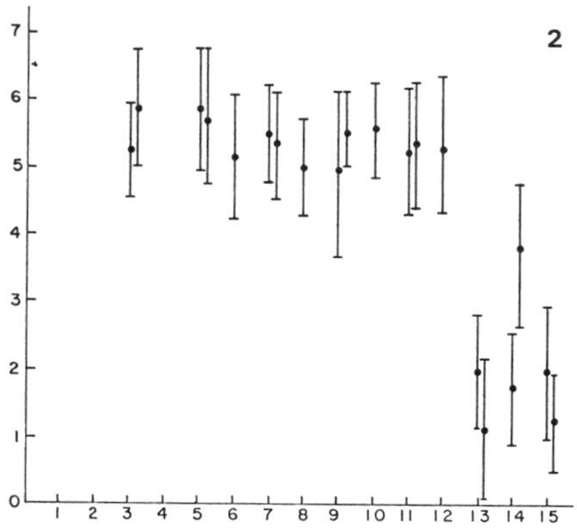

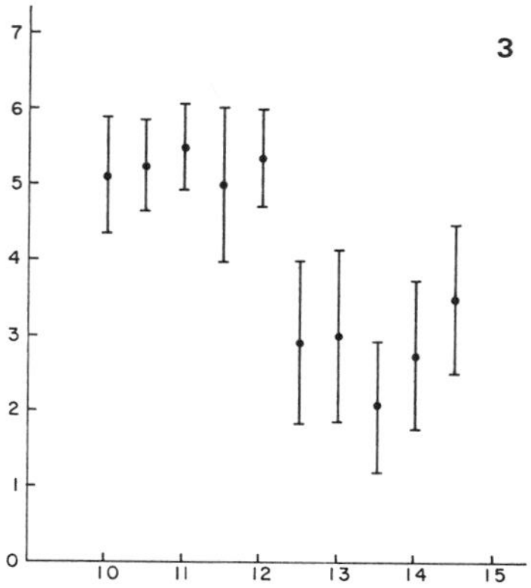

Figs 2, 3. Abscissa: days at 9°C; *ordinate:* mean chiasma frequency per cell.
Mean chiasma frequency and standard deviation per cell at diplotene from racemes sampled daily while being subjected to a temperature of 9°C. Chiasma frequency reduction occurred at day 13 (*fig. 2*) in the first experiment and at day 12½ (*fig. 3*) in the second experiment. Each point represents a sample of 25 cells from one raceme.

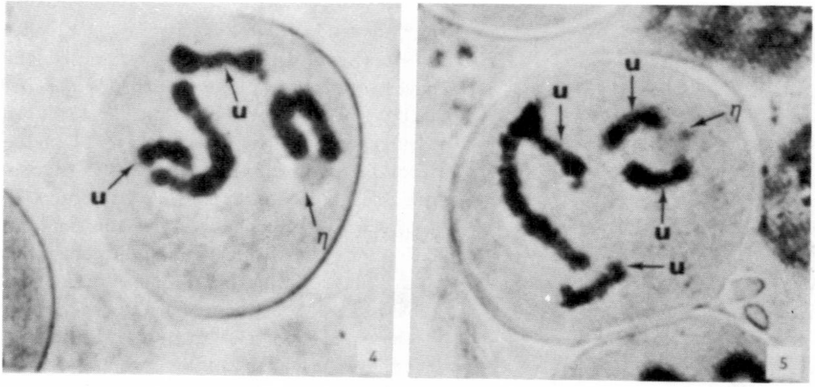

Figs 4, 5. Cells at diplotene showing two univalents and four univalents, respectively. Cells are from racemes sampled on day 13 of treatment at 9°C. *N*, nucleolus; *U*, univalents.

graphs revealed that through the first 12 days of cold treatment the latest stage labelled in any of the timed samples was a prepachytene stage. At day 12½, labelled cells were observed in pachytene, diplotene and metaphase I (fig. 8). It is important to note that the time of arrival of labelled cells in diplotene coincides with the reduction in chiasma frequency and the appearance of univalents at diplotene. Thus, these experiments suggest

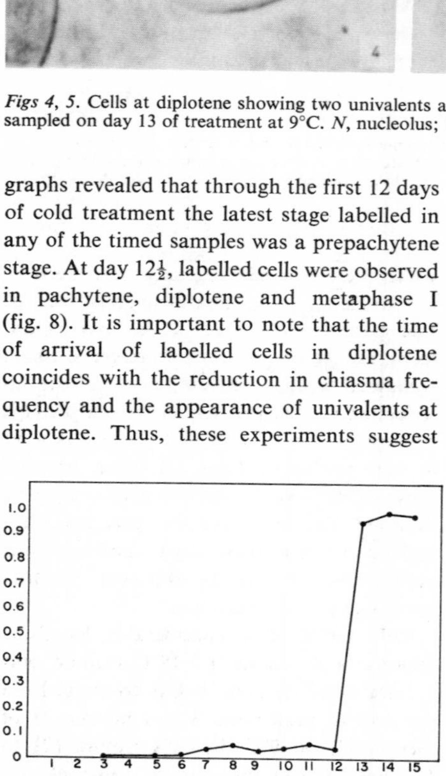

Fig. 6. Abscissa: days at 9°C; *ordinate:* proportion of cells in metaphase containing univalents.

Proportion of cells in metaphase which contain univalents after being subjected to 9°C for various periods of time. Each point represents a sample of 100 cells.

that the cells in which chiasma frequency reduction occurred were most likely in the premeiotic DNA synthetic period(s) or G2 when the cold treatment was begun. Those cells which were at later stages of meiosis when the cold treatment was started showed no interruption of chiasma formation. Two exceptions were noted. The inflorescence sampled on day 10 in the first experiment contained labelled cells in diplotene which showed no reduction in chiasma frequency. In the second experiment, the inflorescence sampled on day 12½ showed both labelled and unlabelled cells in diplotene, yet all cells in diplotene demonstrated chiasma frequency reduction.

DISCUSSION

These results suggest that the lowered temperature is disrupting an event or set of events occurring during premeiotic DNA synthesis, premeiotic G2 or both. These events must be essential for normal chiasma to form. Those cells which showed complete disruption of chiasma formation were in premeiotic DNA synthesis or G2 when the cold

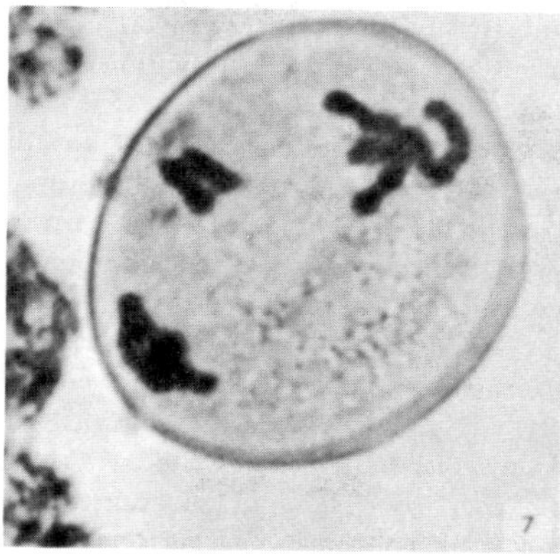

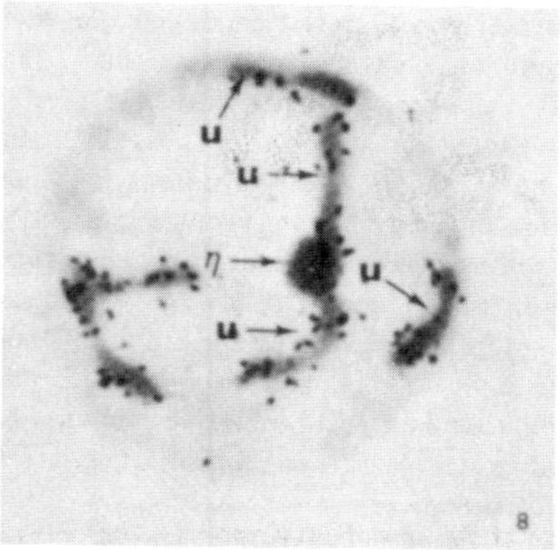

Fig. 7. Cell at anaphase I showing univalent. Cell is from raceme sampled on day 7 of treatment at 9°C.
Fig. 8. Autoradiograph of cell at diplotene from raceme sampled 13 days after the beginning of cold treatment. The cell contains four univalents. The cell was in contact with ^{3}H-thymidine for 12 h just prior to cold treatment. *N*, nucleolus; *U*, univalent.

treatment was begun. This is shown by the fact that diplotene cells (with one exception) which displayed univalents, also contained tritium label. We cannot rule out G2 or possibly early leptotene being the cold-sensitive period. The cells were in contact with ^{3}H-thymidine for 12 h at room temperature before being placed at 9°C. Thus, some labelled cells could have left the S period and entered the G2 period when cold treatment was begun. Then too, since we sampled every 12 h, we may have missed cells with chiasma disruption which were not labelled. Hence, we can say that the cold-sensitive period is either premeiotic DNA synthesis, G2 or early leptotene. Since we do not know the duration of the G2 period (if, indeed, any exists) we cannot estimate the possibility of the cells being at early leptotene when treatment was begun. The exceptional inflorescence observed in the second experiment (fig. 3) perhaps suggests that the cold sensitive period is the G2 period since only part of the diplotene cells that showed chiasma reduction were labelled. Those diplotene cells which were not labelled may have been at G2 when the cold treatment was started. The exception observed in the first experiment, i.e. the one 10 day inflorescence which contained labelled cells in diplotene with no reduction in chiasma frequency, can perhaps be explained by noting that the plants were in part seedlings. Thus, variation between some of the sexually derived plants might be expected. This may have been a variant form with a shorter than usual post-synthesis period and a slightly different reaction syndrome to cold treatment.

Cold temperature considerably lengthens meiosis in *O. virens*. At 18°C meiotic cells require approximately 4 days to travel from the end of premeiotic S to anaphase II of meiosis [3]. At 9°C the cells require $12\frac{1}{2}$ to 13 days to pass through the same phases. Thus, the Q_{10} at this temperature range is approx. 3.

Previous investigations have attempted to determine the time period during meiosis when genetic recombination or chiasma formation occurs. Different results have been obtained for different organisms and different experimental procedures. Hence, genetic recombination or chiasma formation has been suggested to occur prior to meiosis [11], or at a prepachytene stage of meiosis in maize [12] at premeiotic DNA synthesis in *Drosophila melanogaster* [6], at zygotene in a locust, *Schistocerca gregaria* [4], at pachytene in a grasshopper, *Goniaea australieasiae* [14] and in *Lilium* [5]. Rossen & Westergaard [15] and Chiang & Sueoka [1] present evidence that recombination is a post-replicative event in the ascomycete, *Neottiella*, and an alga, *Chlamydomonas reinhardi*, respectively. Lawrence [7, 8] has suggested that there are two or more time intervals during meiosis when chiasma formation can be disrupted by gamma irradiation in *Tradescantia* and *Lilium*. Recombination in *Chlamydomonas* is sensitive to gamma radiation during prophase and premeiotic interphase [9, 10]. Westerman [17] has demonstrated two X-ray sensitive periods when chiasma frequency may be altered during meiosis in the locust, *Schistocerca gregaria*. Chiasma formation in the grasshopper, *Melanoplus femur-rubrum*, can be altered by heat during zygotene and X-rays during leptotene [2], and heat treatment affects crossover frequency at both premeiotic and meiotic stages in *Neurospora* [13].

These apparently conflicting results can be reconciled by viewing chiasma formation as a culmination of a series of events which occur throughout meiosis and premeiosis; disruption of any single set of events can lead to modified chiasma frequency. Stern & Hotta [16] have outlined these events as they occur in *Lilium*. They suggest that in order for a cell to undergo meiotic determination a suppression of the replication of 0.3% of the chromosomal DNA occurs during premeiotic S. In addition to this, a set of events occurs during G2 which is important for meiotic determination as well as pairing determination. During early zygotene, pairing is initiated and chiasma formation occurs during late zygotene or early pachytene. The authors point out that disruption of any of these events could interrupt chiasma formation. If this general outline can be shown to apply to all meiotic systems, then it appears that cold treatment when applied to *O. virens* may be specifically altering pairing determination events which occur during late S or G2.

On the other hand, our results could also be explained by assuming that genetic recombination and hence chiasma formation, is occurring during the premeiotic S period or the G2 period in *O. virens*. Based on observations in *O. virens* alone this interpretation seems more likely. *Ornithogalum* is similar to *Lilium* and *Trillium* (viz. [16]) in that a small quantity of DNA is synthesized during zygotene–pachytene. This is shown by introducing a DNA synthesis inhibitor, deoxyadenosine, to the pollen mother cells during zygotene–pachytene. Fragmented chromosomes are produced when the cells reach anaphase (Church, unpublished). If this late DNA synthesis is related to chiasma formation, we have been unable to demonstrate it by other manipulations. For example, in *Lilium*, treatment with cycloheximide during pachytene will disrupt chiasma formation [5]. This is not true with *O. virens* (Church, unpublished). Experiments have been performed using a spectrum of concentrations of cycloheximide on all stages of meiosis and we have not been able to induce univalents. Disruption of chromosome movement occurs when cycloheximide is applied at anaphase I or II, but no effects are observed when it is applied during zygotene or pachytene. In addition to cycloheximide we have tried high and low temperature treatment during zygotene–pachytene with no disruption of chiasma formation. Based on this negative evidence we can only conclude that if chiasma formation is occurring during zygotene or pachytene in *O. virens*, it is very resistant to experimental manipulation. On the other hand, we can conclude (based on experiments reported herein) that chiasma formation may be readily disrupted by the application of cold temperature to a preprophase stage of meiosis in *O. virens*.

This work has been supported in part by a USPHS grant (GM11702), in part by a PHS research career development award (GM8465) and in part by PHS training grants (GM373) and (GM01433), all from the National Institute of General Medical Sciences.

REFERENCES

1. Chiang, K S & Sueoka, N, J cell physiol 70, suppl. 1 (1967) 89.
2. Church, K & Wimber, D E, Can j genet cytol 11 (1969) 209.
3. — Ibid 11 (1969) 573.
4. Henderson, S A, Nature 211 (1966) 1043.
5. Hotta, Y, Parchman, L G & Stern, H, Proc natl acad sci US 60 (1968) 575.
6. Grell, R & Chandley, A C, Proc natl acad sci US 53 (1965) 1340.
7. Lawrence, C W, Heredity 16 (1961) 83.
8. — Rad bot 1 (1961) 92.
9. — Nature 206 (1965) 789.
10. — Genet res 9 (1967) 123.
11. McGuire, M P, Proc natl acad sci US 55 (1966) 44.
12. — Genetics 60 (1968) 353.
13. McNelly-Ingle, C, Lamb, B C & Frost, L C, Genet res 7 (1966) 169.
14. Peacock, W J, Replication and recombination of genetic material (ed W J Peacock & W Brock) p. 242. Australian Acad Sci, Canberra (1968).
15. Rossen, J M & Westergaard, M, Comp rend lab Carlsberg 35 (1966) 233.
16. Stern, H & Hotta, Y, Genetics 61, suppl. 1, part 2 (1969) 27.
17. Westerman, M, Chromosoma 22 (1967) 401.

CYTOPHOTOMETRY OF DNA AND HISTONE IN MEIOSIS OF *PYRRHOCORIS APTERUS*

E. N. ANTROPOVA and YU. F. BOGDANOV

It is known that the cells, when entering into prophase of mitosis, have already completed their DNA and histone synthesis and have 4C of both DNA and histone [1, 4, 18]. By comparison, spermatocyte nuclei of *Crillus domesticus* at early prophase I (leptonema–zygonema) contain only 3C of histone stainable with fast-green at pH 8.2, while the DNA is already doubled before these stages (i.e. is 4C). Doubling of the histone amount in the nucleus terminates at pachynema only [6].

It was supposed that some deficiency of histone at the early meiotic prophase I may be one of the causes of the change of the chromosome structure, as compared with the prophase of mitosis and may be one of the causes of homologous chromosome pairing in zygonema [6]. If these considerations are true, we can expect to observe the same phenomenon of delayed doubling of nuclear histone in meiotic prophase I in all objects with typical meiosis.

This paper presents some evidence for the delayed doubling of the nuclear histone amount in meiotic prophase I in Hemipteran *Pyrrhocoris apterus*.

MATERIAL AND METHODS

Testes of the larvas of *Pyrrhocoris apterus* (*Hemiptera*, Insecta) were fixed in 10 % neutral formalin for 12 h, washed in running tap water for 24 h, squashed in a drop of 45 % acetic acid and dried in air after removal of the coverslips on dry ice. The slides were then stained according to Feulgen or with fast-green FCF (G. Gurr, Michrome stains, London) at pH 8.1–8.2 [2] with the following modifications: (1) pH 8.1–8.2 was adjusted by the use of 0.0035 M borate buffer; (2) ethanol was substituted for tertiary butanol to dehydrate the slides.

The amount of bound dye was estimated by double wavelength cytophotometry [8, 9] with the cytophotometer constructed by Dr A. I. Sherudilo [14]. Light with 540 nm and 514 nm wavelength was used for the measurement of intensity of the Feulgen reaction and 598 nm and 579 nm for the measurement of fast-green amount.

Calculation of the amount of the light-absorbing substance was made according to a modernized method of Sherudilo [15].

In spite of long washing of the slides, some traces of formaline always remained within the cells, which caused a slight non-specific Feulgen reaction. This formaldehyde-bound Schiff reagent was estimated in cytoplasm using one of the small optical plugs of the cytophotometer. The average value of such

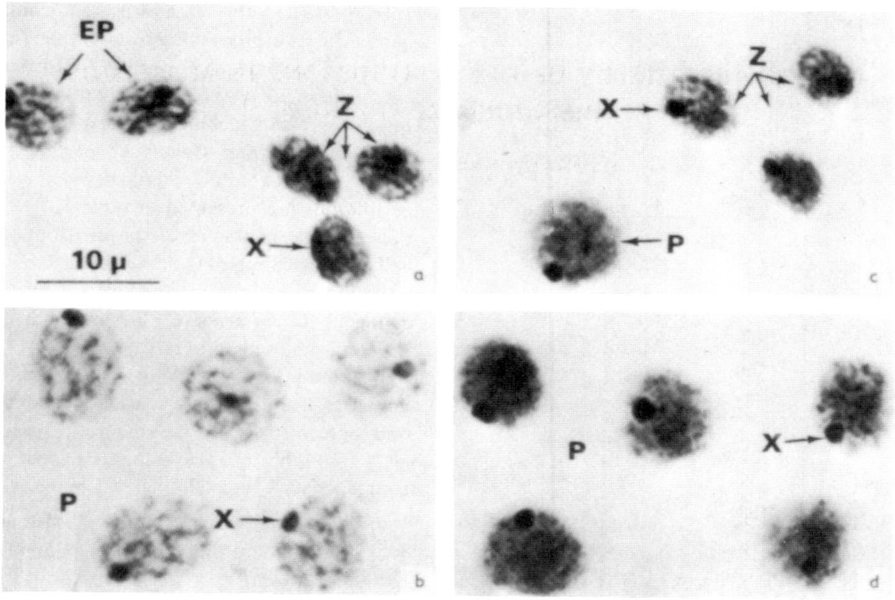

Fig. 1. Morphology of spermatocytes I of *Pyrrhocoris apterus* after fixation with neutral formalin, squashing in 45 % acetic acid and staining according to Feulgen (*a*, *b*) or with fast-green FCF at pH 8.1–8.2 (*c*, *d*). *Z*, zygonema; *EP*, early pachynema; *P*, middle pachynema; X, univalent of X-chromosome.

estimations for 30 cells of a given meiotic stage was recalculated for the areas of larger plugs used for measurement of the dye amount in the nuclei or sets of chromosomes at that stage. The correction obtained was subtracted from the value of dye amount estimated for the whole cell, giving the amount of dye per nucleus or set of chromosomes.

For different cell types these corrections are from less than 1 % for the spermatogonia and leptonema–zygonema to 4 % for the metaphase I. The whole random error of the cytophotometric methods used (including cytochemical procedures, measurements and calculations) was estimated as a coefficient of variation (CV) of the amount of a substance measured in the chromosome set of metaphase I and being 10 % for the Feulgen reaction and 16 % for the fast-green. Other details of the methods are described elsewhere [6].

RESULTS

In spite of some difference in morphology of the pachytene nuclei after Feulgen (fig. 1 *a*, *b*) and fast-green (fig. 1 *c*, *d*) procedures, this and the previous leptotene–zygotene stages may be identified with confidence by means of the nuclear diameter and other general indications which do not change after these procedures as long as the cells are dried on slides after formalin fixation. There is no difficulty in identifying other meiotic stages in *P. apterus* after the two procedures used.

The histograms (fig. 2) show the distributions of the nuclei by the amount of DNA (left vertical column) and of histone (right vertical column) plotted in arbitrary units. Only one histogram differs significantly (by the chi-square test) from the histograms of the normal distribution.

The exceptional histogram, i.e. that of histone amount in spermatogonial nuclei differs significantly by chi-square test from

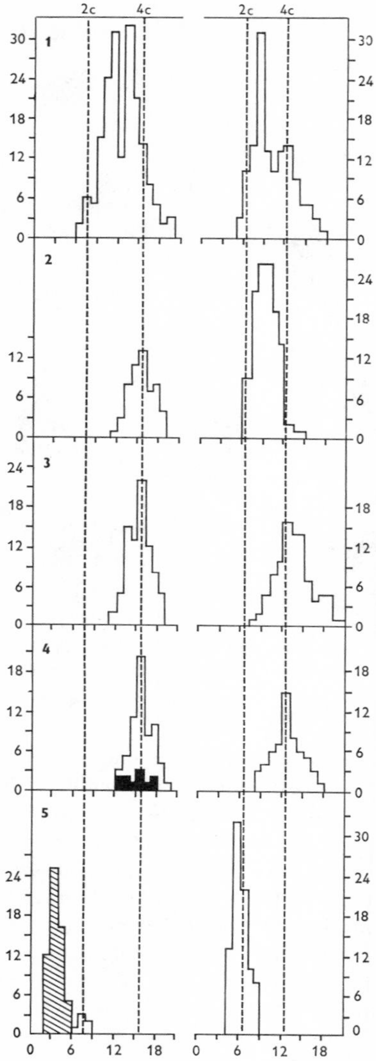

Fig. 2. Abscissa: amount of the substance in arbitrary units; *ordinate:* number of cells measured.

Cytophotometrical data for DNA (left vertical column and histone (right vertical column) amount in the nuclei at various stages of spermatogenesis in the bug *Pyrrhocoris apterus*. *1*, Spermatogonia; *2*, leptozygonema; *3*, pachynema; *4*, metaphase I (□) and diakinesis (■); *5*, telophase I (□) and early spermatides (hatched). The figures along abscissa are ten times reduced compared with those in table 1.

the normal distribution at 0.99 probability level ($\chi^2 = 19.17$; df = 11) and is significantly asymmetric at a probability level of 0.95 (coefficient of asymmetry $A = 0.552 \pm 0.214$; Student's criterion of significance $t = 2.6$; df = 12). This histogram may be subdivided into two normal distributions. Calculated parameters [17] of these two new "daughter" normal distributions are presented in table 1.

The normal character of the distributions justify an evaluation of the range of variability in the amount of DNA and histone by the coefficients of variation (CV). All the statistical data are presented in table 1.

The amounts of DNA in metaphases I, telophases I and in early spermatides are in good agreement with the theoretical ratio 4:2:1 (see table 1). The amounts of histone in metaphases I and telophases I agree with the ratio 4:2 (see table 1). Thus we have the series of classes of the amount of DNA and histone to evaluate the amount of these substances at other stages of spermatogenesis. It is necessary to note that a comparatively high CV of DNA amount in early spermatides (20 %) is conditioned by the XO formula of sex chromosomes in *P. apterus*.

Analysing the histograms and table 1 we arrive at the following conclusions.

(1) The range of variability of the interphase spermatogonial nuclei by the amount of DNA and of histone surpasses the random error of the method (see Material and Methods). This means that there are nuclei in the population of spermatogonia that differ significantly from each other by the amount of both DNA and histone. The range of variability of the spermatogonia is from 3C to 4C, i.e. from S to G2 phases of mitotic cycle.

Table 1. *Results of cytophotometry of DNA and histone*

Cell type	Number of cells measured	DNA amount mean ± S.E. (arbitr. units)	S.D.	CV ± S.D. (%)	Classes of ploidy (C)
DNA					
Spermatogonia	180	126 ± 2	26	21 ± 1	3–4
Lepto-zygonema	55	153 ± 2	17	11 ± 1	4
Pachynema	82	151 ± 2	17	11 ± 1	4
Metaphase I	62	157 ± 2	16	10 ± 1	4
Telophase I (one pole)	6	75 ± 3	8	11 ± 3	2
Early spermatide	58	37 ± 1	8	21 ± 2	1
Histone					
Spermatogonia	131 (93[1] / 38[1])	102 ± 2 (87 ± 2[1] / 139 ± 2[1])	28 (15[1] / 15[1])	27 ± 2 (17 ± 1[1] / 11 ± 1[1])	3–4
Lepto-zygonema	120	92 ± 2	16	17 ± 1	3
Pachynema	93	135 ± 3	27	20 ± 1	4
Metaphase 1	52	127 ± 2	21	16 ± 2	4
Telophase 1	86	60 ± 1	12	20 ± 1	2

[1] The characteristics of two normal theoretical distributions, reconstructed [17] on the grounds of coefficients of asymmetry and excess of the empirical distribution of histone amount in spermatogonia (cf fig. 2).

(2) In leptonema–zygonema, pachynema and diakinesis studied, the nuclei have 4C DNA. Thus DNA synthesis is completed before the beginning of leptonema. This is true only within the limits of resolution of the method, i.e. 10 %.

(3) The amount of histone stained at pH 8.1–8.2 after hydrolysis in hot TCA at leptotene–zygotene stages reaches 3C and will reach the 4C level only at pachytene stage (see table 1).

Thus in meiosis of the bug *P. apterus* we found the same picture as in the case of Ortopteran *G. domesticus* [6]: partial uncoupling of nuclear histone doubling from that of DNA, when the time sequence is considered; and postponed completion of nuclear histone doubling in early meiotic prophase I.

DISCUSSION

The time of meiotic DNA synthesis is studied by various methods in different organisms, including cytophotometry [6]. One of the conclusions of the present work, i.e. concerning the termination of DNA synthesis before zygotene state, coincides with the conclusions of other authors working with other objects.

The number of cytophotometrical studies concerning the determination of the amount of histone in meiosis in the classes of ploidy could well be scored [6]. The main question which remains unanswered is whether the alkaline fast-green procedure allows comparison of the quantity of the total histone in cells with different chromosome morphology, i.e. with a different organization of a nucleohistone and perhaps also with a different type of histone itself [3]. There are some reasons for fearing the possible loss of part of the histone after hot TCA hydrolysis in spite of formaline fixation [11]. It may be fI histone, since it is so very labile a part. At the same time, all we know about the histone which remains after hot TCA hydrolysis and is stained with fast-green at alkaline pH (let us call it as "TCA-stable histone") gives evidence that this histone is a constant part of the chromosome loci in their different physiological and morphological states [16]. In the present work this conclusion is supported by the fact that the amount of the histone in the meiocytes with different levels of nuclear RNA synthesis [12] and chromosome morphology (pachynema, metaphase I, telophase I and others in the previous study [6]) may be expressed in classes of ploidy and remains equal to 4C at all stages of meiosis except leptotene–zygotene, when it is equal to 3C.

This circumstance permits us to state that the doubling of the "TCA-stable histone" is not completed at this stage but is completed later—at the pachytene stage, when the ratio of the classes of ploidy histone to DNA = I, resumes (see table 1).

Two recent findings must be taken into consideration, when discussing the results of the present work. Bloch & Teng [5] showed that in the late-replicating X chromosome of the premeiotic spermatocytes of *Rehnia spinosus* the synthesis of total histone minus f1 histone occurs simultaneously with DNA synthesis, but the amount of f1 histone remains constant. The authors assume that it is synthezised later, i.e. in prophase I.

Our colleagues [10] confirmed the fact of increase of the nuclear histone amount in meiotic prophase I in cricket [6] by interference microscopical measurements of a dry mass of the nuclei before and after extraction of histone at pH 0.7. However, the difference in the dry mass of histone between late pachynema and leptonema found by them exceed ten-fold that discovered by cytophotometry [6]. These authors suppose that their measurement may be overestimated because of the extraction of f3 histone with the acetic-alcohol fixative (3:1) used [7]. This brings us back to the problem of some changes in histone quality during meiotic prophase I [3, 13].

The authors are very much indebted to Dr A. Sherudilo for the design and accomplishment of the microphotometer. We acknowledge the various kinds of aid by the work of professor A. Prokofyeva-Belgovskaya, Mrs N. Liapunova, Dr L. Lomakina, Dr A. Zelenin and Dr G. Ginzburg.

REFERENCES

1. Alfert, M, Exptl cell res, suppl. 6 (1959) 227.
2. Alfert, M & Geschvind, J, Proc natl acad sci US 39 (1953) 991.
3. Ansley, H R, J biophys biochem cytol 4 (1958) 59.
4. Bloch, D P & Goodman, G C, J biophys biochem cytol 1 (1955) 17.
5. Bloch, D P & Teng, Ch, J cell sci 5 (1969) 321.
6. Bogdanov, Yu F, Liapunova, N A, Sherudilo A I & Antropova, E N, Exptl cell res 52 (1968) 59.
7. Dick, C & Johns, E W, Exptl cell res 51 (1968) 626.
8. Ornstein, L, Lab invest I (1952) 250.
9. Patau, K, Chromosoma 5 (1952) 341.
10. Poletayeva, T P, Liapunova, N A & Malenkov, A G, Tsitologia (in Russian). In press.
11. Ringertz, N R & Zetterberg, A, Exptl cell res 42 (1966) 243.
12. Rocchi Brasilello, A, Exptl cell res 53 (1968) 252.
13. Scheridan, W E & Stern, H, Exptl cell res 45 (1967) 323.
14. Sherudilo, A I, Izvestia Sibirskovo Otdelenia Academii Nauk SSSR Ser med biol (in Russian) 12 (1964) 145.
15. Scherudilo, A I, Biophysica (in Russian) 13 (1968) 741.
16. Swift, H, The nucleohistones (ed J Bonner & P Ts'o) p. 169. Holden Day, San Francisco (1964).
17. Urbakh, V J, Biometrical methods (in Russian), pp 415. Nauka, Moscow (1964).
18. Woodard, Y, Rash, E & Swift, H, J biophys biochem cytol 9 (1961) 445.

SEQUENTIAL FORMS OF ATPase ACTIVITY CORRELATED WITH CHANGES IN CATION BINDING AND MEMBRANE POTENTIAL FROM MEIOSIS TO FIRST CLEAVAGE IN *R. PIPIENS*

G. A. MORRILL, ADELE B. KOSTELLOW and JANET B. MURPHY

One of the more intriguing phenomena in developmental biology is the mitotic inertness of the primary oocyte and the stepwise release of this inertness at maturation and fertilization. In the vertebrate, the oocyte becomes blocked in meiotic prophase during maternal embryogenesis and remains blocked until maturation begins. Maturation, followed by ovulation, is induced by progesterone or a progesterone-like hormone [1, 2] which removes the prophase block and meiosis proceeds to the metaphase of the second maturation division. The primary hormone action appears to be at the egg surface [3]. Fertilization, or an artificial stimulus, must follow to complete meiosis and permit cleavage to begin.

Previous studies have shown that the egg cortex undergoes a sequence of electrical [4] and ion permeability [5, 6] changes which are correlated with changing metabolic states at maturation, fertilization and first cleavage.

During maturation the egg cortex changes from a K^+ to a Na^+ selective system and the egg cytoplasm becomes positive relative to the external medium. At fertilization the egg cortex undergoes a transient hyperpolarization with the rising phase due to increasing Cl^- efflux [5]. The egg cytoplasm remains positive until just before furrow formation when the cortex slowly repolarizes and the cytoplasm again becomes negative relative to the external medium [4, 7].

The present study examines the relationship between these electrical and ion permeability changes and the changes in ATPase activity and cation binding during meiosis. The results suggest that hormonal stimulation during maturation produces a marked increase in cortical Ca^{2+} and Mg^{2+}. The rise in divalent cation in turn inhibits the cortical, ouabain sensitive, "Na-pump" ATPase with a concomitant appearance of a Ca-activated, ouabain insensitive, ATPase. The inhibition of the Na^+ transport system allows an accumulation of Na^+ by the egg, and this results in a depolarization of the egg cortex. Fertilization triggers the release of Ca^{2+} and Mg^{2+}, accompanied by a transient positive hyperpolarization with the rising phase due to Cl^- efflux and the falling phase to efflux of Na^+ and K^+. "Na-pump" ATPase activity reappears by the completion of meiosis and the egg cortex slowly repolarizes by first cleavage. Preliminary reports of this work have been presented [8, 9].

MATERIALS AND METHODS

Materials

Individual fully grown *Rana pipiens* oocytes were freed from the ovarian membranes of a pithed frog. Oocytes were collected from late September to early May. Ovulation was induced by an intraperitoneal injection of macerated frog pituitaries and the eggs were either collected from the oviducts (unfertilized) or expressed from the oviducts into a sperm suspension (two minced testes in 50 ml. 0.1 strength Ringer solution).

Electrical measurements

For measurements of the membrane potential, oocytes or eggs were placed in 2 mm depressions in a paraffin-bottomed lucite vessel and impaled with a KCl filled glass micropipette as described previously [4]. Oocyte membrane potentials were measured in isotonic Ringer solution; ovulated eggs and subsequent stages were measured in 0.1 strength Ringer solution. All recordings were made with the electrode tip 0.1 to 0.2 mm below the surface of the animal pole. The ovulated, unfertilized eggs were either expressed into a sperm suspension in 0.1 strength Ringer solution and immediately impaled, or were artificially activated by pricking with a glass needle. Continuous recordings were then made through first cleavage. All experiments were carried out at room temperature (18 to 22°C).

Cell fractionation

Ovulated eggs and subsequent stages were chemically dejellied [10] before being homogenized with 0.24 M sucrose solution at 5°C with a Potter-Elvehjem homogenizer and a Teflon pestle. The homogenate was centrifuged at 300 g for 10 min. The pellet was resuspended in 0.24 M sucrose, recentrifuged at 300 g for 10 min and the combined supernatants were centrifuged successively, with one washing at each step, at 1 700 g for 10 min, 11 000 g for 16 min, and 105 000 g for 60 min. Each pellet was resuspended in ion-free water and an aliquot taken to dryness for dry weight and analysed for Na^+, K^+, Ca^{2+}, and Mg^{2+}. The dry weight of the final combined supernatant was corrected for the dissolved sucrose.

Cation analysis

Whole eggs or aliquots of cell fractions were digested with fuming nitric acid for 12 h at room temperature. The digests were diluted with ion-free water and were analysed for Na^+, K^+, Ca^{2+}, and Mg^{2+} by atomic absorption spectroscopy.

Assay of ATPase activity

The incubation system (4.5 ml) contained 30 mM Tris-HCl (pH 7.4), 5 mM Tris-ATP, Na^+, K^+, Ca^{2+} or Mg^{2+} chlorides as indicated, and 10 to 50 mg wet weight of oocytes or eggs. The reaction mixture was incubated at 25°C and 1.0 ml aliquots were taken at 10 sec and 10, 20, and 30 min intervals. Each aliquot was immediately mixed with an equal volume of cold 10% perchloric acid, centrifuged, and the liberated inorganic phosphate was determined on the supernatant by the method of King [11]. Inorganic phosphate liberation was also measured in duplicate samples to which no ATP has been added and the apparent ATPase activity was corrected for endogenous phosphatase activity. Under these conditions, liberation of inorganic phosphate from ATP was linear for at least 60 min, indicating that the ATP concentration was not rate-limiting. Crude ATPase preparations can be stored at -50°C for at least 6 months without loss of activity but the egg homoge-

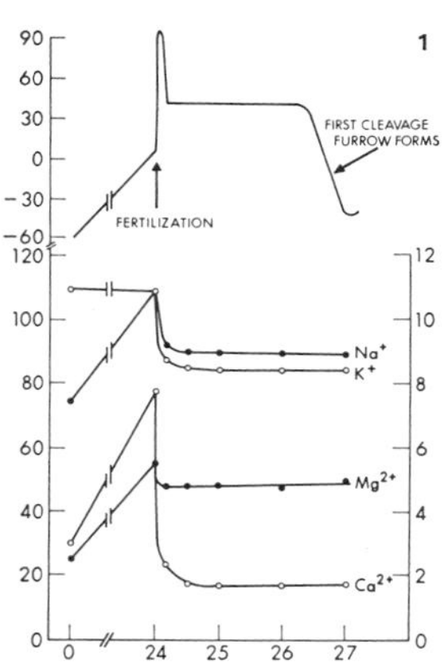

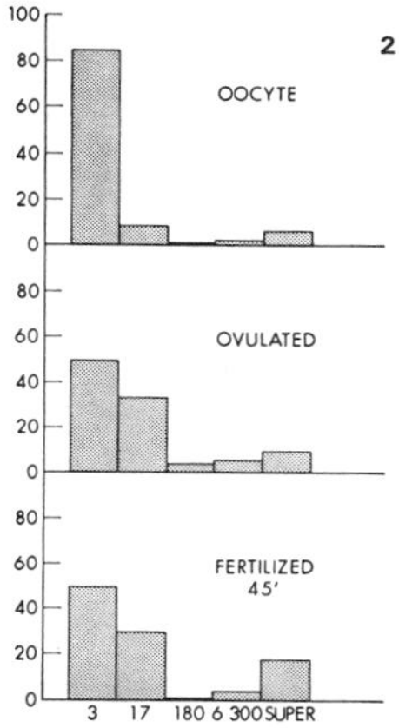

Fig. 1. Abcissa: hours; *ordinate:* (*left*) (EmV Na^+, K^+or Mg^{2+} (mmoles/kg dry wt); (*right*) Ca^{2+}(mmoles/kg dry wt).

Changes in membrane potential and intracellular Na^+, K^+, Ca^{2+} and Mg^{2+} concentrations between meiotic prophase (zero time on abscissa) and first cleavage. As shown here, the eggs are fertilizable 24 h after pituitary injection. This is a representative time interval and would vary with the number of pituitaries used and the season. For a description of the sequence of the morphological events during maturation and ovulation, see [30]. The potential differences and cation concentrations shown are average values for oocytes from five frogs and the subsequent stages were typical sequential samples of ovulated and fertilized eggs from the same frog. Variations between eggs from the same and different frogs have been reported previously [4, 6, 12].

Fig. 2. Abscissa: $g \times min \times 10^{-3}$; *ordinate:* subcellular distribution of dry wt (% of total).

Distribution of dry weight between five subcellular fractions prepared from 0.24 M sucrose homogenates of oocytes, ovulated eggs and 45 min fertilized eggs. The examples shown are typical of three or more fractionations of each developmental stage. The ovulated and fertilized egg fractions shown were prepared from eggs from the same frog.

nates were generally assayed immediately after preparation.

Determination of intracellular inorganic phosphate

Eggs were homogenized at ice-bath temperatures in 30 mM Tris-HCl (pH 7.4) or in distilled water and an aliquot of the homogenate immediately mixed with an equal volume of cold 10% perchloric acid. The mixture was centrifuged and the supernatant analysed for inorganic phosphate by the method of King [11].

RESULTS

Changes in intracellular cation concentrations: a correlation with changes in membrane potential

The sequential changes in cortical membrane potential at ovulation, fertilization, and first cleavage and the corresponding changes in intracellular Na^+, K^+, Ca^{2+} and Mg^{2+} are shown in fig. 1. These potentials have been described previously for the oocyte, ovulated, fertilized and cleaved *R. pipiens* eggs [4]; the present plot shows a typical continuous recording of the changes from ovulation to first cleavage. A 30% increase in egg hydration occurred by ovulation [12]; subsequent changes in egg hydration cannot be readily measured because of jelly coat swelling immediately following fertilization. The dry weight per egg and major chemical constituents (protein, carbohydrate and lipid) remain unchanged from meiotic prophase to first cleavage. Thus the ion concentration and enzyme activity changes reported here

Table 1. *Changes in the subcellular distribution of Na^+, K^+, Ca^{2+} and Mg^{2+} in the amphibian egg from meiotic prophase to first cleavage*

Developmental stage	Yolk platelet fraction				Pigment granule fraction				6.3×10^6 g min supernatant			
	Na	K	Ca	Mg	Na	K	Ca	Mg	Na	K	Ca	Mg
	mmoles/kg dry weight											
Prophase (oocyte)[a]	25	23	1.35	10.6	2	2	0.33	5.4	47	82	1.3	8.0
2nd Metaphase[b] (ovulated)	11	4	0.47	2.5	4	15	1.29	28.6	91	87	6.0	23.0
Fertilized 45 min[b]	4	3	0.27	0.88	2	5	0.45	17.6	83	76	1.1	30.5
First cleavage[b]	25	18	0.48	23.5	15	20	0.93	25.0	49	46	0.4	1.5

[a] Average values of fractions prepared from oocytes from five frogs; S.E. ± 2 mmoles or less for Na and K, ± 0.3 mmoles or less for Mg, and ± 0.1 mmoles or less for Ca.
[b] Values are from sequential samples of eggs from the same frog and are typical of four experiments. Average relative standard deviation less than ± 2 % based on three analyses of each sample.

are expressed in terms of kg dry weight of egg to allow comparison between the different developmental stages.

As shown in fig. 1, intracellular Na^+, Ca^{2+} and Mg^{2+} concentrations per egg rose during the depolarization before fertilization with no significant change in intracellular K^+ concentration. Upon addition of spermatozoa to freshly ovulated eggs there was a transient hyperpolarization accompanied by a rapid efflux of all four cations from the egg. New steady-state levels of both membrane potential and intracellular cation concentrations were achieved by the completion of meiosis (15 to 20 min). In the case of two cations, K^+ and Ca^{2+}, the levels in the fertilized egg fell below that of the oocyte. The Na^+ and Mg^{2+} levels of the fertilized egg approached but did not return to preovulation levels.

The rapid efflux of cations following fertilization is a response to egg activation and not to the transfer to a hypotonic environment. Eggs expressed directly from the oviducts into ion-free water did not lose significant amounts of Na^+, K^+, Ca^{2+} or Mg^{2+} for several hours.

Disaggregation of subcellular particulates at ovulation

Four particulate fractions and a final supernatant were obtained by differential centrifugation of 0.24 M sucrose homogenates and were compared for the distribution of total dry weight in the oocyte, ovulated unfertilized, and fertilized egg (fig. 2). In the oocyte, over 80 % of the total dry weight sedimented as a yellow-black pellet at a very slow speed (300 g for 10 min). However, centrifugation of ovulated and fertilized egg homogenates at 3 000 g min sediments only the creamy platelets and a small amount of the pigment granules. The pigment granules and small platelets sedimented in the second fraction (17 000 g min) and successive centrifugations at 180 000 and 6.3×10^6 g min produced a white pellet and a translucent red pellet, respectively. Thus an apparent subcellular disaggregation occurs during ovulation. No further major change occurred after fertilization, with the exception of a further release of material into the soluble portion of the cell. The disaggregation could in large part be reproduced by the addition of 10 mM EDTA to the homogenizing medium for the ovarian oocyte, suggesting that divalent cations play an important role in the particulate aggregation in the oocyte and the disaggregation during ovulation.

Changes in cation binding during meiosis and first cleavage

Appreciable quantities of all four cations sedimented with the yolk platelet (3 000 g min) and pigment granule (17 000 g min) fractions. The changes in the distribution of the four cations between these two fractions and the high speed supernatant is shown from meiotic prophase to first cleavage in table 1. These cations cannot be released by washing the isolated particulates with ice-cold 0.24 M sucrose solution and may thus be sequestered by the cell organelles and/or bound to specific chemical constituents. Since the alkali metal and alkaline earth cations can bind to nucleic acids [13–15] and phospholipids [16, 17] as well as proteins [18], bound cations are expressed per kg dry weight of egg.

In meiotic prophase (oocyte) the bulk of all four bound cations were recovered in the low speed platelet fraction. However, as noted above (fig. 2), in the oocyte this fraction contained most of the pigment granule material that would appear in the 17 000 g min fraction after ovulation. The fact that three of the four cations, K^+, Ca^{2+} and Mg^{2+}, appear primarily in the pigment granule fraction after ovulation suggests that these cations may be associated with the pigment granules in the oocyte. One cation, Na^+, was concentrated in the platelet fraction both before and after ovulation and may thus be largely associated with the yolk platelets. During maturation and ovulation there is a two-fold increase in the total particulate bound Mg^{2+} with no major change in the total bound Ca^{2+}. Total bound Na^+ and K^+, on the other hand, decreased 25 and 50 %, respectively, by ovulation. As can be seen in the last four columns, the released cations were recovered in the high speed supernatant. Following fertilization there was a loss of K^+, Ca^{2+} and Mg^{2+} from the pigment granule fraction, and a loss of Na^+ from the yolk platelet fraction. Thus, during meiosis Na^+ appears to be primarily associated with the yolk platelets, and K^+, Ca^{2+} and Mg^{2+} with the pigment granules. By first cleavage (3 h), 80 and 97 % of the Ca^{2+} and Mg^{2+}, respectively, and 45 % of the Na^+ and K^+ is recovered in the combined yolk platelet and pigment granule fractions.

Changes in $[P_i]_i$, adenosine triphosphatase (ATPase) and phosphoprotein phosphatase (PPPase) activity during meiosis and first cleavage

The levels of endogenous inorganic phosphate, ATPase and PPPase activity were compared for the oocyte, ovulated unfertilized, fertilized and cleaved egg (fig. 3). Enzyme activity is expressed as mmoles P_i released per kg dry weight of egg/min for comparison with intracellular inorganic phosphate and cation concentration changes. Since *R. pipiens* eggs contain 111 ± 3 (S.E., 7 frogs) mg protein N/g dry weight, the enzyme activities in figs 3–6 can be converted to μmoles P_i mg protein N/min.

In the oocyte, ATPase activity was at a low level but increased five-fold by ovulation. After fertilization, the endogenous ATPase activity decreased slowly but was still three times greater than that of the oocyte by first cleavage. These changes in ATPase activity during early development, are, at least in part, in agreement with those recently reported by Haaland & Rosenberg [19] for hen's egg. As with frog, hen's egg Mg-dependent ATPase activity increased by ovulation, but unlike that of frog egg,

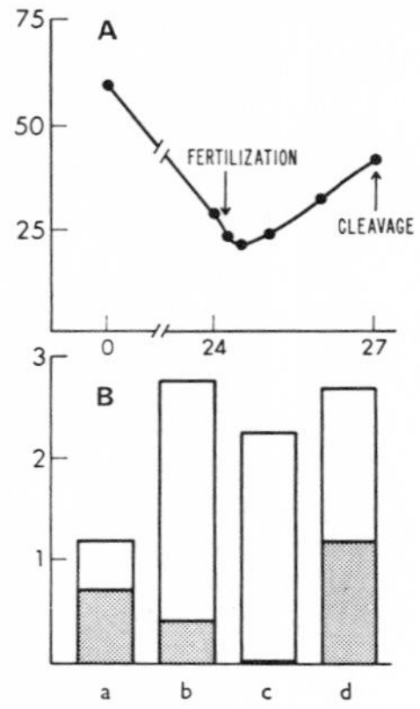

Fig. 3. Abscissa: A, a, oocyte; *b,* ovulated; *c,* fertilized; *d,* cleaved; *B.* hours; *ordinate: A,* P_i released (mmoles/kg/m); *B,* $[P_i]_i$ (mmoles/kg dry wt). ☐, ATPase; ▩, PPPase.

A comparison of adenosine triphosphatase (ATPase), phosphoprotein phosphatase (PPPase), and intracellular inorganic phosphate in the oocytes, ovulated egg, 15 min fertilized and cleaved egg. Endogenous ATPase and PPPase activities were measured on total homogenates at pH 7.4 with no additional cations added to the reaction mixture. Values for oocytes are average values for preparations from five frogs. Ovulated, fertilized and cleaved preparations are from sequential samples from the same frog and are typical of analyses from five frogs.

increased further at fertilization. However, as noted by these authors, this increase may be a result of sperm tails since polyspermy has been reported in hens.

Phosphoprotein phosphatase (PPPase) activity is found primarily in the yolk platelet fraction [20]. It decreased markedly by ovulation, disappeared during fertilization and rose prior to cleavage. The intracellular inorganic phosphate levels paralleled the changes in PPPase activity. As shown in the upper portion of fig. 3, the oocyte contained about 60 mmoles of inorganic phosphate per kg dry weight, making inorganic phosphate one of the principal intracellular anions. By the end of the depolarization at ovulation, inorganic phosphate had decreased two- to three-fold, with an additional small loss immediately following fertilization. Inorganic phosphate levels rose during the plateau phase of the membrane potential and approached the level of the oocyte by first cleavage.

Effect of Mg^{2+} and Ca^{2+} on ATPase activity

There are marked differences between the oocyte ATPase and ovulated egg ATPase. As shown in fig. 4, the oocyte ATPase could be activated several fold by the addition of 2 mM Mg^{2+} but was inhibited at higher Mg^{2+} concentrations. It could be further activated by the addition of Na^+ and K^+ ions, and the Mg-dependent activity was 80% inhibited by 10^{-4} M ouabain. The ovulated egg enzyme, on the other hand, was maximally active without added Mg^{2+}. Increasing Mg^{2+} concentrations had no inhibitory effect. The addition of Na^+ and K^+

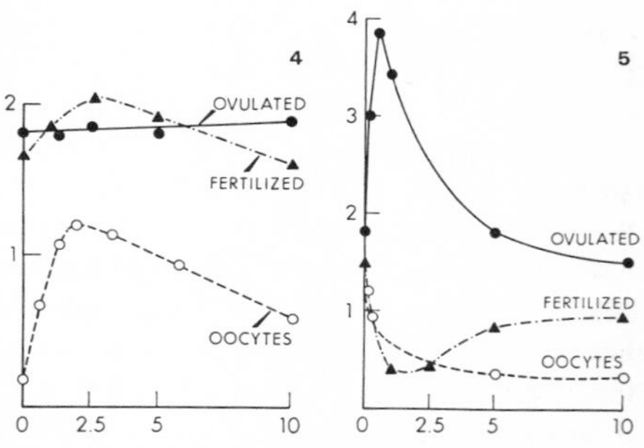

Fig. 4. Abscissa: $[Mg^{2+}]$mM; *ordinate:* P_i released (mmoles/kg/m).

Effect of increasing Mg^{2+} concentrations on ATPase activity of homogenates of oocytes, ovulated and fertilized eggs. ATPase activity was measured at pH 7.4 as described under Methods. Values for oocytes, ovulated and fertilized eggs are as described under fig. 3.

Fig. 5. Abscissa: $[Ca^{2+}]$mM; *ordinate:* P_i released (mmoles/kg/m).

Effect of increasing Ca^{2+} concentrations on ATPase activity of homogenates of oocytes, ovulated, and fertilized eggs. All ATPase analyse were carried out in the presence of 2 mM Mg^{2+} at pH 7.4 (see Methods). Values are averages or typical values as described under fig. 3.

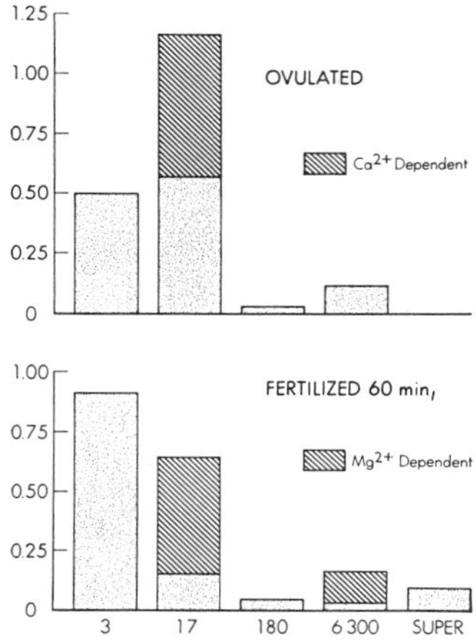

Fig. 6. Abscissa: g × min × 10^{-3}; *ordinate:* P_i released (mmoles/kg/m).

Subcellular distribution of ATPase activity in ovulated and 60 min fertilized eggs. Ovulated and fertilized egg ATPase activity was measured at pH 7.4 with no additional ions added (stippled bars) and with the addition of: (1) 2 mM Mg^{2+} or (2) 2 mM Mg^{2+} and 0.5 mM Ca^{2+} (crosshatched bars). Both preparations were from the same frog and are typical of three experiments.

had no further activating effect, and the addition of up to 1 mM ouabain produced no detectable inhibition. After fertilization the endogenous ATPase activity decreased (as noted above), and the Mg^{2+} requirement returned. By the completion of meiosis the ATPase was again partially inhibited (35 %) by 10^{-4} M ouabain.

The Mg-activated oocyte ATPase was inhibited by the addition of Ca^{2+} ion; 50 % inhibition occurring at a Ca:Mg ratio of 1:2 in the incubation medium (fig. 5). The ATPase of the ovulated egg was, in contrast, activated by low concentrations of Ca^{2+} (0.1 to 0.5 mM), although higher concentrations were inhibitory. This Ca^{2+} activation could be demonstrated only when Mg^{2+} was present. After fertilization the Ca^{2+} activated enzyme disappeared and the system again behaved like the oocyte ATPase.

Subcellular distribution of ATPase activity in the ovulated and 60 min fertilized egg

The subcellular distribution of ATPase activity is compared (fig. 6) for the ovulated and 60 min fertilized egg. A similar ATPase distribution cannot be shown for the prophase oocyte, since as described above (fig. 2), the subcellular particulates of the oocyte cannot be readily separated by differential centrifugation. In the fractionations shown in fig. 9, more than 95 % of an enzyme associated with the yolk platelets, phosphoprotein phosphatase, appeared in the platelet (3 000 g min) fraction, indicating only minor platelet contamination of the pigment granule (17 000 g min) fraction.

The ATPase present in the yolk platelet fraction of both the ovulated and fertilized egg was not significantly activated or inhibited by the addition of Mg^{2+} and/or Ca^{2+}. In contrast the ATPase associated with the pigment granule fraction of the ovulated egg was activated by Ca^{2+} in the presence of Mg^{2+}. After fertilization the Ca-dependence disappeared and the pigment granule ATPase became at least in part, Mg-dependent and Ca-inhibited. This latter Mg-dependent ATPase can be further activated by the addition of Na^+ plus K^+ or inhibited by ouabain. A second Mg-dependent ATPase was present in the 6.3×10^6 g min fraction of both the ovulated (not shown) and fertilized egg and does not appear to change during meiosis.

DISCUSSION

The pattern emerging from these studies indicates that the alkali and alkaline earth metal ions are an important if not key factor in the sequential changes during

Table 2. *Cortical changes during meiosis*

Developmental stage	Phase of meiosis	Membrane[a] potential (mV)	Membrane[b] ion selectivity	ATPase activity Ca (0.5 mM)	Mg (2 mM)	Ouabain (1 mM)
Oocyte	Prophase	−60	K^+	inhibits	activates	inhibits (80 %)
Ovulated (pond water)	Second metaphase plate	+40	Na^+	activates	no effect	no effect
Fertilized						
0–2 min		+40 to +80	Cl^-	....	....	....
2–15 min	Completes meiosis	+80 to +40	Na^+, K^+	inhibits	activates	inhibits (35 %)

[a] Potential of cell cytoplasm relative to the external medium.
[b] See [4, 5, 6].

earliest development. The general findings are outlined in table 2.

In the oocyte, the principal ATPase is Mg-activated and ouabain sensitive, and thus appears similar to that of the Na^+ "pump" or transport ATPase. This is consistent with the negative membrane potential and the K^+ selectivity of the oocyte cortex [6]. Furthermore, this pump ATPase appears to be present in the pigment granules. These granules are primarily localized near the periphery of the animal hemisphere, in close proximity to the egg plasma membrane.

After hormonal stimulation, the ouabain sensitive "Na pump" ATPase disappears and the cortical ATPase becomes Ca-activated. Inhibition of the Na-pump would account for the rise in cytoplasmic Na^+ and the accompanying depolarization during maturation (fig. 1).

Following the hyperpolarization at fertilization, the endogenous ATPase activity decreases and the enzyme is again activated by Mg^{2+} and inhibited by Ca^{2+} or ouabain. A comparison of tables 1 and 2 and fig. 3 indicates that the increase in endogenous ATPase activity (and disappearance of the Mg-requirement) by ovulation coincides with the increase in both Ca^{2+} and Mg^{2+} binding to the pigment granule fraction containing the ATPase activity; likewise, the decrease in total ATPase activity, and the reappearance of the Mg-requirement and ouabain sensitivity coincides with the release of bound Ca^{2+} and Mg^{2+} following fertilization. Thus it appears that a primary action of progesterone may be to increase Ca^{2+} and Mg^{2+} binding to specific cortical structures. This is consistent with the finding of Smith & Ecker [3] that the site of action of progesterone in inducing maturation is extranuclear, possibly at the egg surface. Similarly, a primary action of fertilization may be to release bound Ca^{2+} and Mg^{2+} from these same cortical structures.

The cortical changes outlined above appear to be responsible for a number of metabolic changes during meiosis. Major changes in phosphate metabolism are evident from the decrease in intracellular inorganic phosphate during meiosis and the subsequent rise prior to first cleavage. Preliminary chemical analyses of the principal organic phosphate components indicate that the labile phosphate is primarily associated with the phosphoproteins. The decrease in $[P_i]_i$ during maturation (fig. 3) parallels the inhibition of phosphoprotein phosphatase (PPPase), and the subsequent rise in $[P_i]_i$ following fertilization again parallels the reappearance of endogenous PPPase. These results suggest that a protein kinase similar to that described by Rabinowitz & Lipmann for hen's egg [21] may be activated during the depolarization phase of meiosis. It is worth noting that the depolarization of the neuronal membrane is accompanied by an increased turnover of protein-bound phosphorylserine groups in the membrane [22] and the intracellular release of cyclic AMP [23]. Cyclic AMP has been shown to stimulate the protein kinase of brain [24, 25] as well as the protein kinase from a wide variety of tissues [26]. Castaneda & Tyler [27] had reported earlier that an activation of adenyl cyclase occurs after fertilization in sea urchin eggs, suggesting that the production of cyclic AMP may be one of the sequential steps leading to first cleavage. Since protein kinases preferentially phosphorylate histones, the changes in $[P_i]_i$ may reflect the control mechanism for the synthesis of heterogenous RNA during maturation [28] and the replication of DNA between the end of meiosis and first cleavage.

The ion binding changes that occur during meiosis should alter the net charge and hydration, and thus the configuration, solubility, biological activity, etc., of the intracellular macromolecules. These physical changes may, in part, account for the massive subcellular disaggregation reported here, for the progressive cytological changes [29] and for the marked increase in protein synthesis [30] during maturation. Physical and chemical changes have also been observed in certain protein fractions of invertebrate eggs during meiosis (for review see Monroy [31]). For example, in sea urchin eggs there is a release of particulate bound glucose-6-phosphate dehydrogenase at fertilization [32] and this enzyme may be artificially released by an increase in ionic strength in the medium. Tyler & Monroy [33] suggest that the increase in exchangeable K^+ that occurs at this stage in sea urchin may be the releasing mechanism. It should also be mentioned that many years ago Mazia [34] observed a release of bound Ca^{2+} in sea urchin eggs during the first 10 min following fertilization, suggesting that similar mechanisms may be operating in both vertebrate and invertebrate eggs. However, a direct comparison is complicated by the differing relationships between maturation and fertilization in diverse species.

It should be emphasized that the binding described here may involve only that cation which is tightly bound or sequestered in cell organelles. Disruption of mammalian cells in aqueous media is known to result in the loss of cations from the cell nucleus. This can be avoided, at least in part, by the use of non-aqueous techniques for nuclear isolation [35]. However, unlike the readily exchangeable Na^+ in most mammalian cells, only about 10 to 15 % of the total Na^+ in the fully grown oocyte is exchangeable with extracellular Na^+ [12, 36–38]. More recently, Dick & Lea [39] have estimated that in *Bufo bufo* oocytes about 48 % of the internal Na^+ is either bound to proteins or sequestered in cell organelles. As shown in table 1, about 35 % of the internal Na^+ is recovered with the yolk platelet and pigment granules in *R. pipiens* oocytes. Thus, there may be cation "pools" with varying degrees of lability, and the pool(s) reported here may be considered to be the more tightly bound components.

Finally, the membrane potential changes shown in fig. 1 and the cation selectivity changes reported previously [6] indicate that the egg cortex undergoes a sequence of changes similar to that seen during excitation in nerve and the excitation-contraction coupling in muscle. One principal difference between the egg and muscle or nerve is that the duration of excitation is only milliseconds in nerve and muscle but lasts from minutes to hours in eggs. The relatively long time periods that are required for egg activation may make it possible to study the molecular changes which must underlie excitation both in nerve and muscle and in meiosis.

The authors thank Dr Amir Ascari for valuable discussions.

This investigation was supported in part by USPHS Research Grant GM-HD-10757. G. A. M. was a Career Scientist (I-632) of the Health Research Council of the City of New York.

REFERENCES

1. Schuetz, A W, J exptl zool 166 (1968) 347.
2. Masui, Y, J exptl zool 166 (1968) 365.
3. Smith, L D & Ecker, R E, Develop biol 19 (1969) 281.
4. Morrill, G A & Watson, D E, J cell physiol 67 (1966) 85.
5. Maeno, T, J gen physiol 43 (1959) 139.
6. Morrill, G A, Rosenthal, J & Watson, D E, J cell physiol 67 (1966) 375.
7. Woodward, D J, J gen physiol 352 (1968) 509.
8. Morrill, G A & Kostellow, A B, Biophys j 9 (1969) A-186.
9. Morrill, G A & Murphy, J B, Fed proc 29 (1970) 802abs.
10. Morrill, G A & Kostellow, A B, J cell biol 25 (1965) 21.
11. King, E J, Biochem j 26 (1932) 292.
12. Morrill, G A, Exptl cell res 40 (1965) 664.
13. Naora, H, Naora, H, Mirsky, A E & Allfrey, V G, J gen physiol 44 (1961) 713.
14. Wacher, W E C & Vallee, B L, J biol chem 234 (1959) 3257.
15. Morrill, G A & Reiss, M M, Biochim biophys acta 179 (1969) 43.
16. Manery, J F, Fed proc 25 (1966) 1804
17. Abood, L G, Neurosci res (ed S Ehrenpreis & O C Solnitzky) vol. 2, p. 41. Academic Press, New York (1969).
18. Klotz, I M, Trends in physiology & biochem (ed E S G Barron) p. 427. Academic Press, New York (1952).
19. Haaland, J E & Rosenberg, M D, Nature 223 (1969) 1275.
20. Flickinger, R A, J exptl zool 131 (1956) 307.
21. Rabinowitz, M & Lipmann, F, J biol chem 235 (1960) 1043.
22. Trevor, A J & Rodnight, R, Biochem j 95 (1965) 889.
23. Kakiuchi, S, Rall, T W & McIlwain, H, J neurochem 16 (1969) 485.
24. Miyamoto, E, Kuo, J F & Greengard, P, Science 165 (1969) 63.
25. Weller, M & Rodnight, R, Nature 225 (1970) 187.
26. Kuo, J F & Greengard, P, Proc natl acad sci US 64 (1969) 1349.
27. Castaneda, M & Tyler, A, Biochem biophys res comm 33 (1968) 782.
28. Brown, D D, Differentiation and development. Proc of symposium sponsored by New York Heart Association, p. 101. Little, Brown & Co., Boston (1964).
29. Brachet, J, Nature 208 (1965) 596.
30. Smith, L D, Ecker, R E & Subtelny, S, Proc natl acad sci US 56 (1966) 1724.
31. Monroy, A, Comprehensive biochemistry (ed M Florkin & E H Stolz) vol. 28, p. 1. Elsevier, Amsterdam (1967).
32. Isono, N, J fac sci univ Tokyo sect 4 10 (1963) 37.
33. Tyler, A & Monroy, A J, J exptl zool 142 (1959) 675.
34. Mazia, D, J cell comp physiol 10 (1937) 291.
35. Itoh, S & Schwartz, I L, Am j physiol 188 (1957) 490.
36. Abelson, P H & Duryee, W R, Biol bull 96 (1949) 205.
37. Naora, H, Naora, H, Izawa, M, Allfrey, V G & Mirsky, A E, Proc natl acad sci US 48 (1962) 853.
38. Dick, D A T & Lea E J A, J physiol 174 (1964) 55.
39. — Ibid 191 (1967) 289

*PROTEIN SYNTHESIS DURING MEIOSIS**

BY YASUO HOTTA, L. G. PARCHMAN, AND HERBERT STERN

Studies on the behavior of chromosomes during the extended prophase of meiosis have been largely in the province of cytogenetics. Biochemical analyses have been few in number and these have centered on the properties of deoxyribonucleic acid (DNA). The significance of the meiotic history of DNA to chromosome pairing and crossing-over is patent, but the adequacy of such history in accounting for these two major phenomena is rendered improbable by the complexity of chromosome organization during meiotic prophase. The pioneering autoradiographic studies of Taylor[1] demonstrated the occurrence of protein synthesis during meiotic prophase, and his conclusions were subsequently confirmed in a general way by biochemical analysis.[2] Other studies of meiotic proteins have been more specifically concerned with histones.[3, 4] Taken as a whole, the attempts to define a relationship between proteins and the behavior of meiotic chromosomes have fallen short of their goal. Past studies only reveal that unidentified proteins are synthesized during meiotic prophase and that these proteins are located in both nucleus and cytoplasm.

This report furnishes limited evidence for the selective synthesis of certain nuclear proteins during meiotic prophase and for their possible relationship to synapsis and crossing-over. The evidence is limited in the sense that no attempt has been made to survey all of the proteins synthesized during meiotic prophase nor to purify those proteins which appear to have a distinctive functional role. The purpose of these experiments has been to establish some temporal correlations between the synthesis of particular nuclear proteins and the behavior of chromosomes during meiotic prophase.

Methods.—Meiotic cells obtained from two horticultural varieties of lily, "Cinnabar" and "Bright Star," were cultured *in vitro* as described in an earlier publication.[5] Protein synthesis was followed by the addition of uniformly labeled C^{14}-leucine or H^3-leucine (labeled in 4 and 5 positions) to the culture medium at concentrations of 0.25 μc/ml or 2.5 μc/ml, respectively. H^3-thymidine (methyl label) was used at a concentration of 10 μc/ml in studies of deoxyribonucleic acid (DNA) synthesis. The procedures used for harvesting the cells were the same as those previously described.[6]

Nuclei were isolated in glycerin-sucrose media.[7] The nuclear fraction was homogenized in a glass tissue grinder with 25–50 vol of 0.1 *M* K-Na phosphate buffer (pH 8.0). The pestle was held in the chuck of a $^1/_4$-in. electric drill controlled by a voltage regulator. The homogenate was allowed to stand in an ice bath for 20 min and then centrifuged for 10 min at 30,000 $\times$ *g*. The precipitate was re-extracted once and the combined supernatant fluids were dialyzed against 0.005 *M* phosphate buffer (pH 8.8) for 2 hr with two changes of external medium. This extract will be referred to as the "alkaline" extract. The residue remaining after alkaline extraction was resuspended in 0.1 *M* phosphate buffer (pH 6.0). The suspension was centrifuged at 30,000 $\times$ *g* for 10 min after standing in an ice bath for 20 min. The resulting supernatant fluid will be referred to as the "pH 6.0" fraction. The residue was resuspended in 1.0 *M* NaCl containing 0.01 *M* phosphate buffer (pH 7.0). The suspension was shaken overnight at 2°C and then centrifuged for 10 min at 10,000 $\times$ *g*. The supernatant fluid is designated as the "1.0 *M*" fraction. The residue remaining will be referred to as such.

The alkaline extract was fractionated on a column of *O*-(diethylaminoethyl) (DEAE) (0.8 $cm^2 \times 25$ cm), which was previously washed with 0.005 *M* phosphate buffer (pH 8.8). A linear gradient of increasing sodium chloride concentration (0–1.0 *M*) in the presence of 0.005 *M* phosphate buffer was used to elute the adsorbed material. The gradient was obtained by use of two cylinders, one containing 80 ml of 0.005 *M* phosphate buffer and the other containing 80 ml of 1.0 *M* NaCl–0.005 *M* phosphate buffer. A flow of 10 ml/hr was maintained and fractions of 1.6 ml were collected in tubes. Individual fractions or portions of the various extracts were mixed with trichloroacetic acid to a final concentration of 10% (w/v). The precipitates were collected on glass filter paper and the filter disks, after being dried, were placed in vials containing 10 ml of scintillation fluid.

A solution of crystalline deoxyribonuclease (5 μg/ml) buffered with 0.01 *M* tris(hydroxymethyl)aminomethane (Tris) (pH 6.8) and containing 0.001 *M* $MgCl_2$ was used to remove DNA. Ribonuclease was used at a concentration of 40 μg/ml and was dissolved either in 0.15 *M* NaCl:0.015 *M* Na citrate:0.01 *M* Tris (pH 7.0) or in 1.0 *M* NaCl:0.01 *M* ethylenediaminetetraacetic acid:0.01 *M* Tris (pH 8.0). DNA was purified with the use of "Pronase" and chloroform:amyl alcohol for deproteinization and of isopropanol for precipitation.[8]

Results.—Distribution of radioactivity among nuclear fractions: The purpose of these experiments was to determine whether there were any appreciable differences in specific activities between the four fractions of proteins obtained from isolated nuclei. A mixture of cells at various stages of meiotic prophase was cultured for two days in the presence of C^{14}-leucine at 19°C. Very few of the cells thus cultured were initially in late pachynema so that virtually none of the cells advanced beyond meiotic prophase during the interval of exposure. The combined results of three such experiments are summarized in Table 1. Our attention was drawn to the alkaline extract which accounted for 15 per cent of the protein and for 48 per cent of the radioactivity. Although the pH 6.0 fraction had an even higher specific activity, the amount of protein present was too small for further analysis.

All subsequent experiments were directed at the alkaline extract. The selec-

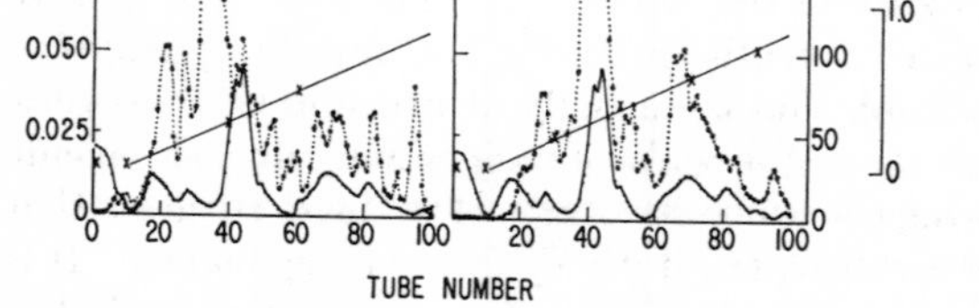

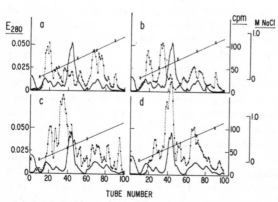

FIG. 1.—Elution profiles of "alkaline extracts" from cells at various meiotic stages cultured in the presence of H^3-leucine for 2 days at 19°C. The extracts were prepared from isolated nuclei and adsorbed on DEAE-columns as described under *Methods*.

Solid curves represent optical density; dotted curves, radioactivity. The straight line indicates the concentration of sodium chloride in eluate.

The figures from (*a*) to (*d*) represent progressively later stages of meiosis: (*a*) Leptotene-zygotene; (*b*) zygotene-pachytene; (*c*) pachytene; (*d*) diplotene to completion of meiosis.

Premeitic interphase is not included in this figure but, as is pointed out in the text, the optical density is the same as that of other stages, and the radioactivity profile more or less tracks the optical density.

TABLE 1. *Distribution of radioactivity in nuclear proteins of cells in meiotic prophase.*

Nuclear fraction	Protein content (% of total)	Specific activity (cpm/μg protein)	Radioactivity (% of total)
Alkaline (pH 8.0)	15.2	37.6	48.2
pH 6.0	0.6	42.7	2.1
1.0 *M* NaCl	16.3	12.8	17.2
Residual	67.9	6.8	33.4

Cells in the zygotene and pachytene stages of meiosis were cultured for 2 days at 19°C in the presence of 0.25 μc of C^{14}-leucine/ml. The nuclei were isolated and fractionated as described under *Methods*. Attention is drawn to the alkaline fraction that contains about half of the radioactivity. All subsequent experiments reported here are concerned with this fraction.

tion was necessarily arbitrary, since no criteria were available for assigning significance to any particular protein or group of proteins in relation to chromosomal behavior during meiotic prophase. The value of this selection for an understanding of meiotic events will become apparent from the results.

Protein synthesis in relation to meiotic stage: A satisfactory, but by no means complete, resolution of the proteins present in the alkaline extract was obtained by the use of DEAE-columns. With a linearly increasing concentration of sodium chloride as eluant, several peaks of protein concentration were observed (Fig. 1). The elution pattern was the same for all stages of meiosis and also for the premeiotic cells. Thus, no gross changes in composition of the alkaline extract were observed during the progression of the cells from premeiotic interphase to completion of meiosis. The apparent constancy in protein composition was not reflected, however, in the profiles of radioactivity. Only in the case of cells exposed to labeled amino acids during the interval of premeiotic mitosis did the profile of radioactivity track that of optical density.

The respective profiles of radioactivity in proteins obtained from cells at different meiotic intervals are shown in Figure 1. A striking feature of the profiles is the elution of a major peak of radioactivity at a salt concentration of about 0.2 *M*. This peak was found only in cells that had been exposed to isotope during the zygotene-pachytene intervals. This peak was pronounced in extracts from pachytene cells (Fig. 1*c*). Another peak of radioactivity eluting at 0.08 *M* salt was present in extracts from cells in early zygotene through pachytene (Fig. 1*a–c*) but not in postpachytene cells. Other differences in radioactivity profiles may also be noted but the actual differences between successive meiotic stages are probably blurred to some extent because of the overlapping of stages between the groups tested. The most important inference that may be drawn from these results is the occurrence of a progressive change in patterns of protein synthesis paralleling the progress of the cells through meiosis. The change is marked by a distinctive pattern of synthesis during the zygotene-pachytene interval. The pattern is reproducible and is virtually identical in the two varieties of lily tested. The fact that the major peaks of radioactivity do not coincide with those of optical density during the prophase stages (Fig. 1*a–c*) suggests that the principal syntheses occur in certain quantitatively minor components. It is during these same stages that pairing and chiasma formation occur and that quantitatively minor components of DNA are synthesized.[6, 9]

Relationship between DNA and protein synthesis during meiotic prophase: The

TABLE 2. *Distribution of protein and DNA between alkaline and other nuclear fractions in zygotene-pachytene cells.*

Fraction	Extraction method	Protein content (%)	DNA content (%)	H^3-DNA (total cpm)	H^3-DNA (cpm/μg)
"Alkaline"	"Complete"	15.2	5.5	3134	111
	"Mild"	3.25	0.7	1983	302
Remainder	"Complete"	84.8	94.5	3732	—
	"Mild"	96.75	99.3	4663	—

The distribution of total protein and DNA was determined in three separate experiments. For "mild" extraction, the nuclear suspension was homogenized with 5 strokes of the pestle at a setting of 50 v (see *Methods*). For "complete" extraction, three successive extractions were made, each with 20 strokes of the pestle and at a setting of 90 v. This latter method gave similar results to passage of the nuclear suspension through a French press. The low contents of DNA and protein in the mild alkaline extract showed appreciable variations between experiments (2.5–5% for protein and 0.25–1.0% for DNA), but the average of these is recorded in the table. Radioactivities of proteins under conditions of complete extraction are shown in Table 1. In this table only the data on H^3-thymidine incorporation have been included. In the actual experiment zygotene and pachytene cells were exposed simultaneously to C^{14}-leucine and H^3-thymidine for 2 days at 19°C.

coincidence of nuclear protein and DNA syntheses in zygotene-pachytene cells points to the possibility that the two syntheses might be functionally interrelated. To test this possibility, zygotene and pachytene cells were incubated in the presence of C^{14}-leucine and H^3-thymidine. Nuclei were then isolated and the distribution of each of the labels among the nuclear fractions was determined. The results of this experiment are summarized in Table 2. Of major interest is the observation that about half of the DNA synthesized during zygotene and pachytene stages is extracted at pH 8.0, even though about 95 per cent of the total DNA remains unextracted. The precise significance of the actual proportion of labeled DNA that is extracted in the alkaline medium is unclear. Neither labeled protein nor DNA can be extracted from the nuclei prior to their disruption. As illustrated in Table 2, the yield of labeled material partly depends upon the degree of nuclear disruption. It is reasonably clear, however, that the DNA synthesized during meiotic prophase is selectively extracted by the same procedure that selectively extracts certain nuclear proteins with high specific activity. Although milder disruption of nuclei yields lower amounts of labeled DNA, the specific activity of that DNA is much higher. More vig-

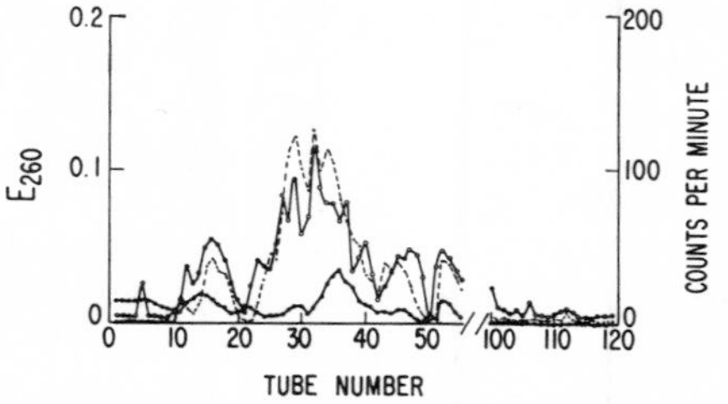

FIG. 2.—Elution profile of alkaline extract from zygotene-pachytene nuclei. Cells were cultured in the presence of H^3-thymidine and C^{14}-leucine at 19°C. and the extract adsorbed on a DEAE-column. Conditions of elution are described under *Methods.* Solid curve with solid circles represents optical density. Solid curve with open circles represents H^3-DNA activity, and dotted curve represents C^{14}-leucine activity. The optical density profile is different from that shown in Fig. 1 because readings are recorded at 260 rather than 280 mμ. Profile of eluate between tubes 55 and 100 has been omitted because no DNA counts have been found in that region; H^3-counts do occur but they are entirely due to RNA. In this particular fractionation 1.3-ml rather than 1.6-ml samples were collected.

FIG. 3.—Distribution of labeled DNA and protein from alkaline extracts of zygotene-pachytene nuclei in a CsCl gradient following centrifugation for 60 hr at 73,500 × *g*. The regression lines represent the density (gm/cc) of the fractions in the collecting tubes. The solid curve is the optical density profile. Dashed curve is profile of H^3-DNA activity and dotted curve is profile of C^{14}-protein activity. (*a*) Native extract; (*b*) extract adjusted to pH 13.0 and allowed to stand for 20 min at room temperature prior to addition of CsCl solution and subsequent centrifugation. The shaded portion represents the profile of native DNA that was added to the mixture just prior to centrifugation.

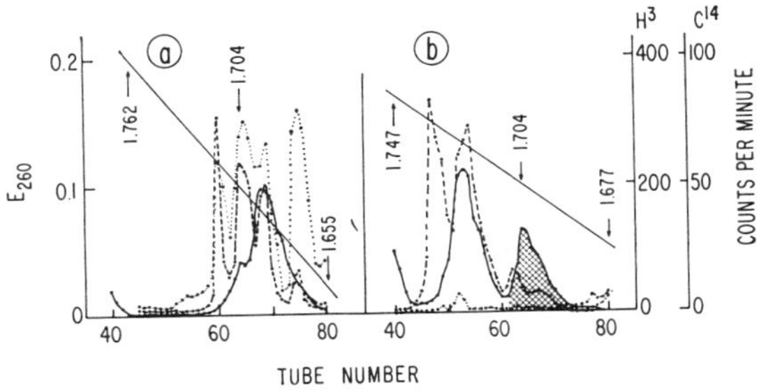

orous disruption of nuclei renders a higher proportion of the unlabeled DNA extractable at pH 8.0.

In order to determine whether any association exists between the prophase-labeled DNA and protein, the extracts were examined by column chromatography and by centrifugation in solutions of CsCl. The elution profile of the doubly labeled material is shown in Figure 2. Most of the labeled DNA is collected in tubes 20–40 and the curve of tritium activity tracks that of C^{14}-leucine. This is the region of the labeled protein profile which is most prominent in extracts of pachytene nuclei (Fig. 1*b* and *c*). A high peak of tritium activity (not shown in the curve) occurs between tubes 50–100 but it is not due to DNA. It is unaffected by DNase and is completely removed by RNase. This latter result is in line with other unpublished observations in our laboratory on the incorporation of label from thymidine into RNA. The nature of the association between RNA and nuclear protein has not been pursued in these studies. Of direct relevance to the present study is the fact that those nuclear proteins that are distinctively synthesized during the zygotene-pachytene interval behave on a DEAE-column as though they were associated with the DNA that is also distinctively synthesized during that same interval.[6]

Centrifugation of the alkaline extract in the presence of CsCl also reveals the presence of a complex between labeled protein and DNA (Fig. 3*a*). However,

FIG. 4.—Profile of radioactivity of DNA purified from alkaline extract obtained from zygotene-pachytene nuclei cultured in the presence of H^3-thymidine and C^{14}-leucine. DNA was prepared as described under *Methods*. Solid curve represents the optical density profile of marker DNA added just prior to centrifugation. Dashed curve represents H^3-activity. The dots along the bottom line represent C^{14}-activity and indicate the absence of labeled protein in the purified preparation.

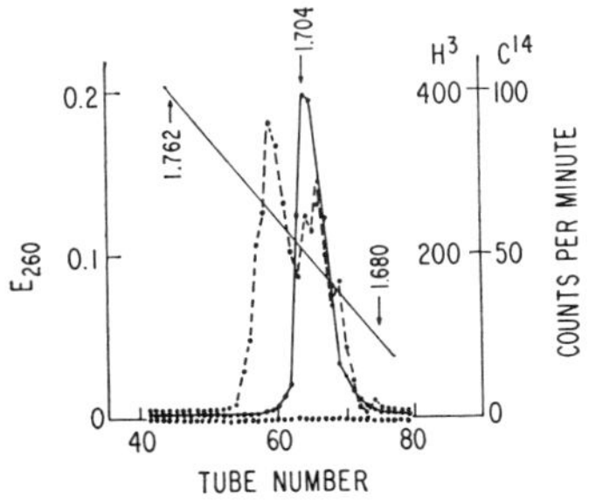

TABLE 3. *Effect of cycloheximide (CHI) on synthesis of nuclear protein during zygotene-pachytene stages.*

Fraction	Specific Activity (cpm/μg) −CHI	+CHI	Per cent inhibition
Alkaline	37.6	13.1	65.2
pH 6.0	42.7	2.0	53.6
1.0 *M* NaCl	12.8	13.9	52.4
Residual	6.8	52.5	23.5

Conditions of culture were the same as those described under Table 1. The columns marked "+CHI" represent the values obtained for cells that had been cultured in the presence of 0.5 μg/ml of cycloheximide and C^{14}-leucine.

the proportion of labeled protein apparently complexed with the labeled DNA in the CsCl gradient is much less than that observed with column chromatography. About 30 per cent of the protein count is found in the DNA region of the gradient. Four peaks of protein radioactivity match the positions of the four peaks of DNA radioactivity. If the extract is denatured with alkali prior to centrifugation, the protein counts are lost from the DNA region and the DNA peaks are found in positions expected for the denatured product (Fig. 3*b*). The amount of protein associated with each of the native DNA peaks may be estimated by comparing Figure 3*a* with the profile of DNA that has been purified from the alkaline extract (Fig. 4). These estimates, however, are unlikely to have much value except for indicating that the protein content is of the order of 0–12 per cent, and that each of the peaks probably has a different protein content. For present purposes it is sufficient to point out that the radioactive profile of the DNA prepared from the alkaline extract is the same as that reported for DNA prepared from whole zygotene-pachytene cells,[6] and that this DNA remains associated with protein synthesized during the zygotene-pachytene interval. The physical association may imply a functional relationship.

Significance of protein synthesis to meiotic development: The relevance of protein synthesis to meiotic development was examined by culturing meiotic cells in the presence of cycloheximide. At concentrations of the drug exceeding 2 μg/ml, protein synthesis is virtually abolished and meiotic development is completely arrested. The responses of the cells to concentrations in the range of 0.2–1.0 μg/ml are more interesting. At these lower concentrations of cycloheximide, inhibition is incomplete and, as will be described elsewhere,[10] meiotic development is only partially suppressed. The inhibition of protein synthesis appears to be selective. Generally, the proteins in the alkaline fraction are more sensitive than those of the other fractions to 0.5 μg/ml of cycloheximide (Table 3). Moreover, a selective effect is also apparent within the alkaline fraction. The peaks of radioactivity that are characteristic of zygotene-pachytene cells are abolished by 0.5 μg/ml of cycloheximide. The remaining radioactivity tracks the optical density profile in a pattern that is more or less characteristic of interphase nuclei (Fig. 5). The cytological consequences of partially inhibiting protein synthesis depend upon the particular stage at which the cells are exposed to the inhibitor and the duration of exposure. By appropriately timing the interval of exposure it can be shown that selective inhibition of protein synthesis at the end of zygonema causes a failure of chiasma formation.[10] It would thus

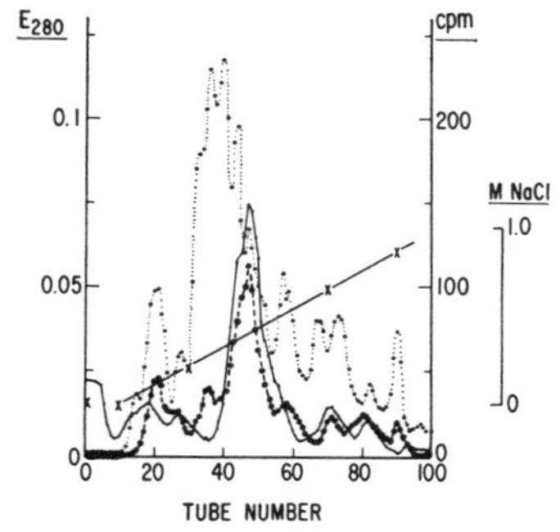

FIG. 5.—Effect of cycloheximide on synthesis of nuclear protein in alkaline extract from zygotene-pachytene cells. Conditions of culture are the same as those described under Fig. 1. Tritiated leucine was used in these experiments. The alkaline extract from the isolated nuclei was adsorbed on a DEAE-column and eluted as described. The straight line indicates molarity of NaCl in eluate. Solid curve represents optical density. Dotted curve represents radioactivity profile of control and dashed curve represents radioactivity profile of extract from cells cultured in the presence of 0.5 μg/ml of cycloheximide.

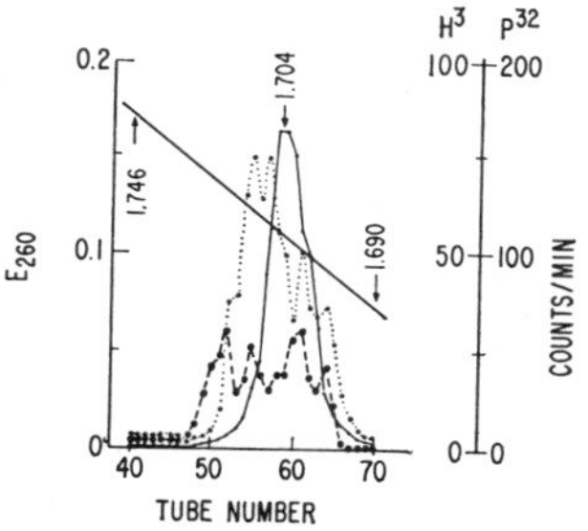

FIG. 6.—Effect of cycloheximide on DNA synthesis in zygotene-pachytene cells. One group of cells (*dotted curve*) was cultured from the zygotene stage for 5 days in the presence of 0.2 μg/ml of cycloheximide. At the end of this period the medium was replaced with standard medium containing H^3-thymidine and cultured for 2 more days. The second group of cells (*dashed curve*) was cultured for 7 days in the presence of P^{32}-phosphate and 0.2 μg/ml of cycloheximide. DNA was prepared from each of the groups as described under *Methods*, and the two preparations were combined for analysis in a CsCl gradient. Solid curve represents optical density. Regression line represents density (gm/cc) of fractions.

appear that the proteins which are distinctively synthesized during the late zygonema-early pachynema are essential to formation of chiasmata.

A prominent biochemical consequence of exposure to cycloheximide is the arrest of DNA synthesis. Data pointing to this conclusion are shown in Figure 6. Concentrations of cycloheximide as low as 0.2 μg/ml markedly suppress DNA synthesis. The dashed curve in Figure 6 shows the pattern of synthesis in zygotene-pachytene cells during a seven-day exposure to P^{32}-phosphate and cycloheximide. Much the same pattern of synthesis is observed if the cells are harvested following a three-day exposure. DNA synthesis is thus brought to a halt at least within three days of exposure to cycloheximide. The inhibition is entirely reversible as illustrated by the dotted curve. The pattern of DNA synthesis in cells that had been exposed for five days to cycloheximide and then cultured for two days in a normal medium is the same as in untreated cells.

These observations permit the conclusion that the synthesis of DNA during the zygotene-pachytene stages requires the simultaneous synthesis of certain nuclear proteins. Whether the required proteins are those that are physically associated with the DNA is open to conjecture. One obvious implication of this requirement is that those processes which have been shown to depend upon prophase DNA synthesis[9] must, at least indirectly, depend upon prophase protein synthesis.

Discussion.—The experiments reported here represent no more than an arbitrary entry into the general problem of protein functions during meiotic prophase. The results are gratifying, inasmuch as they do offer convincing evidence that synthesis of certain nuclear proteins is essential to the synthesis of DNA and that the combination of these syntheses that occur during zygonema and pachynema is essential to chromosome pairing and chiasma formation. A variety of speculative schemes is made possible by these observations, but their pursuit is probably of little value in the absence of more precise information concerning the nature of the proteins synthesized. The apparent physical association between the protein and DNA synthesized during meiotic prophase is no more than a pointer to a possible structural relationship between them. The principal conclusion that may be drawn from the present studies is that the respective syntheses of DNA and protein are functionally interrelated.

* This work was supported by a grant from the National Science Foundation (NSF-GB-5173x) and by supplementary assistance from the Institute for Studies in Developmental Biology (USPHS-HD03015 and NSF GB 6476).

[1] Taylor, J. H., *Am. J. Botany*, **46**, 477–484 (1959).
[2] Hotta, Y., and H. Stern, *J. Cell Biol.*, **19**, 45–58 (1963).
[3] Ansley, H. R., *J. Biophys. Biochem. Cytol.*, **4**, 59–62 (1958).
[4] Sheridan, W. F., and H. Stern, *Exptl. Cell Res.*, **45**, 323–335 (1967).
[5] Ito, M., and H. Stern, *Devel. Biol.*, **16**, 36–53 (1967).
[6] Hotta, Y., M. Ito, and H. Stern, these PROCEEDINGS, **56**, 1184–1191 (1966).
[7] Hotta, Y., and H. Stern, *Protoplasma*, **60**, 218–232 (1965).
[8] Hotta, Y., A. Bassel, and H. Stern, *J. Cell Biol.*, **27**, 451–457 (1965).
[9] Stern, H., and Y. Hotta, in *The Control of Nuclear Activity*, ed. L. Goldstein (New York: Prentice Hall, 1967), pp. 47–76.
[10] Parchman, L. G., and H. Stern, manuscript in preparation.

AUTORADIOGRAPHIC STUDY OF RIBONUCLEIC ACID SYNTHESIS DURING SPERMATOGENESIS OF *ASELLUS AQUATICUS* (CRUST. ISOPODA)

ANGELA ROCCHI BRASIELLO

In order to study the metabolism of ribonucleic acid in the spermatogenetic cells and nurse cells of the testicle of *Asellus* (*Asellus*) *aquaticus* and to observe the existence of a possible relationship, with regard to this acid, between these two kinds of cells, use has been made of autoradiographic technique, after injection of ^{3}H-uridine.

The nurse cells in *Asellus* are large-sized cells with a probably polyploid nucleus; cyclic variations, as regards their form, have been described by Montalenti, Vitagliano and De Nicola [16]. These authors, by means of histochemical stains, have described a passage of ribonucleic acid from the nurse cells to the germinal cells. In an experiment performed on *Asellus coxalis* [20], using ^{3}H-thymidine as a labeled precursor, it was possible to show that these cells either divide very rarely or encounter processes of endomitosis.

The use of tritiated uridine to clarify further aspects of the problem seemed very interesting.

Autoradiographic data on the metabolism of RNA in meiosis and in spermatogenesis already exist for certain plants and animals. Taylor [21] has carried out a study of this kind on *Lilium longiflorum*, Monesi [14] on the mouse, Utakoji [23] on *Cricetulus griseus*, Block and Brack [2], Muckenthaler [17], Henderson [7, 8], Das and others [3] on various species of Orthoptera, Meusy [12] on an amphipod Crustacean *Orchestia gammarella*, Olivieri and Olivieri [18] on *Drosophila melanogaster*, Berlowitz [1] on the mealy bug.

MATERIALS AND METHODS

The specimens used for the experiments were males of an isopod Crustacean *Asellus* (*Asellus*) *aquaticus* collected *in natura*. ^{3}H-uridine (spec. act. 2.73 c/m*M*, concentration 500 μc/ml) was injected into each individual, by means of a glass micropipette, dorso-laterally. After the treatment, *Aselli* were put back into water and kept at a constant temperature of 18°C. All the individuals were sacrificed at various intervals of time, from 10 min to 16 days after the injection. The testicles were in part fixed in Bouin, enclosed in paraffin and cut to a thickness of 5 μ; in part they were squashed, after fixing and staining in acetic carmine. In order to remove all traces of precursors not incorporated, a certain number of slides were treated with 5 per cent TCA at 4°C for 5 min, but these slides showed no differences compared with the others. All the slides were covered with Kodak NTB_2 emulsion and after developing they were stained in Mayer's haematoxylin. Three experiments were carried out, in which a total of about 120 individuals were examined.

RESULTS

The gonads of *Asellus* each consist of three lobes that lead, at different points, into a duct deferens. The primary spermatogonia occupy the base position of each lobe. As maturation proceeds the cells are driven, without any seriation being created, into the lumen of the follicle: that is to say, they do not become established in fixed positions or come to form part of cysts as occurs in many animals; all this makes it very difficult to distinguish all the interkinetic stages from one another.

In order to be able to recognize the cells and cellular structures that synthesize RNA directly, in cases such as this in which the labeled precursor is supplied *in vivo* to the individual by means of an injection, it is necessary to observe the preparations made by sacrificing the animals at a short interval after administration, since it is considered that the radioisotope remains available for some hours in the body of the animal to be incorporated in the RNA molecules that are in process of synthesis. We considered as labeled by direct synthesis the cells labeled within the first hour.

Cells of the germ line

Five to ten min after administration of the ^{3}H-uridine the cells that incorporate are labeled exclusively in the nucleus. The mitotic stages that incorporate the radioisotope actively are the prophases and prometaphases, which appear with the chromosomes heavily labeled (Figs 1, 2) while the metaphases and anaphases are not labeled at all (Fig. 5). The stages of the meiotic prophase almost all appear labeled, though with very different degrees of intensity; the labeling increases from the leptotene and zygotene stages, in which it is very low, to the pachytene stage and becomes intense in the advanced pachytene stage; it then decreases again to the diplotene and diakinesis stages and disappears at metaphase (Fig. 4). Neither the early nor the late spermatids show any labeling (Fig. 3), nor do the spermatozoa.

The preparations made 30 min and 1 h after the injection show no substantial differences from the previous preparations. Two hours later the metaphase (Fig. 6) and the mitotic anaphase have begun to show the labeling in the cytoplasm. After 5–7 h the late pachytene stage is still heavily labeled in the nucleus, and the nuclei of the gonial prophases are also heavily labeled; in addition, a slight labeling of the cytoplasm begins to appear in all the types of cells (Fig. 8).

24–28 h after the administration of the radioactive precursor, all the labeled cells continue to be, especially in the nucleus. After 3–4 days, slightly labeled spermatids appear (Fig. 10). After 6 days the lepto-zygotene stages are labeled only in the cytoplasm, while certain cells in interkinesis or in mitotic prophase and pachytene are very highly labeled, especially in the nucleus. We know that the pachytene stage is one that lasts some days [20] and we suppose that all the other cells still labeled in the nucleus are cells that have not divided.

After 7 days the tails of the sperms appear slightly labeled; after 10–14 days their labeling is much more intense (Fig. 12), and the spermatids are also highly labeled.

Nurse cells

A few minutes after administration of ^{3}H-uridine part of the nurse cells reveal a heavily labeled nucleus (Fig. 3). After 30 min, some of the numerous nucleoli of the nurse cells appear more intensily labeled than the rest of the nucleus. However, we cannot exclude the possibility that some of those that we have called nucleoli may really be karyosomes.

Two hours after the injection, in the nucleus intensely labeled of some

---

Fig. 1.—Labeled gonial mitotic prophase and interkinetic cell, 15 min after injection of ^{3}H-uridine. ×640.

Fig. 2.—Labeled gonial mitotic prometaphase and sperms devoid of labeling, 15 min after injection of ^{3}H-uridine. ×640.

Fig. 3.—Spermatids devoid of labeling and labeled nurse cells, 15 min after injection of ^{3}H-uridine. ×400.

Fig. 4.—Meiotic metaphase I, meiotic anaphase I and meiotic metaphase II, all devoid of labeling, and labeled interkinetic cells, 15 min after injection of ^{3}H-uridine. ×400.

Fig. 5.—Gonial mitotic metaphase devoid of labeling and labeled gonial mitotic prophase, 30 min after injection of ^{3}H-uridine. ×400.

Fig. 6.—Gonial mitotic metaphase in which the labeling is passing from the nucleus to the cytoplasm, 2 h after injection of ^{3}H-uridine. ×640.

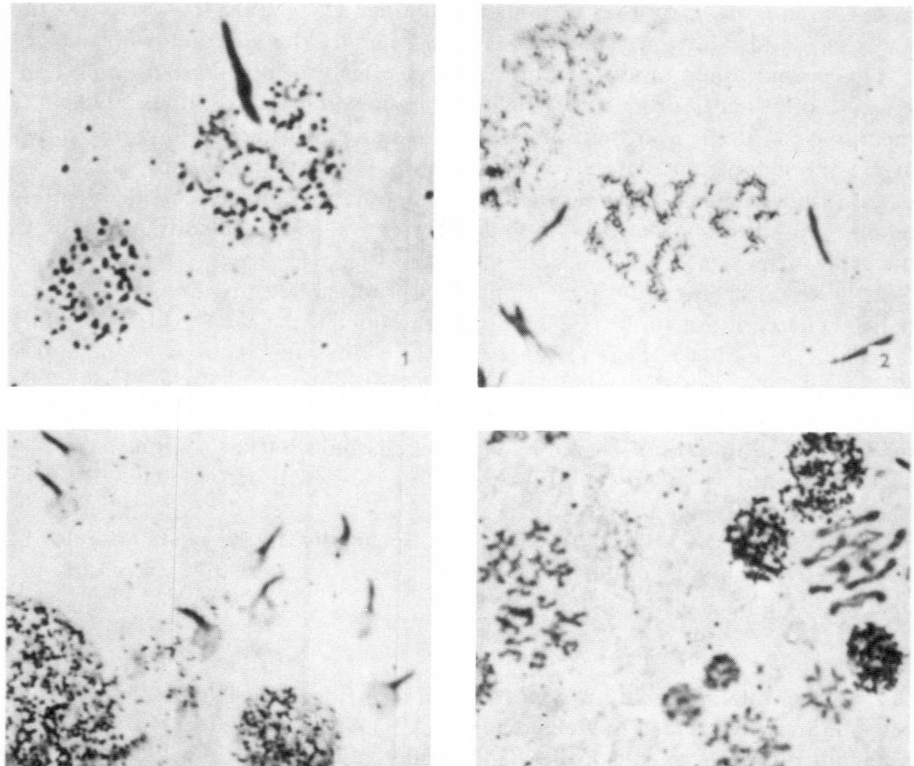
1
2
3
4
5
6

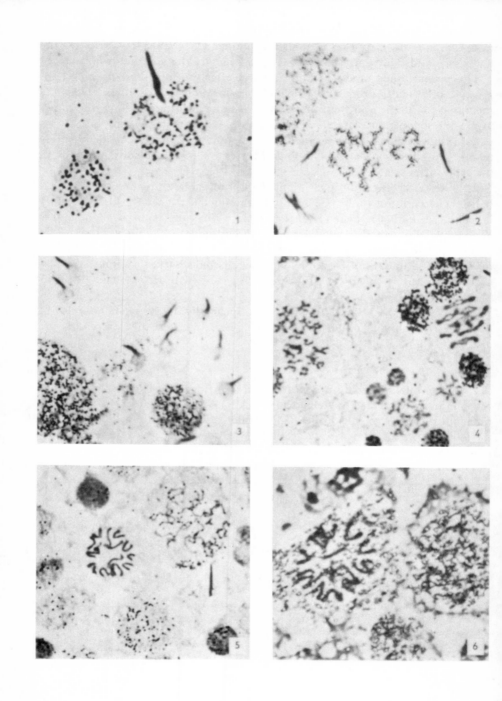

nurse cells, we were able to observe an empty space, as regards both the labeling and the staining (Fig. 7). The phenomenon was clearly observed but it occurs rather rarely.

After 5–7 h the nucleus of the nurse cells is still heavily labeled while a slight cytoplasmatic labeling begins to appear (Fig. 8). After 3–4–7 days the nucleus is still more intensely labled than the cytoplasm and continues to be so until 16 days after.

DISCUSSION

It may be concluded that, as regards the spermatogonia, RNA synthesis occupies practically the whole cellular cycle, excluding the time during which the cells are in metaphase and anaphase; no labeling is observed in these stages during the first hour after administration of the radioactive precursor.

The same holds for the prophase of spermtocytes I; the whole meiotic prophase actively incorporates ^{3}H-uridine, though with a different intensity in the various stages. The leptotene and zygotene stages immediately appear slightly labeled, the labeling increases up to a maximum at the advanced pachytene and then decreases in diplotene, diakinesis becoming nil at metaphase. All the authors that have studied RNA synthesis in meiosis have observed a different intensity of synthesis in the various stages of prophase; they are not in agreement, however, as to which are the stages that synthesizes most.

Monesi [14] has observed, in the mouse, a pattern similar to ours; he indicates the late pachytene stage as the most active one, as does also Utakoji [23] Henderson [7, 8], in Orthoptera, has observed an almost constant incorporation in the whole meiotic prophase and a decrease during diakinesis; Muckenthaler [17], in *Melanoplus differentialis*, has noted a decrease in the labeling during zygotene and an intense synthesis in diplotene and in diakinesis; Hotta and Stern [9, 10] have observed in *Trillium erectum* a situation somewhat resembling that of *Asellus*.

Fig. 7.—Nurse cells showing an empty space devoid of labeling 2 h after injection of ^{3}H-uridine. ×640.

Fig. 8.—Nurse cells in which part of the labeling is passing into the cytoplasm, 5 h after injecton of ^{3}H-uridine. ×640.

Fig. 9.—Zygotene and early pachytene stages, labeled in nucleus, 16 h after injection of ^{3}H-uridine. ×400.

Fig. 10—Labeled spermatids, 4 days after injection of ^{3}H-uridine. ×400.

Fig. 11.—Interkinetic cells, intensely labeled in nucleus, 16 days after injection of ^{3}H-uridine. ×400.

Fig. 12.—Spermatozoa with tails labeled, 14 days after injection of ^{3}H-uridine. ×250.

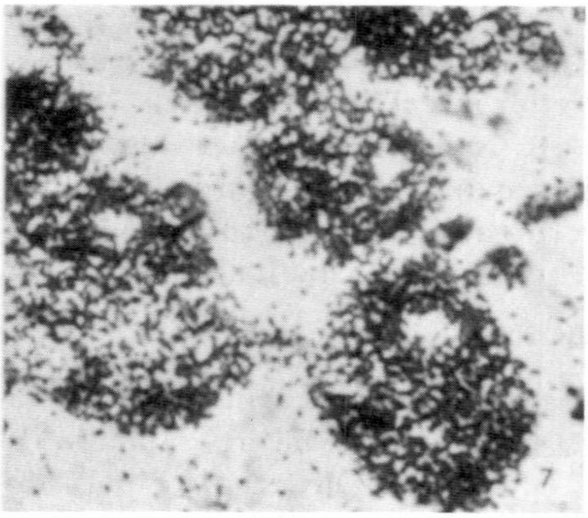
7
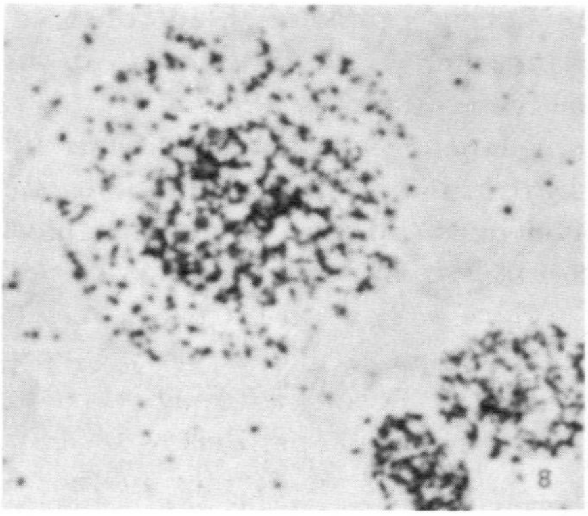
8
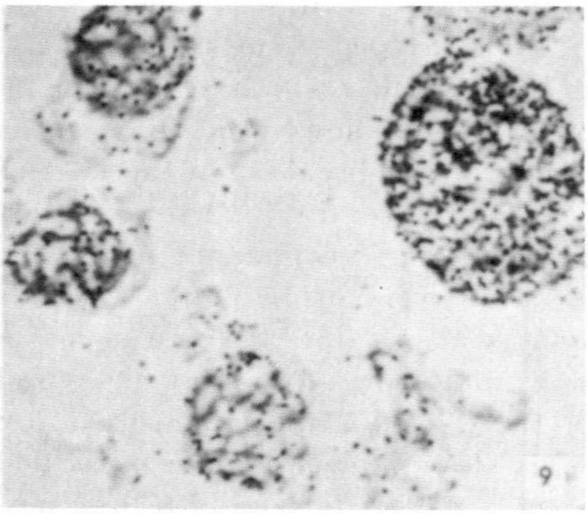
9
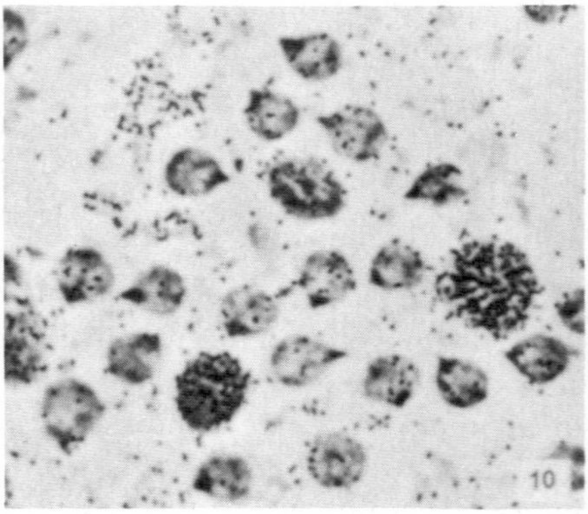
10
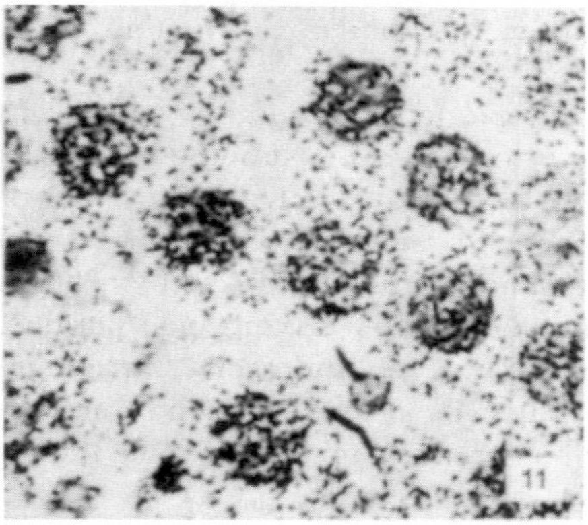
11
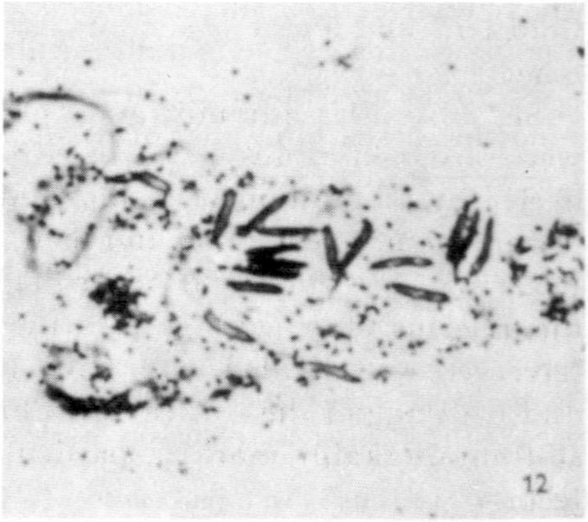
12

Another point on which authors are not in agreement is the presence of active synthesis in the spermatids. In *Asellus* it was not possible to observe any labeling on this type of cells before the 4th–6th day after administration of radioisotope, when the spermatids may have derived through maturation by the spermatocytes that have incorporated. Monesi [13] observed, in the mouse, a few granules in a small fraction of young spermatids, only after 60 days of exposure; Utakoji [23] found a distinct but low synthetic activity in telophase II and in early spermatids; both Muckenthaler [17] and Henderson [8] observed young labeled spermatids a few hours after administration of radioisotope; Berlowitz [1] has not observed any incorporation in the spermatids of mealy bug, neither have Olivieri and Olivieri [18] in Drosophila.

Cytoplasm probably does not incorporate actively uridine. In the first 2 h after administration of isotope, none of the cell types shows cytoplasmatic labeling; after 2 h, the gonial cells going into mitotic metaphase show a labeled cytoplasm, while the chromosomes which showed labeling up to prometaphase are now perfectly empty: probably on breaking of the nuclear membrane, the chromosomes release in the cytoplasm the RNA that was associated with them. In any case, about 5–7 h after injection, the cytoplasm of all the cells begins to be slightly labeled; it is very probable that, besides the mass passage of RNA into the cytoplasm at metaphase, there occurs a slow passage from the nucleus to the cytoplasm, during all cellular stages and in all types of cells. Monesi and Crippa [15] arrived at the same conclusion upon somatic cells of mice in culture. This slow passage has been verified by a great number of authors with the most varied material and techniques, e.g. Taylor [21] on *Cricetulus griseus*, Zolokar [24] on *Neurospora crassa*, Goldstein and Plaut [6] in nuclear grafting in *Amoeba proteus* and many others.

Six days after administration of the isotope, leptotene and zygotene cells were observed exclusively labeled in the cytoplasm. It is supposed that they derive by maturation from the gonia that have incorporated in the nucleus during the first few hours and then, following division, have shed the labeled RNA into the cytoplasm. Accordingly we should expect to find still labeled in the nucleus after several days the cells which, after having incorporated, have not yet divided and have therefore not yet undergone metaphase and breaking of the nuclear membrane, the cells at the pachytene stage are still unmistakably marked, particularly on the nucleus, 7 days after the injection.

Some authors [4, 5, 11] have shown that the uridine is transformed, to a small extent, into a precursor for the synthesis of DNA; that is, it is incorporated, in a small degree, in the DNA; however, it does not seem to us that this can provide an explanation for the heavy nuclear labeling that we still observe, after so many days, in some types of cell, nor for the fact that, also after several days, no mitotic or meiotic metaphase are observed with labeled chromosomes.

The nurse cells show, in great number, a heavy nuclear labeling from the first few minutes after the administration of the radioisotope and maintain it for 16 days; from a few hours after the administration, also their cytoplasm appears slightly labeled. These cells possess large nucleoli and some karyosomes; about 1 h after the injection, some nucleoli appear more intensely labeled than the rest of the nucleus; no nucleoli labeled before the rest of the nucleus have been found. After about 2 h, in a few cells, an empty space was observed, with no labeling and no staining; it might be left by a nucleolus that has completely dissolved. It now seems clear that the precursors of the ribosomes are synthesized in the nucleolus; Painter and Biesele [19] state that certain types of glandular and nurse cells undergo processes of endomitosis also in order to have available, with the increased number of chromosomes, a greater number of nucleolar organizers and consequently more nucleoli and more polyribosomes available for secretory activity of the large cells. In the case of *Asellus* the function of the large nurse cells is not clear; however, it is certain that they have a very intense metabolism of ribonucleic acid, contrary to what has been observed by Meusy [12] in *Orchestia gammarella*, and it therefore seems obvious that they also have a considerable secretory activity.

Concerning the hypothesis of Montalenti, Vitagliano and De Nicola [16], the present observations do not exclude the possibility that there is a passage of ribonucleic acid from nurse to germinal cells; however, with our technique we have not been able to observe it; soon after the injection both types of cells are already labeled.

SUMMARY

By means of autoradiography, a study has been made of the incorporation of ^{3}H-uridine in the cells of the testicle of *Asellus aquaticus*. RNA synthesis has been observed in all the cells during the whole mitotic cycle except in metaphase and in anaphase. The meiotic prophase incorporates RNA with different intensities in the various stages, with a maximum in late pachytene. Meiotic metaphase and anaphase are not incorporated at all.

The first labeled spermatids are observed only four days after the administration of the radioisotope. The spermatozoa show labeled tails from the 7th day onwards. The nurse cells show, in great number, a heavy nuclear labeling from the first few minutes and maintain it until the 16th day. About 7 h after the injection, the cytoplasm of all the cells begins to be slightly labeled. It is probable that there exists, in addition to a massive passage of RNA from the nucleus to the cytoplasm, at metaphase a slow passage during all stages of the cellular cyle.

REFERENCES

1. BERLOWITZ, L., *Proc. Natl Acad. Sci.* **53**, 68 (1965).
2. BLOCK, D. P. and BRACK, S. D., *J. Cell Biol.* **22**, 327 (1964).
3. DAS, N. K., SIEGEL, E. P. and ALFERT, M. J.*Cell Biol.* **25**, 387 (1965).
4. FEINENDEGEN, L. E., BOND, V. P. and HUGHES, W. L., *Exptl Cell Res.* **25**, 627 (1961).
5. FRIEDKIN, M. and KORNBERG, A., *in* W. D. MCELROY and B. GLASS (eds), The chemical basis of heredity, p. 609. Johns Hopkins Press, Baltimore, 1957.
6. GOLDSTEIN, L. and PLAUT, W., *Proc. Natl Acad. Sci.* **41**, 874 (1955).
7. HENDERSON, S. A., *Nature* **200**, 1235 (1963).
8. —— *Chromosoma* **15**, 345 (1964).
9. HOTTA, Y. and STERN, H., *J. Cell Biol.* **16**, 259 (1963).
10. —— *ibid.* **19**, 45 (1963).
11. MCMASTER-KAYE, R. and TAYLOR, J. H., *J. Biophys. Biochem. Cytol.* **5**, 461 (1959).
12. MEUSY, J.-J., *Arch. Anat. Microscop. Morphol. exptl.* **4**, 287 (1964).
13. MONESI, V., *J. Cell Biol.* **22**, 521 (1964).
14. —— *Exptl Cell Res.* **39**, 197 (1965).
15. MONESI, V. and CRIPPA, M., *Atti soc. Ital. Anat.* **71** (1963).
16. MONTALENTI, G., VITAGLIANO, G. and DE NICOLA, M., *Heredity* 4, 75 (1950).
17. MUCKENTHALER, F. A., *Exptl Cell Res.* **35**, 31 (1964).
18. OLIVIERI, G. and OLIVIERI, A., *Mut. Res.* **2**, 366 (1965).
19. PAINTER, T. S. and BIESELE, J. J., *Proc. Natl Acad. Sci.* **56**, 1920 (1966).
20. ROCCHI BRASIELLO, A., *Acc. Naz. Lincei* **42**, 264 (1967).
21. TAYLOR, J. H., *Am. J. Bot.* **46**, 477 (1959).
22. —— *Ann. N.Y. Acad. Sci.* **90**, 409 (1960).
23. UTAKOJI, T., *Exptl Cell Res.* **42**, 585 (1966).
24. ZALOKAR, M., *Nature* **183**, 1330 (1959).

The Inhibition of Protein Synthesis in Meiotic Cells and its Effect on Chromosome Behavior

L. G. PARCHMAN and HERBERT STERN

Introduction

The purpose of this paper is to report on a series of experiments concerning the role of protein synthesis in meiosis. The occurrence of protein synthesis during meiotic prophase was first demonstrated autoradiographically by TAYLOR (1959). Biochemical analyses by HOTTA and STERN (1963) supported this conclusion. In neither these studies nor others, however, was any evidence adduced for specific functions of proteins during meiotic prophase. The recent development of a technique for culturing meiotic cells has made possible a more thorough investigation of the role of protein synthesis (ITO and STERN, 1967). The autoradiographic and inhibitor studies reported in this paper together with the biochemical studies to be reported elsewhere (HOTTA, PARCHMAN, and STERN, 1968) provide unambiguous although limited evidence concerning the function of proteins during meiotic prophase. Two principal conclusions may be drawn from the data presented: (1) Protein synthesis during prophase stages is generally essential to meiotic development;

(2) Interference with protein synthesis at certain intervals of prophase results in cytological abnormalities which are characteristic for the particular interval at which such interference occurs.

Materials and Methods

Microsporocytes in various stages of meiotic prophase were explanted from the hybrid lily variety, "Cinnabar", and cultured at 20° C, as described by Ito and Stern (1967).

For autoradiographic studies of amino acid incorporation, tritiated leucine (45 c/mM) was included in the basic medium at a concentration of either 10 or 100 μc/ml. Cells were usually precultured for a few hours in isotope-free medium, to avoid incorporation of label by cells which died during the initial phase of explantation. At the end of the exposure period, the filaments of cells were washed in chilled, basic medium containing no isotope and fixed in FAA (5% formalin: 60% ethyl alcohol: 5% acetic acid: 30% water). For pulse-chase experiments, the isotopic medium was replaced by fresh medium containing 0.008 M leucine.

Initially, the standard dry-ice squash technique (Darlington and La Cour, 1962) was used for autoradiography with Kodak AR-10 stripping film. Since the squash technique gave poor resolution between nucleus and cytoplasm, in later experiments the tissue was embedded in paraffin and sectioned to a thickness of 5 μ before autoradiography. Quantitation of grain density was obtained by counting the number of grains over a unit area measured on enlarged photomicrographs of the labeled cells stained with methyl green-pyronin.

Cycloheximide (CH) was found to be the most suitable of several protein inhibitors examined and it was tested at concentration ranging from 3.5×10^{-7} to 1.7×10^{-5} M. At different times, samples of cultured filaments of cells were taken, squashed in propionic-orcein, and scored cytologically for stage and types of abnormalities. The effect of the inhibitor on amino acid incorporation was also noted.

Results

1. Autoradiographic Analysis

These experiments were designed to examine the relative rates of protein synthesis at different intervals of meiotic prophase, the stability of proteins synthesized, and the distribution of newly formed proteins between nucleus and cytoplasm. We also wished to determine whether the changes in chromosome morphology during meiotic prophase were correlated with distinctive periods of protein synthesis and/or distinctive localizations of the proteins formed. The data obtained show that proteins synthesized during meiosis are present in all parts of the cell and that the rate of synthesis is approximately constant through the prophase stages. These conclusions are consistent with earlier autoradiographic studies (Taylor, 1959).

Rates of protein synthesis at different meiotic stages were determined in two ways. Cells at a particular stage were explanted into culture medium containing radioactive leucine and samples were re-

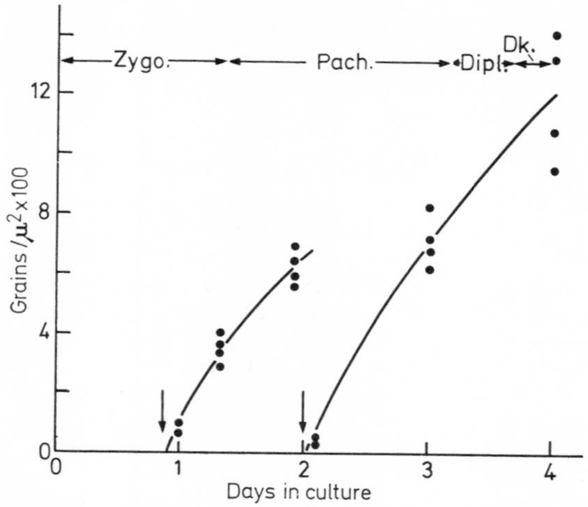

Fig. 1. Incorporation of tritiated leucine into cultured microsporocytes. Cells were explanted into isotope-free culture medium at about mid-zygonema (zero time). Arrows indicate the time at which labeled leucine was added to the cultures. The radioactivity was adjusted to 10 μcuries/ml. Grain counts were made from squash preparations exposed to the emulsion for 3 days. Meiotic stages are shown at the top of the figure. The cells removed after 4 days in culture were at metaphase I or later stages. In this experiment the cells were cultured at 25°C. Abbreviations: *Zygo.* zygonema; *Pach.* pachynema; *Dipl.* diplonema; *Dk.* diakinesis

moved for analysis after the cells had progressed through one or more additional stages. Alternatively, cells were explanted into non-radioactive medium and labeled leucine was added after the cells had reached the desired stage. In this way, any effects of preculturing on rates of amino acid incorporation could be noted. The results of one set of experiments are summarized in Fig. 1. For cells which were explanted in late leptonema, the rates of protein synthesis were similar and almost constant over 1—2 days of incubation, regardless of whether isotope was introduced in mid-zygonema or mid-pachynema. In these experiments the cells labeled in zygonema were followed to pachynema and those labeled in pachynema were followed to first metaphase. Thus, excluding the possibility of short term oscillations, the rate of protein synthesis appears to be approximately constant from zygonema to metaphase I.

The stability of the proteins formed during meiotic prophase was examined by exposing cells for 10 hours to radioactive leucine and then transferring them to a cold medium in which they were allowed to complete meiosis. Samples were removed periodically for analysis. The results of these experiments are shown in Fig. 2. Cells at leptotene, zygotene, and pachytene stages were explanted into radioactive medium.

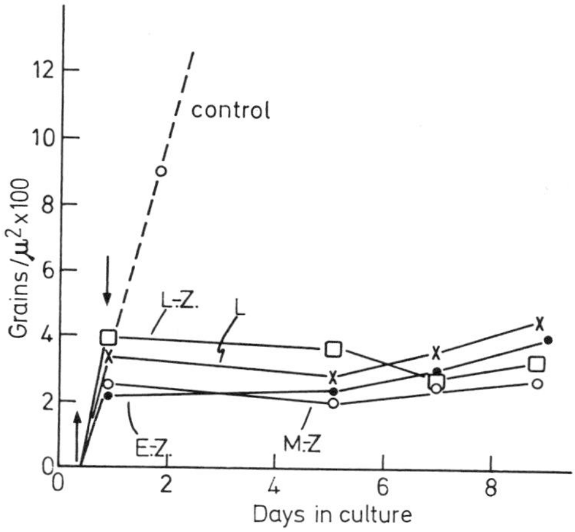

Fig. 2. Radioactivity of microsporocytes cultured in isotope-free medium following brief exposure to tritiated leucine. Cells at different meiotic stages were explanted at zero time and after 12 hours were transferred to a medium containing H^3-leucine (100 μcuries/ml). After 10 hours of exposure the cells were again transferred to a medium containing 0.008 M unlabelled leucine. Samples were taken at the indicated times and the cells embedded in paraffin, sectioned, and exposed to emulsion for 3 days. Cultures were incubated at 20° C. The upward pointing arrow indicates the time which the cells were exposed to isotope and the downward pointing one indicates the time at which the cells were transferred to cold leucine. Abbreviations: *L* leptonema; *E.Z.* early zygonema; *M.Z.* mid zygonema; *L.Z.* late zygonema. The cells which were explanted at late zygonema reached metaphase I by nine days. The rate of development is somewhat slower in the presence of leucine at the concentration used for the chase

and then transferred to isotope-free medium following 10 hours of exposure. Regardless of the stage at which the cells were exposed to isotope, no major change in radioactivity of the cells occurred after five days of growth in cold medium. The "chase" is clearly effective insofar as transfer to cold medium immediately arrests the steady accumulation of label which otherwise occurs when cells are continuously exposed to isotope (Fig. 2). The retention of label by cells under "chase" conditions is however, not necessarily a consequence of protein stability. Protein turnover with reutilization of breakdown products would result in the same labeling behavior. The question of protein stability during meiosis thus remains unresolved.

The chemical studies described elsewhere (Hotta, Parchman, and Stern, 1968) indicate that different kinds of protein are synthesized during different meiotic periods but these studies too provide no evidence on stability.

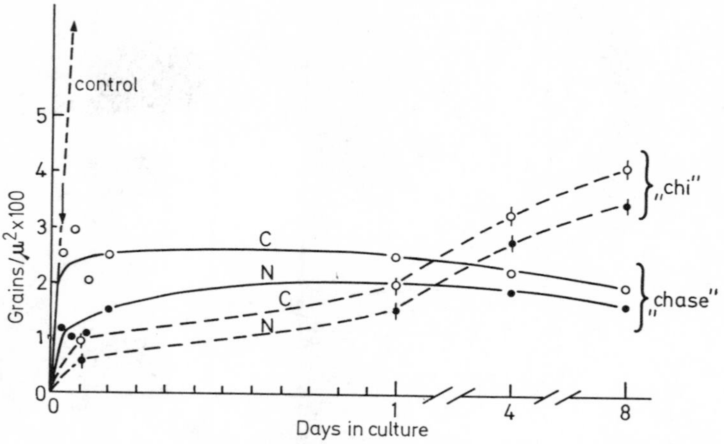

Fig. 3. The incorporation and intracellular distribution of H^3-leucine under conditions of "chase" or of exposure to cycloheximide ("Chi"). Cells were precultured in isotope-free medium for 2 hours. In the "chase" experiment the cells were exposed to H^3-leucine (10 μcuries/ml) for 40 minutes and then transferred as described under Fig. 2. In the "Chi" experiment, the cells were exposed to 3.5×10^{-6} M cycloheximide prior to the addition of radioactive leucine. Grains were counted from sectioned material. The arrow indicates the time at which cells were transferred to isotope-free medium in the "Chase" experiment. "C" and "N" represent cytoplasm and nucleus respectively. In the case of cells at metaphase or anaphase, "N" was determined by counting grains over the chromosomes

The proteins synthesized during meiotic prophase do not accumulate in any particular region of the cell. The evidence for this conclusion is shown in the curves of Fig. 3 and in the photomicrographs of Figs. 4 and 5. After exposure of cells to isotope for 40 minutes the concentration of label in the nucleus is approximately half of that in the cytoplasm (Figs. 3, 4a). This difference disappears if the cells are exposed continuously for several hours or if the cells which had been exposed for 40 minutes are transferred to cold medium (Figs. 3, 4b). Whereas the concentration of label falls slightly in the cytoplasm after 2—4 hours, that in the nucleus rises slightly. The significance of the shift in label from cytoplasm to nucleus is unclear. The important conclusion which can be drawn from these experiments is that newly synthesized proteins appear throughout the cell during meiotic prophase.

Fig. 4. a Distribution of label in pachytene cells after exposure to H^3-leucine (10 μcuries/1 ml) for 40 minutes. The autoradiograph was allowed to develop for 30 days. The concentration of grains over the cytoplasm is approximately twice that over the nucleus (*n*) × 640. b Distribution of label in meiotic cells after long term culture. Microsporocytes in late zygonema were exposed for 40 minutes to H^3-leucine (10 μcuries/ml) and allowed to progress to metaphase in the presence of

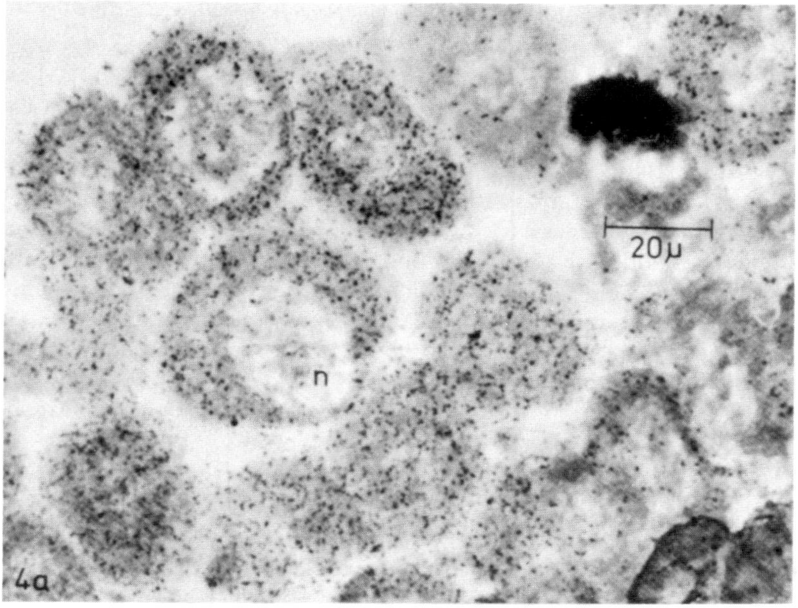

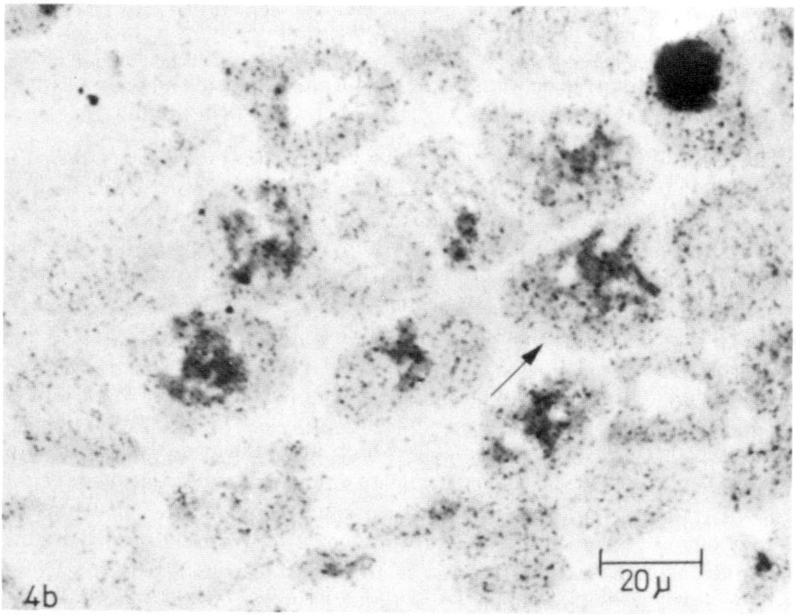

0.008 M leucine chase medium. Label is evenly distributed throughout the cells. This result is obtained with either continuous or "pulse-chase" labeling. The arrow is pointing to a cell in metaphase. Other stages are also shown. ×580

2. *Effects of Cycloheximide on Protein Synthesis and Meiotic Development*

In order to examine the function of the proteins synthesized during meiotic prophase, the responses of cells to inhibitors of protein synthesis were studied. Of the various inhibitors tested, cycloheximide (SIEGEL and SISSLER, 1964) proved to be the most effective. This reagent had a rapid effect on protein synthesis as evidenced by the fact that the degree of inhibition measured was the same regardless of whether cycloheximide was added to the culture medium one hour prior to or simultaneously with the addition of radioactive leucine. At the end of one hour, control cultures had accumulated 5—10 times as much label as those incubated in the presence of 3.5×10^{-6} M cycloheximide (Fig. 3). By 24 hours the difference was 20-fold. No evidence was obtained for a selective inhibition of synthesis of either nuclear or cytoplasmic proteins.

The physiological response of meiotic cells to cycloheximide matches the pattern of protein inhibition. In the presence of the drug, the rate of meiotic development is markedly reduced and cytological abnormalities appear. The severity of the effect is a function of drug concentration but, as will be seen, the type of abnormality produced is also dependent upon the stage of treatment. The relationship between meiotic rate and cycloheximide concentration is clearly shown in Table 1 (p. 307). At a drug concentration of 3.5×10^{-7} M, zygotene cells require about 50% more time to reach metaphase; at 1.8×10^{-6} M, the cells take three times longer to reach metaphase than do the controls. At the highest concentration used (3.5×10^{-6} M), meiotic development is virtually arrested. Thus, regardless of the specific functions of the newly synthesized proteins, it is clear that some or all of these proteins are essential to meiotic development and are not synthesized for some later post-meiotic function.

This latter conclusion is strengthened by observations of the cytological abnormalities induced by cycloheximide. As previously stated, these abnormalities are both concentration and stage dependent. The stage dependence is apparent from response of the cells to the highest concentration of cycloheximide at which development still occurs at an appreciable rate (1.8×10^{-6} M). Cells which are continuously exposed to inhibitor beginning at stages earlier than mid zygonema do not develop beyond the prophase stage. Eventually, such cells die in culture after a period of about 25 days. Mid zygotone cells, however, do reach metaphase but chromosome morphology is abnormal (Table 2, p. 308). These effects are in marked contrast to those observed in cells treated with a DNA synthesis inhibitor beginning at the same interval. Inhibition of DNA synthesis leads to chromosome fragmentation (ITO, HOTTA, and

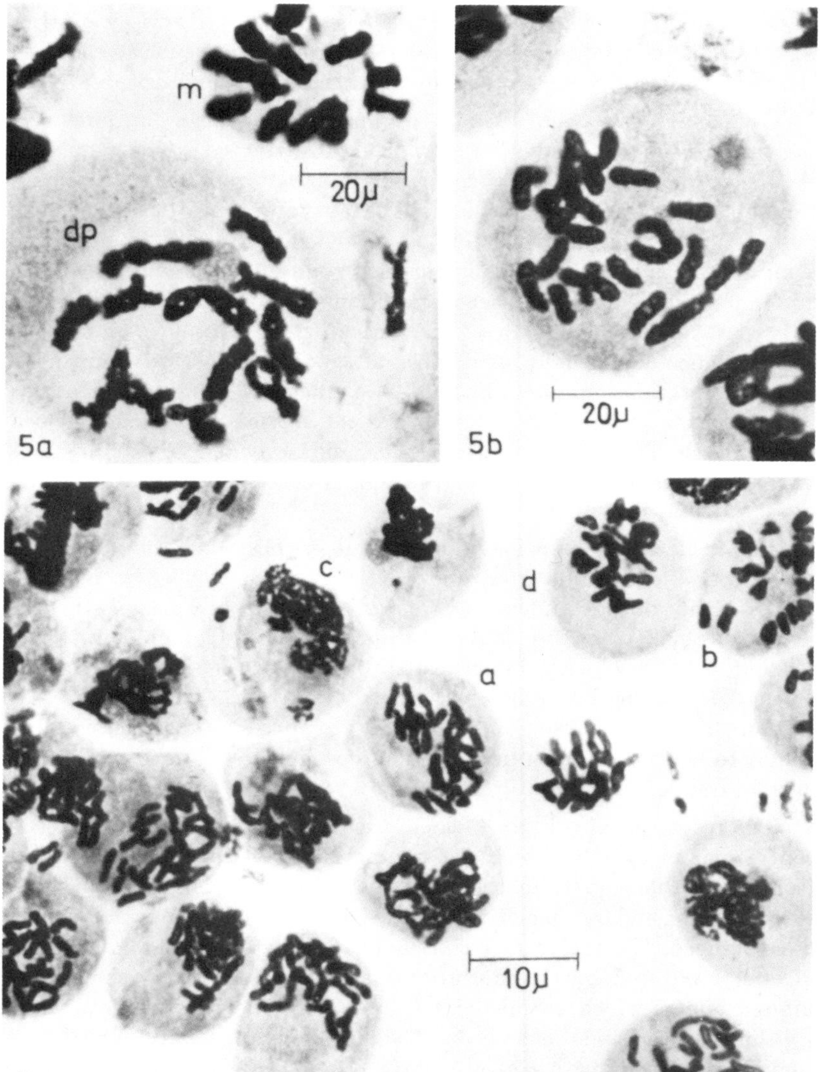

Fig. 5. a Control cells cultured from late zygonema for 4 days in basic medium attain the normal configuration for late diplonema (*dp*) and metaphase (*m*). Twelve bivalents are evident at both stages. b An achiasmatic metaphase cell which resulted from explanting late zygotene cells into 1.8×10^{-6} M cycloheximide and culturing for 16 days. There are 24 univalents present. c Achiasmatic cells in a 16 day culture in cycloheximide. Cell *a* is in metaphase; in cell *b* the chromosomes are beginning to separate to yield 48 chromosomes; in cell *c* the chromosomes have already uncoiled to yield several groups of chromatin; cell *d* is a partial achiasmatic since there are at least 2 bivalents present with normal repulsion of homologous centromeres. a and b $\times 640$, c $\times 320$

Stern, 1967). Inhibition of protein synthesis causes a variety of alterations in the general morphology of the chromosomes such as stickiness and supercontraction but fragmentation has not been observed.

Cells exposed to cycloheximide beginning either in late zygonema or early pachynema respond differently from those treated at mid zygonema. Such cells develop normally except for their failure to form chiasmata (Fig. 5). Metaphase chromosomes have a typical meiotic morphology but are present as univalents in the diploid number (24). In some cells a mixture of univalents and bivalents is found. The univalents do not undergo normal anaphase separation, but frequently the chromatids fall apart and thus give rise to cells with a tetraploid number of chromosomes. Eventually an abnormal cytokinesis occurs resulting in cells with various numbers and sizes of nuclei. The failure in normal segregation is probably not due to defects in the spindle since in cells which have a mixture of bivalents and univalents, the bivalents undergo normal anaphase segregation, whereas the univalents are randomly distributed.

The achiasmatic response is not found in cells which have been exposed to cycloheximide from mid pachynema on. Generally, however, the bivalents do not undergo anaphase separation and frequently they do not desynapse but fuse. There is thus a critical stage in meiotic development beyond which inhibition of protein synthesis has no effect on chiasma formation but still has a variety of effects on chromosome association and segregation. Even cells as advanced as diakinesis fail to complete the second meiotic division when cultured in the presence of cycloheximide.

The responses of cells at various stages of meiosis to inhibitory concentrations of cycloheximide are summarized in Table 2. What is clear from this table and from the preceding discussion is that continuous synthesis throughout meiotic prophase is essential to normal meiotic development. The major shortcoming of these results is the multiplicity of effects induced by the inhibition and the lack of a few sharply delimited effects which would point to a few specific functions of protein synthesis in the meiotic process. It should be emphasized nevertheless, that the observed disturbances in chromosome behavior are not generally linked to a deterioration in cell viability. Moreover, as will be shown below, cells exposed to the drug at certain stages of development may be transferred to inhibitor-free medium in which they develop normally. The main difficulty in the interpretation of these experiments is the distinction between primary and secondary effects. In view of the fact that newly synthesized proteins are located throughout the meiotic cell, the primary effects of inhibiting protein synthesis cannot be assigned to the chromosomes even though the visible effects of such inhibition are manifest in the chromosomes.

3. Protein Synthesis and Chiasma Formation

The conclusion that the proteins synthesized during meiotic prophase are essential to meiotic development is based upon three lines of evidence: (1) The continuous incorporation of labeled amino acids into protein during the prophase stages. (2) The strong inhibitory effect of cycloheximide upon such incorporation. (3) The correspondingly strong effect of cycloheximide upon the behavior of meiotic cells. It should be noted, however, that even at the strongest concentration tested (3.5×10^{-6} M), the proteins of the exposed cells accumulate label at 5% the rate of the controls (Fig. 3). At lower concentrations of inhibitor, the accumulation rate is greater, and it is at these concentrations that the abnormalities described above have been observed. Two possibilities exist with respect to such partial inhibition. The inhibition may be random and thus produce a uniform decrease in rate of synthesis of all cellular proteins. Alternatively, it may be selective and thus produce a situation in which certain proteins are synthesized whereas others are not. The correspondence between inhibitor concentration and the degree of deceleration of meiotic development (Table 1) are consistent with the first alternative.

Table 1. *Effect of cycloheximide concentration on meiotic rate*

Cycloheximide concentration (molarity)	Duration of prophase (days at 20° C)
0	6
3.5×10^{-7}	8—9
7×10^{-7}	12—13
1.8×10^{-6}	15—17
3.5×10^{-6}	20+

Filaments of cells determined to be in mid zygonema were explanted into medium containing inhibitor at the indicated concentrations. The time required to reach metaphase was noted as a cytological measure of the degree of inhibition. Many abnormalities were observed in addition to the rate decrease.

However, biochemical analyses (HOTTA, PARCHMAN, and STERN, 1968) show that the synthesis of certain proteins is particularly sensitive to low concentrations of cycloheximide. Cytological abnormalities, whatever their source, could therefore be attributed to a failure in coordination of protein synthesis. If this is true, then cells which have been exposed for limited intervals of time to low concentrations of cycloheximide and subsequently returned to inhibitor-free medium should reflect the transient failure in coordination.

Generally, cells at all meiotic stages resume normal development if exposed to cycloheximide for one day and then returned to normal medium. The lack of any persistent cytological defect cannot be attributed to a delay in cycloheximide action since inhibition of protein synthesis occurs within less than 4 hours. If, therefore, biosynthetic processes become uncoordinated during this 24 hour interval, the defect

Table 2. *Cytological effects of cycloheximide on meiotic cells*

Initial stage	Cytological abnormality		
	Type	Stage[a]	Frequency (%)
leptonema	arrest	early prophase	90+
early zygonema	[b]		
early-mid zygonema	arrest	late prophase	80+
mid zygonema	chromosome	metaphase I	40—70
late zygonema	achiasmata	metaphase I	40—80
early pachynema	achiasmata	metaphase I	20—60
early-mid pachynema	chromosome and disjunction	anaphase I	30—50
mid pachynema	[b]		
late pachynema	chromosome and disjunction	diakinesis and metaphase I	60—80
early diplonema	chromosome and disjunction	metaphase I and anaphase I	50—80
late diplonema	arrest	anaphase I through prophase II	70—80
diakinesis	arrest and disjunction	prophase II and metaphase II	80—90

Cells were explanted at the stages indicated into culture medium containing 1.8×10^{-6} M cycloheximide. The frequency is a measure of the percentage of living cells which showed the abnormality typical for that initial stage. Other abnormalities occur as well and very few cells are completely normal. „Arrest" means the cells failed to develop beyond the stage indicated in the Stage column. "Chromosome" abnormalities indicate stickiness or fusion of bivalents and/or contraction of metaphase arms more or less than normal. "Disjunction" abnormalities refer to failures in segregation of otherwise normal chromosomes.

[a] The approximate stage at which the typical morphological abnormality becomes grossly apparent.

[b] Early zygonema and mid pachynema are not included in this analysis because of difficulty in successful explantation of these stages.

must be entirely reversible. A similar result is obtained if cells in mid-zygonema or mid pachynema are exposed for 2—4 days. This is not the case, however, for cells at late zygonema or early pachynema. These cells resume the normal rate of development after being returned to inhibitor-free medium but, on the whole, the cytological abnormalities noted for cells continually exposed to inhibitor also appear in these cells.

From the standpoint of cytogenetics, the most significant abnormality is the failure of cells to manifest chiasmata. The result implies, but does not prove, that inhibition of protein synthesis can inhibit crossing-over. The conditions under which achiasmatic cells develop can be clearly specified. Cells exposed to cycloheximide for 3—4 days at mid zygonema or earlier form chiasmata after being returned to normal culture media. The same is true for cells exposed at stages later than early pachynema. The stage at which a transient exposure to cycloheximide inhibits chiasma formation lies somewhere in the interval spanning late zygonema to early pachynema. The precise duration of that interval has not been determined. In some cases, all the meiotic cells of one anther become achiasmatic; in most cases, the achiasmatics are seen in a variable but high proportion of the cells (Table 2). In all cases, however, the production of achiasmatics by cycloheximide is restricted to the interval from late zygonema to early pachynema. It thus appears as though cycloheximide can prevent chiasma formation if the cells are exposed to it at a time when chromosome pairing is more or less complete but that it cannot do so if cells are exposed prior to the completion of pairing or after the pairs have remained in contact for 2 or more days.

Discussion

The experiments reported here confirm previous findings that protein synthesis occurs during meiotic prophase (Taylor, 1959; Hotta and Stern, 1965). In addition, these experiments have provided substantial evidence for the functional significance of such synthesis to meiotic development. On the whole, however, the evidence is more general than specific in nature. It is clear that almost every aspect of chromosome behavior during meiosis can be affected by disturbing protein synthesis at one stage or another of the prophase interval. It is also clear that the respective consequences of inhibiting DNA and protein synthesis during meiotic prophase differ in one presumably important way. Partial inhibition of DNA synthesis leads primarily to chromosome fragmentation (Ito, Hotta and Stern, 1967) but does not otherwise appear to affect the morphology and movement of the chromosomes. By contrast, partial inhibition of protein synthesis does not primarily affect chromosome integrity but does drastically affect key meiotic events such as pairing, desynapsis, and segregation. The conclusion could be drawn that chromosome integrity is exclusively DNA-dependent and that proteins act by modifying secondary features of chromosome organization. The conclusion, however, would represent no more than a superficial reading of the experiments. The gap between the cytological abnormalities observed and the mechanisms underlying the production of such abnormalities is wide enough to allow for a variety of possibilities.

The one phenomenon examined in these studies which is open to more specific considerations is that of chiasma formation. The factors which might render an otherwise normal meiotic cell achiasmatic have been studied by a number of investigators. One widely used technique has been the application of heat shock to cells in meiotic prophase (HENDERSON, 1966). Studies of this kind have led to the conclusion that crossing-over occurs during meiotic prophase because chiasma formation can be inhibited by subjecting the cells to elevated temperatures during that stage. The precise time at which heat is effective in rendering cells achiasmatic is uncertain, although it is clear that it lies within or close to the zygotene interval. However, perturbations of the meiotic system at stages prior to zygonema can also lead to achiasmatic cells. This was found by ITO (see STERN and HOTTA, 1967) in the course of explanting premeiotic and meiotic cells into culture media. Cells explanted just prior to or in early leptonema are evidently achiasmatic when they reach diplonema some ten days later. This effect is not observed if cells are explanted at mid-leptonema or later.

The experiments reported here on the production of achiasmatic cells are more specific in nature in that they implicate a requirement for protein synthesis at about the time that chromosome pairing is completed. Although the experiments do not reveal the precise role which protein synthesis plays in the formation of chiasmata, they do reveal that a discrete interval exists following the completion of pairing during which chiasma formation may be effectively inhibited by arresting protein synthesis. The inhibition cannot be due to an absence of the synaptinemal complex since the complex is already formed at termination of zygonema. It would appear, therefore, that although the pairing complex is essential to crossing-over, it is not a sufficient condition for it. This interpretation is supported by the unpublished observations of W. J. PEACOCK (personal communication) and of ROTH and PARCHMAN (1968) that achiasmatic cells do have synaptinemal complexes. The major conclusion which may be drawn from these experiments is that, if the achiasmatic condition can be equated with an absence of crossing-over, then crossing-over is a deliberate event which occurs after the completion of chromosome pairing.

References

Darlington, C. D., and L. F. La Cour: The handling of chromosomes, 4th ed. London: George Allen and Unwin, Ltd. 1962.

Henderson, S. A.: Time of chiasma formation in relation to the time of deoxyribonucleic acid synthesis. Nature (Lond.) **211**, 1043—1047 (1966).

Hotta, Y., L. G. Parchman, and H. Stern: Protein synthesis during meiosis. Proc. nat. Acad. Sci. (Wash.) **60**, 575—582 (1968).

—, and H. Stern: Synthesis of messenger-like ribonucleic acid and protein during meiosis in isolated cells of *Trillium erectum*. J. Cell Biol. **19**, 45—58 (1963).

Ito, M., Y. Hotta, and H. Stern: Studies of meiosis *in vitro*. II. Effects of inhibiting DNA synthesis during meiotic prophase on chromosome structure and behavior. Develop. Biol. **16**, 54—77 (1967).

—, and H. Stern: Studies of meiosis *in vitro*. I. *In vitro* culture of meiotic cells. Develop. Biol. **16**, 36—53 (1967).

Roth, T., and L. G. Parchman: Pachytene synaptinemal complexes in meiotic achiasmatic chromosomes (manuscript in preparation).

Siegel, M. R., and H. D. Sisler: Site of action of cycloheximide in cells of *Saccharomyces pastorianus*. II. The nature of inhibition of protein synthesis in a cell-free system. Biochem. biophys. Acta (Amst.) **87**, 83—89 (1964).

Stern, H., and Y. Hotta: Chromosomal behavior during development of meiotic tissue. In: The control of nuclear activity, ed. by L. Goldstein, p. 47—75. Englewood Cliffs, N.J.: Prentice-Hall 1967.

Taylor, J. H.: Autoradiographic studies of nucleic acid and proteins during meiosis in *Lilium longiflorum*. Amer. J. Bot. **46**, 477—484 (1959).

RAPIDLY-LABELED NUCLEAR RNA IN CHINESE HAMSTER TESTIS

M. MURAMATSU , T. UTAKOJI AND H. SUGANO

In the course of meiosis in the primary spermatocytes of some mammals the cells in late pachytene stage were found to be most active in ribonucleic acid (RNA) synthesis [6, 15]. At 4 h after intraperitoneal injection of ^{3}H-uridine, most of the label was found on the autosomal bivalents in these cells, but not on the sex pairs. The nucleoli were not labeled either (Fig. 1 *a*).

It has been almost established that the nucleolus is the site of ribosomal RNA synthesis via its 45S precursor [9–12], although its detailed process is not yet fully understood. On the other hand, "rapidly-labeled" nuclear RNA ("chromosomal" RNA) with DNA-like base composition was found in various kinds of cells including rat liver [8], HeLa cells [4, 17] and avian erythrocytes [1, 13]. The two striking features of these RNAs were their extraordinarily large size [1, 4, 8, 13, 17] and apparent breakdown in the nucleus without coming out to the cytoplasm [2, 3, 14]. Only a small part of these RNAs may be transferred to the cytoplasm as messenger RNA.

In view of the special feature of the synthetic pattern of RNA in the spermatocytes which undergo meiosis, it is interesting to know if the RNAs synthesized in these cells, are similar to or different from those synthesized in other types of somatic cells.

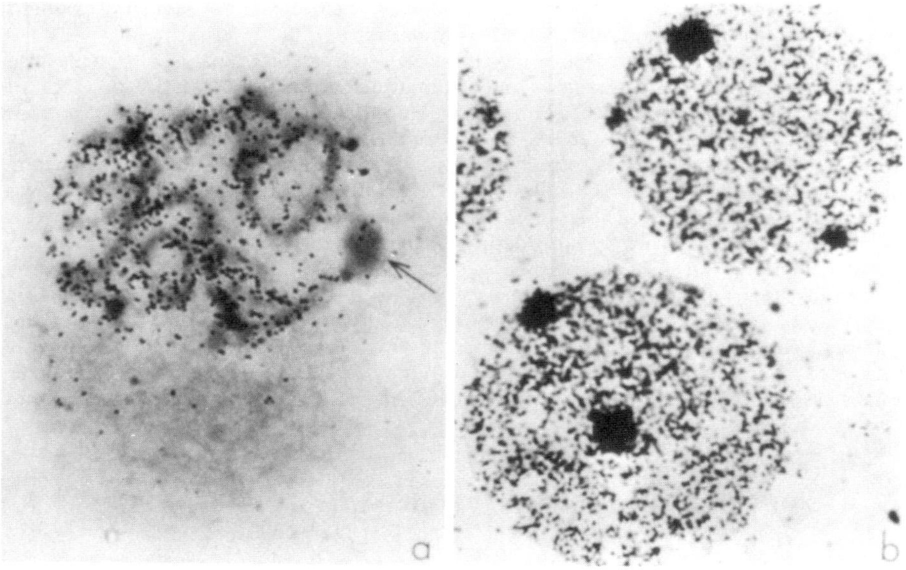

Fig. 1.—Autoradiograms of a primary spermatocyte and liver cell nuclei. The material was obtained from a Chinese hamster at 4 h after intraperitoneal injection of ^{3}H-uridine. The cells were fixed with 50 per cent acetic acid, and Kodak AR 10 stripping film was applied. (*a*), Primary spermatocyte at pachytene stage. The arrow indicates the unlabeled sex vesicle which consists of XY-bivalent and nucleolus. Note the moderately labeled pachytene stage autosomes. Unsquashed preparation. (*b*) Liver cell nuclei. Nucleoli are heavily labeled. Squash preparation.

The present study was undertaken to clarify the molecular nature of the RNA synthesized on the pachytene chromosomes during meiotic stage.

Male Chinese hamsters (*Cricetulus griseus*), approx. 4 months old, weighing 30–40 g, were used throughout the experiments. To analyze the sedimentation characteristics of "rapidly-labeled" testicular RNA, 45 μC of ^{3}H-uridine (8.7 C/mmole) in 0.1 ml saline was injected directly into each testis under ether anesthesia. At specified times the testes were excised and the capsules were removed. The outcoming seminiferous tubules were homogenized with approx. 6–7 vol of 0.25 *M* sucrose containing 20 μg/ml polyvinylsulfate and 0.5 mg/ml bentonite in a loosely fitting glass-Teflon homogenizer. An equal volume of 0.6 per cent sodium dodecylsulfate—0.28 *M* NaCl, 0.02 *M* sodium acetate, pH 5.1, containing 20 μg/ml polyvinylsulfate—was added and the mixture was shaken vigorously for 30 sec. Then, an equal volume (of the above mixture) of phenol–*m*-cresol–water (7:1:2 in volume) mixture [5] added and the RNA was extracted at 65°C as described previously [9, 12].

To determine the base composition of "rapidly-labeled" RNA with ^{32}P, 0.4 mC of carrier-free ^{32}P-orthophosphate was injected into each testis of an individual and the animal was sacrificed at 30 min postinjection. The phenol-prepared RNA was passed through a Sephadex G25 column (2.2 × 30 cm), before sucrose density gradient centrifugation, to remove non-RNA ^{32}P compounds [7].

The method of sedimentation analysis was described previously [8, 9] (see also legends for figures). Fig. 2 presents the sedimentation profiles of testicular RNA at various times after injection of ^{3}H-uridine. The absorbance profiles show a similar pattern to that of the total cellular RNA with 28S, 18S and 4S peaks. The relatively large size of the 18S peak as compared with 28S peak is probably due to the contaminated DNA. At 10 min after injection of ^{3}H-uridine, most of the radioactivity was present in the 4S region of the gradient with some radioactivity appearing in heavier regions than 28S RNA. At 30 min, definite heterogeneous peak of radioactivity was demonstrated in the 45S and higher regions (70–80S), distributing almost to the bottom of the gradient. Similar patterns were obtained for 60- and 120-min labeling experiments. These kinetic patterns are in general agreement with those of other types of cells [1, 4, 8, 13, 17] except for the two differences described below. One difference appeared on the sedimentation profile of RNA at 18 h after injection of the tracer. As seen in Fig. 2*e*, very little radioactivity was associated with the 28S and

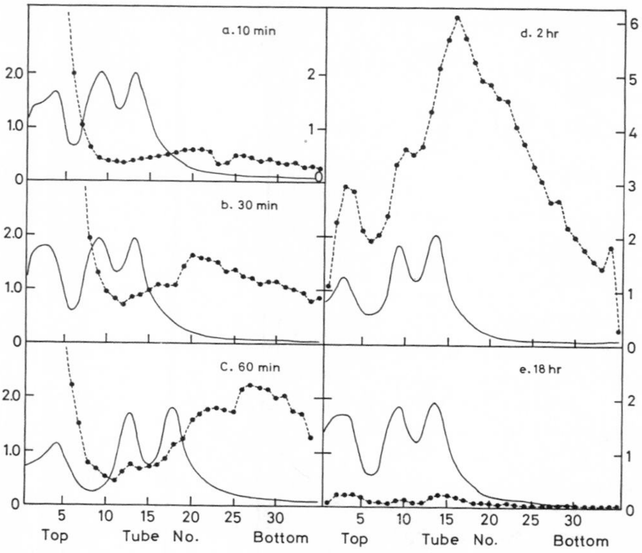

Fig. 2.—Sedimentation profiles of testicular RNA. At specified time intervals after intratesticular injection of 45 μC per testis of ^{3}H-uridine, the animals were sacrified and the testicular RNA was extracted with sodium dodecylsulfate-hot phenol procedure as described in the text. Approx. 800 μg of RNA was layered on a 28 ml of 10–40 per cent linear sucrose gradient containing 0.1 *M* NaCl–1 *mM* EDTA–0.01 *M* sodium acetate, pH 5.1 and centrifuged in a Spinco SW 25.1 rotor at 22,000 rpm for 16 h at 4°C. For 1-h experiment (*c*), the sample was centrifuged at 25,000 rpm for 18 h.

After centrifugation, the sample was fractionated into approx. 1 ml fractions with the aid of an ISCO Density Gradient Fractionator, Model D, with the automatic OD recorder. The radioactivity was determined for each tube in a liquid scintillation system [7].

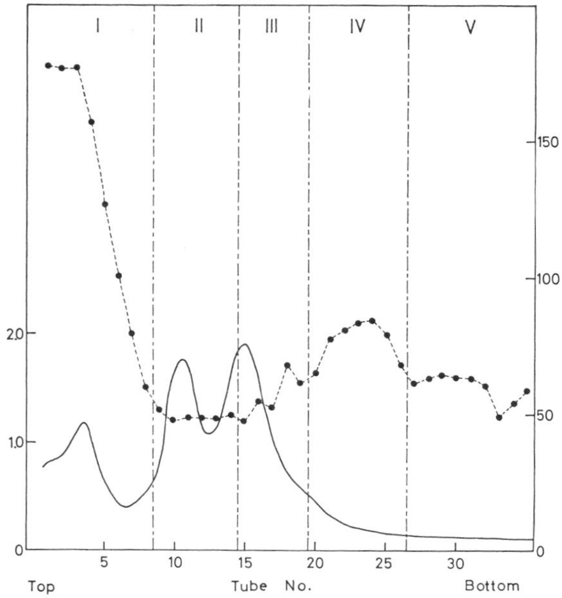

Fig. 3.—Fractionation of "rapidly-labeled" testicular RNA. At 30 min after intratesticular injection of 0.4 mC (per testis) of $^{32}PO_4$, the animal was sacrificed and the testicular RNA was extracted as in Fig. 2. Phenol-prepared RNA was passed through a sephadex G 25 column (2.2 × 30 cm) and centrifuged on the same sucrose density gradient as in Fig. 2. After fractionation with an ISCO apparatus, the fractions were divided into 5 portions according to the scheme shown in this figure and processed for base composition analysis (see Table 1). *Figs 2 and 3: Ordinate:* (left), OD 254 mμ; (right) ^{32}P radioactivity (cpm/tube).

18S absorbance peaks, or anywhere else. This indicates that most of the "rapidly-labeled" RNA found in the testicular tissue is not preserved as ribosomal RNA or relatively stable messenger RNA. The above finding may be consistent with the autoradiographic observation that much of the ^3H-uridine incorporation by testicular cells is due to the primary spermatocytes, especially to the pachytene nuclei, which, in particular, show incorporation of the precursor around the chromosomes but not into the nucleoli (Fig. 1).

Another difference was found in the base composition of these RNSs. To examine the nature of the "rapidly-labeled" RNA in this respect, the RNA labeled with ^{32}P-orthophosphate for 30 min was fractionated on the sucrose density gradient in the same manner as above. The labeling pattern as determined by taking a small aliquot from each fraction was essentially the same as obtained with ^3H-uridine (Fig. 3). The radioactive RNA on the gradient was divided into 5 fractions according to their sedimentation characteristics (I: 4–10 S, II: 16–23 S, III: 28–35 S, IV: $\sim$45 S, V: $>$45 S) and their base composition was determined by the ^{32}P-distribution among the four 3′-mononucleotides. The data shown in Table 1 may be compared to those obtained for the liver nuclear RNA [8], which showed a specific pattern of distribution of AU- and GC-rich RNA throughout the gradient. Of particular interest was the presence of AU-rich RNA in the region of about 18S and >45S both in the liver and the testicular RNA.

However, the ratio A + U/G + C was much higher in the case of testicular RNA than in the liver, resembling rather extranucleolar nuclear RNA of the liver than the total nuclear RNA [8]. This suggests that the contribution by ribosomal RNA precursor in the region of high molecular weight RNA is small in this tissue. It is noteworthy here that the uracil content of the >45S fraction was very high (27–29 per cent) in this tissue, which was also noted in HeLa cells [9] as well as in avian erythrocytes [1, 13]. It should be added that recently the pachytene nuclei were isolated in considerably pure form from these testes [16] and the RNA was examined on these preparations. It was found that the sedimentation profiles were essentially the same as those of the whole testis, indicating that this notion was correct. Details of this study will be presented elsewhere.

In conclusion, the present study gives support to the idea that the RNAs synthesized in pachytene nuclei of Chinese hamster testis are mostly DNA-like, heterogeneous molecules and contain little ribosomal RNA precursors. The former RNAs are synthesized on the extranucleolar chromatin as revealed by autoradiography, and apparently breaks down rapidly without remaining in any form of macro-

TABLE 1. *Base composition* (${}^{32}P$) *of sucrose density fractions of rapidly-labeled testicular RNA*

Fractions pooled acording to the scheme shown in Fig. 3 were precipitated with cold 0.5 *N* perchloric acid after addition of carrier yeast RNA, washed once with the same solution and hydrolyzed with 0.3 *N* KOH at 37°C for 18 h. The hydrolysate was neutralized with perchloric acid and chromatographed on a Dowex 1 formate column, detailed procedures for which were described previously together with the counting procedures [7]. The distribution of radioactivity in the four nucleotides was expressed as percentages of the total incorporation.

Fraction	Approx. S	Testis	Adenylic acid, %	Uridylic acid, %	Guanylic acid, %	Cytidylic acid, %	$\frac{A+U}{G+C}$
I	4–10	r	19.4	25.4	27.6	27.6	0.81
		l	22.8	24.3	24.9	28.0	0.89
II	16–23	r	23.1	27.0	24.8	25.1	1.00
		l	26.3	26.4	24.4	22.8	1.12
III	28–35	r	23.8	25.8	26.4	24.0	0.98
		l	21.2	27.3	25.0	26.5	0.94
IV	~45	r	18.9	27.6	29.1	24.4	0.87
		l	21.7	26.8	25.6	25.9	0.94
V	>45	r	24.0	29.3	23.5	23.2	1.14
		l	24.9	26.9	23.8	24.4	1.07

molecular ribonucleotides in the cell. It may not be surprising if these cells stop synthesizing ribosomal RNA since the ribosomes appear to be lost during spermiogenesis. The fact that the DNA-like RNAs are being synthesized continuously in the testis may indicate that these RNAs have some specific role in the process of spermatogenesis, or else they are essential to more fundamental processes of the cell in general. In any event, the system of RNA synthesis in spermatocytes appears to be an interesting tool in analyzing the "rapidly-labeled" DNA-like RNA in connection with such a phenomenon as specific gene repression or the behaviour of ribosomes in male and female gametogenesis since this is the only system, except for early embryogenesis, in which ribosomal RNA synthesis is specifically suppressed and this particular type of RNA could be obtained in relatively pure form.

REFERENCES

1. Attardi, G., Parnas, H., Huang, M. H. and Attardi, B., *J. Mol. Biol.* **20**, 145 (1966).
2. Harris, H., *Biochem. J.* **73**, 362 (1959).
3. —— *Nature* **201**, 863 (1964).
4. Housais, J.-F., Attardi, G., *Proc. Natl Adad. Sci. U.S.* **56**, 616 (1966).
5. Kirby, K. S., *Biochem. J.* **96**, 266 (1965).
6. Monesi, V., *Chromosoma* **17**, 11 (1965).
7. Muramatsu, M. and Busch, H., *J. Biol. Chem.*, **240**, 3960 (1965).
8. Muramatsu, M., Hodnett, J. L. and Busch, H., *J. Biol. Chem.* **241**, 1544 (1966).
9. Muramatsu, M., Hodnett, J. L., Steele, W. J. and Busch, H., *Biochim. Biophys. Acta* **123**, 116 (1966).
10. Penman, S., Smith, I., and Holtzman, E., *Science* **154**, 786 (1966).
11. Perry, R. P., *Proc. Natl Acad. Sci. US* **48**, 2179 (1962).
12. Scherrer, K. and Darnell, J. E., *Biochem. Biophys. Res. Commun.* **7**, 486 (1962).
13. Scherrer, K., Marcaud, L., Zajdela, F., London, I. and Gros, F., *Proc. Natl Acad. Sci. US* **56**, 1571 (1966).
14. Shearer, R. W. and McCartty, B. J., *Biochem.* **6**, 283 (1967).
15. Utakoji, T., *Exptl Cell Res.* **42**, 585 (1966).
16. Utakoji, T., Muramatsu, M. and Sugano, H. submitted to *Exptl Cell Res.*
17. Warner, J. R., Soeiro, R., Birnboim, C., Girard, M. and Darnell, J. E., *J. Mol. Biol.* **19**, 349, (1966).

Nucleocytoplasmic Interaction at the Nuclear Envelope in Post Meiotic Microspores of *Pinus banksiana*

H. G. DICKINSON AND P. R. BELL

There is accumulating evidence that, during microsporogenesis, the formation of the haploid nucleus is followed by a realignment of the meiocyte cytoplasm from the "diplophase" to a "haplophase" condition. This event is apparently preceded in the pollen mother cells of *Lilium* by the elimination of ribosomes (*4*) and extensive changes in the plastids (*3*) during early meiotic prophase.

In *Lilium* the ribosome population increases again in step with the disintegration of nucleolar material which is synthesized in the dyad nucleus and released from the karyoplasm into the cytoplasm in meiosis II (*2*). Since these events appeared to us of possibly basic significance, we have now investigated microsporogenesis in a gymnosperm, *Pinus banksiana*, to see whether similar phenomena occur at this level of evolutionary complexity. This preliminary report concerns a peculiar form of nucleocytoplasmic interaction in the very young microspores of this species, not so far detected elsewhere in vascular plants.

FIG. 1. Transverse section of microspore tetrad. Note the "inflated" nuclear invaginations (arrows) × 6 250.

FIG. 2. Postmeiotic microspore nuclear envelope showing early membranous invaginations (arrows) into the karyoplasm (*K*). × 52 000.

FIG. 3. Condensation of chromatin around and the accumulation of material within an invagination of the nuclear envelope. Note the terminal sacs (*S*). × 33 000.

FIG. 4. Nuclear invagination surrounded by condensed chromatin and containing granofibrillar material (*M*). × 62 000.

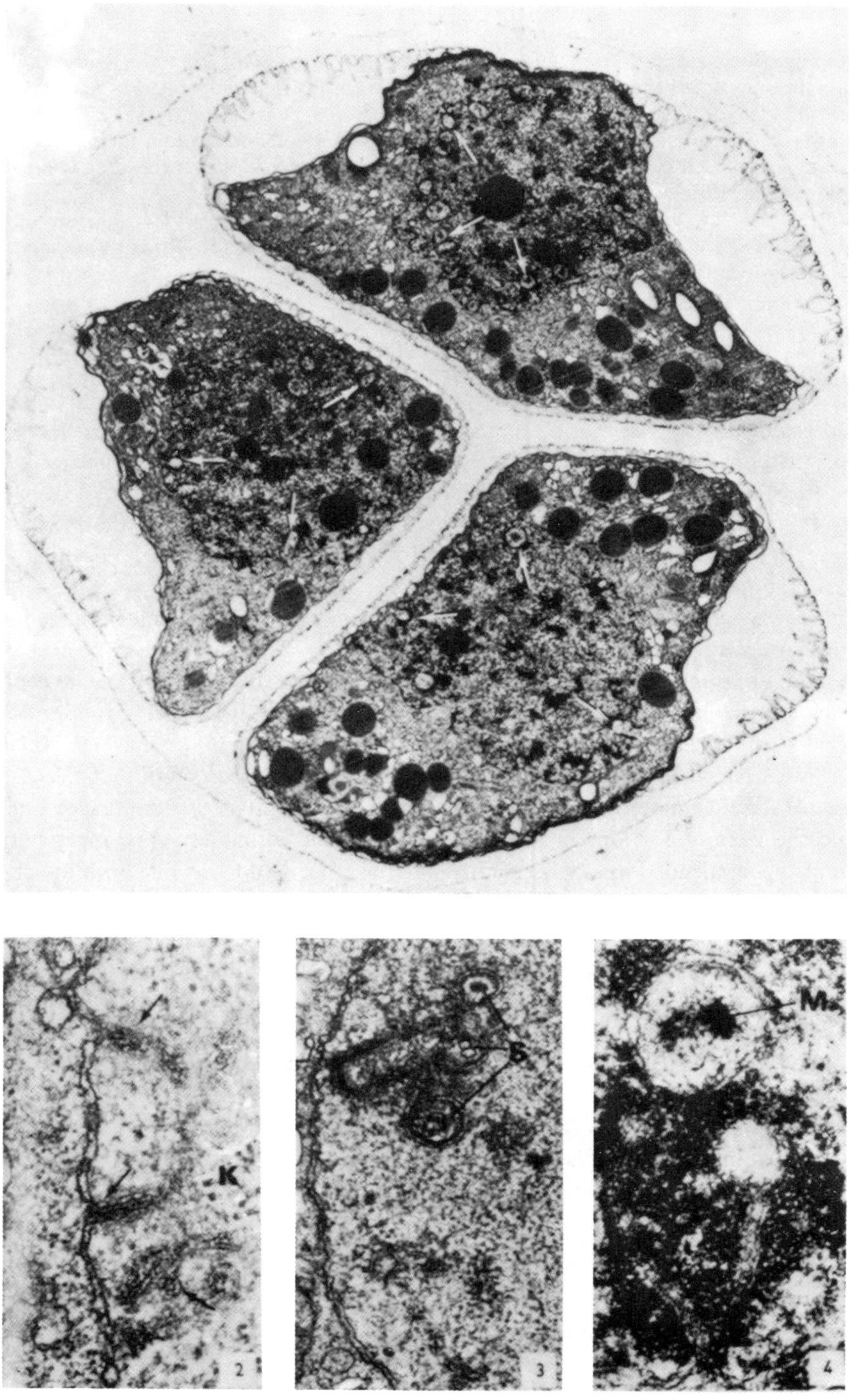
K
2
S
3
M
4

MATERIALS AND METHODS

Sporangia were excised from male cones of *Pinus banksiana* (from the University of London Botanical Supply Unit, Egham, Surrey, England) and plunged into ice-cold 3 % glutaric dialdehyde in 0.05 *M* phosphate buffer (pH 6.8) and fixed for 3–4 hours. After washing in buffer and subsequent postfixation in 2 % aqueous osmic acid for 3 hours, tissues were dehydrated in ethyl alcohols, transferred to 1,2-epoxy propane and embedded in Epikote (Shell, U. K.).

Thin sections for electron microscopy were cut on a Porter-Blum MT 1 ultramicrotome and stained in 7 % aqueous uranyl acetate followed by lead citrate (*5*). Grids were examined in a Siemens Elmiskop I electron microscope operating at 60 kV.

OBSERVATIONS

The envelope that invests the microspore nucleus of *Pinus banksiana* following the final division of meiosis is microscopically indistinguishable from that of a typical nucleus. It consists of two membranes, regular in outline and punctuated only by pores, their diameter being of the order of 0.1 μm. During the subsequent development of the microspores, however, approximately at the stage when the primexine of the pollen wall is being laid down, the form of the nuclear envelope undergoes marked changes.

Over the whole nuclear surface, finger like invaginations, involving both layers of the nuclear envelope, extend into the karyoplasm (Fig. 2). They may branch once or twice, and, although they mostly contain little cytoplasm (the diameter of the central channel is of the order of 0.05 μm), they frequently inflate at their tips giving rise to terminal sacs, which may reach 0.5 μm in diameter, deep inside the nucleus. Nuclear pores have been seen only rarely in the invaginations and sacs.

Electron-opaque material interpreted as chromatin, begins to condense (see Fig. 3) around the narrow portions of the invaginations, and ultimately each invagination is completely embedded in a mass of this material (Fig. 4). Meanwhile both the channels and sacs of the invaginations become filled with an electron-opaque substance, more fibrillar than the chromatin.

As the electron-opaque substance accumulates within the invaginations they swell and eventually become more or less spherical (Fig. 1, arrows). Ultimately it seems that they are resorbed, the material within them being discharged into the cytoplasm (Fig. 1). The nuclear envelope once again becomes regular and the nuclear pores conspicuous.

A similar phenomenon to that described above has been reported as occurring in sporogenous and mitotic cells of *Noctiluca miliaris* by Afzelius (*1*) and later Soyer (*6. 7*). Soyer, while mentioning that these structures often contain a particulate content and may take part in some nucleocytoplasmic interaction, hypothysizes that in *Noctiluca* the invaginations represent reserves of membrane for subsequent utilization by the developing cells. In *Pinus banksiana*, the marked condensation of chromatin around the invaginations, followed by the accumulation of material within their terminal sacs, might indicate that these structures are not membrane reserves, but specialized extensions of the nuclear envelope concerned with the secretion of material into the microspore cytoplasm.

The composition of the granofibrillar substance which accumulates in the sacs of the invaginations is presently unknown. Since it appears in such close proximity to the chromatin, it is conceivable that it is a nucleic acid. There is prima facie evidence, therefore, that a situation similar to that in *Lilium* is also to be found in microsporogenesis in *Pinus*, but that the mechanism of nucleocytoplasmic transfer differs in detail.

REFERENCES

1. Afzelius, B. A., *J. Cell Biol.* **19**, 229 (1963).
2. Dickinson, H. G. and Heslop Harrison, J., *Protoplasma*, in press.
3. —— *Cytobios*, in press.
4. Mackenzie, A., Heslop Harrison, J. and Dickinson, H. G., *Nature (London)* **215**, 997 (1967).
5. Reynolds, E. S., *J. Cell Biol.* **17**, 208 (1963).
6. Soyer, M. O., *J. Microsc. (Paris)* **8**, 569 (1969).
7. —— *ibid.* **8**, 709 (1969).

ORGANIZATION AND ACTIVITY IN THE PRE- AND POSTOVULATORY FOLLICLE OF *NECTURUS MACULOSUS*

R. G. KESSEL and W. R. PANJE

INTRODUCTION

In almost all organisms, the developing oocyte is invested by cells and formed elements collectively designated the follicle envelope. This structure thus represents a morphological and physiological boundary between the oocyte and its environment. Because of the intimate relationship existing between the follicular envelope and developing oocyte, it has long been held that the follicle cells, in particular, play an important role in the nourishment of the oocyte and in mediating the transport of materials into the growing egg. Several different functions have been proposed for the vertebrate follicle envelope. These include: (*a*) the synthesis of nucleic acids and proteins which may then be utilized by the oocyte (see 29, 59); (*b*) the formation of secondary egg membranes (see 69); (*c*) the mediation of the transport of materials from the capillaries in the connective tissue of the follicle envelope to the oocyte (see 89); and (*d*) the biosynthesis of cholesterol and steroid (see 8, 60). Recently, Masui (57) has obtained results with *Rana pipiens* which . . .

"indicate that pituitary gonadotropin acts on the follicle cells to stimulate them to release a hormone that directly acts on the oocyte to induce maturation." Much of the information suggesting that the follicle cells play a role in the synthesis of materials which are then supplied to the oocyte is based on studies of insect ovaries (3, 4, 6, 7, 40, 80, 94). Notwithstanding the fact that amphibian oocytes have been extensively studied at the electron microscope level, few studies have been concerned with variations in structure and activity occurring in the follicle envelope throughout the entire period of oogenesis (see 9, 24, 38, 85, 86, 91). That this is the case probably reflects the technical difficulties of dealing with the follicle envelope separated from its oocyte, and of observing the activity in the follicle envelope over a long period of development.

The present report summarizes information collected by a variety of means about the activity of the follicle envelope during the course of oogenesis in the neotonous salamander, *Necturus maculosus*.

MATERIALS AND METHODS

Electron Microscopy

Necturus oocytes of different sizes, mechanically isolated follicle envelopes (i.e., stripped from oocytes with watchmaker's forceps under a dissecting microscope) from oocytes ranging from 2.5 to 4.5 mm in diameter, and postovulatory follicles were fixed for 2 hr in an ice-cold 1% solution of glutaraldehyde (76) in either 0.1 or 0.05 M Sörensen's phosphate buffer (pH 7.4). In some cases, 0.5 mg of calcium chloride or 0.05 M sucrose was added to the fixative. Similar tissues were also fixed for 2 hr in ice-cold, 1% osmium tetroxide in 0.1 M phosphate buffer (pH 7.4). The fixative, therefore, ranged from approximately 255 to 270 milliosmols. Glutaraldehyde-fixed tissues were subsequently washed in several changes of 0.1 M phosphate buffer containing 0.05 M sucrose for a period of 6–12 hr and postfixed for 2 hr in cold, 1% osmium tetroxide (63) in 0.1 M phosphate (pH 7.4). After rapid dehydration in a series of cold ethanols and treatment with propylene oxide, the tissues were embedded in Epon 812 (55). Sections obtained with a Sorval MT-2 ultramicrotome were stained with uranyl acetate (87) and lead citrate (71) and studied in an RCA EMU-3G electron microscope.

Histochemistry

Schultz's modification of the Liebermann-Burchardt reaction as described by Thompson (84) was used in detecting cholesterol. Whole mounts of the follicle envelope and squashed preparations of the postovulatory follicle were treated and photographed within a 10- to 30-min period. An aqueous Nile blue staining method as described by Thompson (84) was also used in demonstrating the presence of acidic and nonacidic lipids. Both whole mounts and fresh-frozen sections of follicle envelopes and postovulatory follicles were stained at 60°C with Nile blue, rinsed with distilled water at room temperature, and immediately examined. Gomori's (33) method for the detection of alkaline phosphatase was used for the demonstration of this enzyme in fresh-frozen sections of the follicle envelope. Portions of the *Necturus* ovary were fixed in Bouin's or Champy's solution, and paraffin sections were stained with mercuric bromphenol blue (58) or Heidenhain's iron hematoxylin, respectively.

Lipid Extraction–Thin Layer Chromatography

Lipids were extracted from the postovulatory follicles by the method of Folch et al. (31) under an atmosphere of argon. The preparations were examined cytologically prior to extraction, for elimination of possible contamination of the postovulatory follicles with egg debris, atretic follicles, or small oocytes. 5-μl samples of the lipid extract were spotted on activated (110°C for 30 min) Silica Gel G plates (82) and placed into one of six different solvent systems for the specific lipid or steroid separation. Solvent A consisted of 90 parts *n*-pentane, 10 parts diethyl ether, and 1 part acetic acid; solvent B consisted of 65 parts chloroform, 34 parts methanol, and 4 parts water; solvent C, 3 parts benzene, 2 parts ethyl acetate; solvent D, 4 parts benzene, 1 part ethyl acetate; solvent E, 4 parts chloroform, 1 part ethyl acetate; solvent F, 9 parts chloroform, 1 part acetone. The separated lipids and steroid were visually demonstrated by charring the plates with 50% sulfuric acid. The classes of lipids and steroids were identified by their rf values according to Stahl (82) and Randerath (68). In addition, specific color reagents Nos. 123 and 60 were used in detecting the presence of steroids (82).

Soluble Proteins-3 β-Hydroxysteroid Dehydrogenase Activity

A modification of Wattenberg's (88) method for 3 β-hydroxysteroid dehydrogenase (3 β-HSD) localization was used for the demonstration of possible sites of steroid biosynthesis. The technique was applied to frozen sections of postovulatory follicles as well as to soluble proteins separated in polyacrylamide gels. The purpose of using polyacrylamide gel electrophoresis in this portion of the study was to provide a more concentrated enzyme that could be histochemically analyzed for 3 β-HSD activity. In this technique,

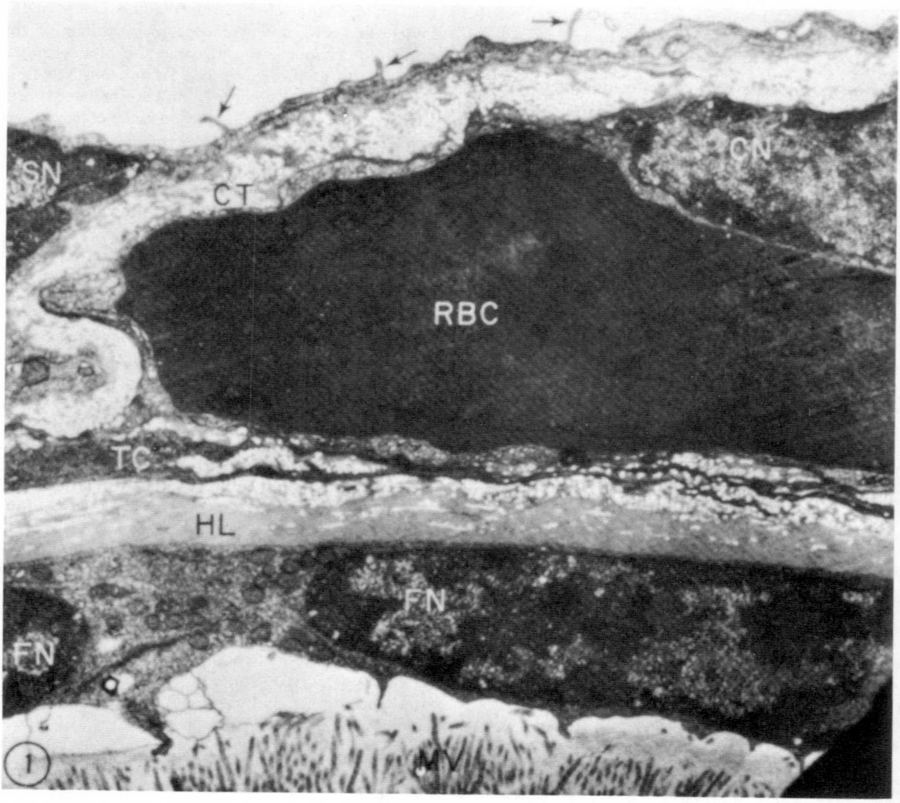

FIGURE 1 Oocyte-follicle complex ~0.75 mm in diameter. A portion of two overlapping follicle cells and their nuclei (*FN*) is illustrated adjacent to the microvilli (*MV*) of the oocyte. The theca contains a capillary with an enclosed erythrocyte (*RBC*). The nucleus (*CN*) of the capillary endothelial cell and a process from a fibroblast cell (*TC*) are also illustrated in the theca. The connective tissue fibers and matrix of the theca are identified (*CT, HL*). The thin serosal cell layer includes a portion of the nucleus (*SN*), and microvilli project from the outer surface of the serosal cells (arrows). Glycogen granules and pinocytotic vesicles are present in the serosal, endothelial, and thecal cells. × 10,200.

approximately 30 postovulatory follicles in each experiment were homogenized at 4°C in 0.58 M sucrose in 0.2 M Tris-HCl buffer, pH 7.54, and the cellular debris was sedimented in an International B-20 refrigerated centrifuge at 20,000 *g* for 1 hr at 4°C. 100-μl samples of the supernatant fraction were loaded onto 7.5% polyacrylamide gels (21) and separated at 3.3 milliamps/tube for 90 min in a Tris-glycine buffer at pH 8.6. The gels were removed and incubated for 20–30 min in the following medium at 37°C: 3 mg of nicotinamide adenine dinucleotide; 0.2 mg of dehydroepiandrosterone (dissolved in dimethyl formamide); 1 mg of nitro-blue tetrazolium salt; 0.03 mg of phenzaine methosulfate; 7 ml of 0.2 M Tris-HCl buffer at pH 7.54; and 3 mg of potassium cyanide. The incubation mixture was then removed, and the gels were washed in tap water and stored in a 7.5% acetic acid solution. For demonstration of unspecific diaphorase reaction, control gels were incubated in the substrate which lacked the dehydroepiandrosterone.

Soluble proteins of the pre- and postovulatory

follicles were demonstrated by treating the electrophoresed gels with a 1% naphthol blue-black stain in a 7.5% acetic acid solution for 30 min. The proteins appeared as blue bands after leaching or electrophoretic destaining.

Radioautography

Small numbers of oocytes with associated follicle envelopes were cultured in 2, 10, or 20 ml of sterile Holtfreter's solution containing uridine-^{3}H (1.71 c/mmole, Schwarz BioResearch, Crangeburg, N.Y.), uridine-5-^{3}H (25 c/mmole, New England Nuclear Corp., Boston), lysine-^{3}H (1.0 c/mmole, Schwarz BioResearch), or L-leucine-4,5-^{3}H (5 c/mmole, New England Nuclear Corp.) in amounts ranging from 10 to 100 μc for durations ranging from 30 min to 22 hr. Adult females were also injected with 250–500 μc of uridine-^{3}H or 150–250 μc of leucine-^{3}H or lysine-^{3}H and maintained for 1–14 days thereafter. In all cases, the oocytes were rinsed briefly in several changes of cold, sterile Holtfreter's solution. They were then fixed in ice-cold, 6% formalin containing 0.5 or 3% trichloracetic acid (TCA) and in some cases the corresponding amino acid in the amount of 0.1 mg/ml of total fixative for 24 hr, or they were fixed in Bouin's solution for 24 hr. The oocytes were embedded in paraffin, sectioned at 5 μ, and processed for radioautography according to the techniques described by Kopriwa and Leblond (50). Deparaffinized and hydrated sections treated with 5% TCA were coated with Kodak NTB-2 emulsion and exposed for periods of 4–8 wk.

RESULTS

Preovulatory Follicle

General

The ovary of *Necturus* is a saccular structure with peritoneum covering the outside and an inner ovarian epithelium lining the inside. Numerous oogonia are dispersed throughout the connective tissue stroma which is interposed between the inner ovarian wall and the peritoneum or serosa. As the oocytes grow, they bulge into the interior of the ovary. During the early prophase of oogenesis, the fibroblasts of the connective tissue differentiate into a layer of cells, the follicle epithelium or granulosa layer, immediately surrounding the oocyte surface. An early stage in the formation of the follicle envelope is illustrated in Fig. 22. In this figure a fibroblast nucleus is closely apposed to the gonial cell, and its contour reflects the curvature of the oogonium. The zona pellucida, which includes the zona radiata (processes from the follicle cells and plasma membrane of the oocyte) and (in urodeles) an acellular or homogeneous layer, is located between the follicle epithelium and the oocyte surface. The follicle epithelium is surrounded by the theca, composed of connective tissue fibers, matrix, fibroblasts, and capillaries. The theca, in turn, is surrounded by the surface epithelium or serosa, constituting the outermost layer of the follicle. These components of the *Necturus* follicle envelope are illustrated in Figs. 1 and 3. In part, the serosa results from the enlargement of the oocyte during its period of tremendous growth so that the oocyte protrudes against the inner ovarian epithelium. Eventually, the inner ovarian epithelium completely surrounds the oocyte, but remains continuous with the peritoneum in a small region which marks the point through which the mature egg will rupture at the time of ovulation (see 91, 92). During the course of oogenesis in *Necturus*, the oocyte and its associated follicle envelope increase in size to approximately 5 mm in diameter at maturity. Previous studies have demonstrated that protein yolk deposition begins in those oocytes 1.1–1.2 mm in diameter (49). Studies dealing with various aspects of the amphibian oocyte and follicle envelope include those by Kemp (45–48), Dollander (24), Wartenberg and Gusek (86), Wartenberg (85), Wischnitzer (91), and Hope et al. (38).

Fine Structure

GRANULOSA: The typical organization of the follicle envelope surrounding oocytes 0.5–1.0 mm in diameter is illustrated in Figs. 1 and 3. The fibrous network associated with the tips of the oocyte microvilli in Fig. 1 represents an early stage in the formation of the vitelline membrane. At this stage of development, the follicle cells are squamous in shape and contain a large, ellipsoidal nucleus. Slender projections of the follicle cell extend between the oocyte microvilli and end in close proximity to the oolemma (Fig. 2). In the area of close contact, the plasma membranes of the two cells appear more dense than in any other region (Fig. 2). The cytoplasm of the follicle cells at this period of development consists of mitochondria, rough-surfaced endoplasmic reticulum, free ribosomes, and one or more Golgi complexes (Fig. 4). In this early period of oogenesis, it is common for adjacent follicle cells to overlap each other for considerable distances as is

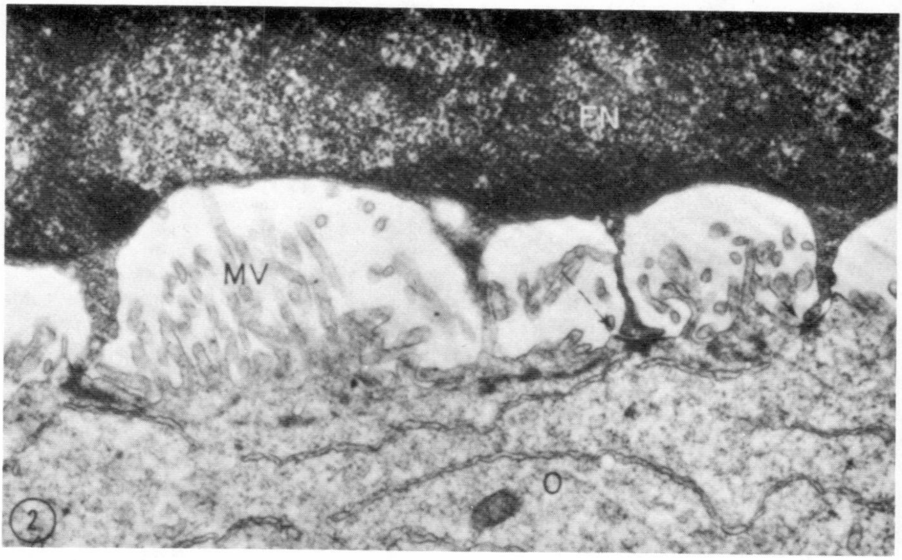

FIGURE 2 Oocyte-follicle complex ~0.5 mm in diameter. The figure illustrates a portion of the oocyte (*O*) and the large, elliptical-shaped nucleus of the follicle cell (*FN*). Processes from the follicle cell extend between the oocyte microvilli (*MV*) and terminate in close proximity to the oolemma (arrows). × 22,000.

apparent in Figs. 1 and 3, and no intercellular spaces are evident between the follicle cells.

As the oocyte-follicle complex increases in size from 1.3 to 2.0 mm in diameter, several changes occur in the follicle cells. For one, the shape of the follicle cell changes from squamous to cuboidal and the nuclei change from ellipsoidal to ovoid in shape (Figs. 5, 6). Further, by the time the oocytes have grown to ~1.3 mm in diameter, overlapping of adjacent follicular cells is uncommon, but large intercellular spaces are apparent which become extensive channels between the follicle cells of larger oocytes (Figs. 7, 8). Moreover, the distance between the follicle cell layer and the oocyte increases as the length of the oocyte microvilli increases and the deposition of the vitelline membrane is completed. At this time, long, slender extensions of the follicle cell extend between the oocyte microvilli and occupy deep but narrow pits of the oocyte surface. Cytoplasmic changes in the follicle cells also occur at this time. An extensive system of branching and anastomosing smooth-surfaced tubules becomes evident, this system representing an elaboration of smooth endoplasmic reticulum (Fig. 10). While free ribosomes are still numerous, the rough endoplasmic reticulum does not appear so extensively developed (Figs. 7–9). Dense, irregularly shaped secretion granules are present in the follicle cells early in development and persist throughout much of oogenesis (Figs. 4, 8, 10). They first appear in the Golgi region (Fig. 4) and later are more concentrated in this area than in any other region of the follicle cell cytoplasm. These secretion granules in the follicle cells appear similar in size, shape, and density to secretion granules in the serosa and theca cells. Although a tubular, smooth endoplasmic reticulum is encountered in those follicle cells fixed in glutaraldehyde as the primary fixative, numerous vesicles of rather uniform size are present in those follicle cells fixed in osmium tetroxide as the primary fixative. Lipid droplets begin to accumulate in the follicle cell cytoplasm during this period of development (Fig. 6). Lipid deposition begins when the oocyte-follicle complex is ~1.3 mm in diameter and increases during the growth to 2.0 mm. Histological sections of the oocyte-follicle cell complex which are suitable for

FIGURE 3 Oocyte-follicle complex ~0.75 mm in diameter. The thin serosal cell layer is identified at *SC*. The extensive theca indicated by the dark vertical line contains fibroblast cells (*FB*) and connective tissue fibers and matrix (*CT*). A portion of two overlapping follicle cells (*FC*) and a follicle cell nucleus (*FN*) are identified. Processes of the follicle cell (*FP*) extend between oocyte microvilli (*MV*). A fibrillar material is associated with the tips of the oocyte microvilli at this stage. Glycogen granules and pinocytotic vesicles are apparent in the theca and serosa cells. × 14,500.

the preservation of lipid indicate that in some of the follicle cells lipid appears packed at 2 mm whereas in other cells lipid accumulation has not progressed to such an extent (Figs. 5, 6). The accumulation of lipid in the follicle cells continues as oogenesis proceeds. A portion of a follicle cell surrounding an oocyte ~4 mm in diameter is illustrated in Fig. 11. The cytoplasm of this cell is nearly filled with lipid. Thus, one major function of the follicle cell during oogenesis in *Necturus* appears to be the synthesis and accumulation of large amounts of lipid.

During the terminal stages of oogenesis (oocytes 4-5 mm in diameter), proteinaceous yolk platelets appear in some, but not all, of the follicle cells (Figs. 12–14). These yolk platelets as visualized in light microscope preparations are similar in size to those in the ooplasm and stain intensely with

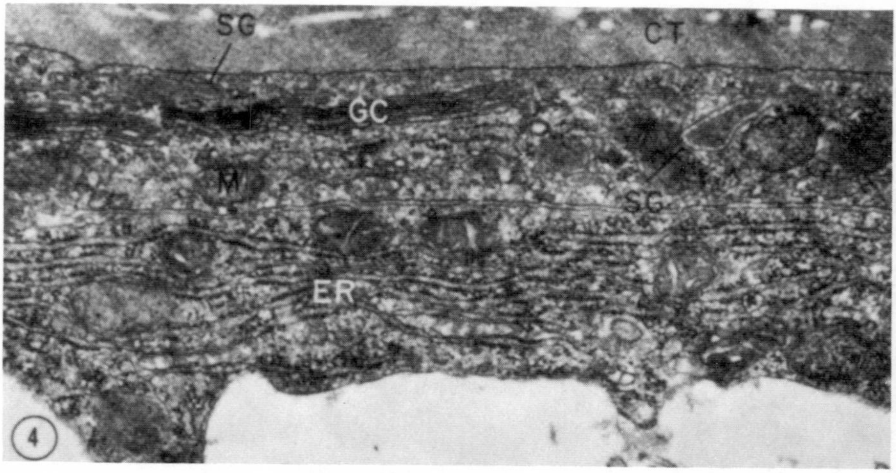

Figure 4 Oocyte-follicle complex ~0.75 mm. The figure illustrates a portion of the cytoplasm of two follicle cells. Mitochondria (*M*), rough-surfaced endoplasmic reticulum (*ER*), Golgi saccules (*GC*), and forming secretory granules (*SG*) are identified. A portion of the connective tissue of the theca is present at *CT*. × 30,000.

mercuric bromphenol blue (Fig. 13). Electron micrographs of the follicle cells around an oocyte nearing maturity confirm the presence of yolk platelets in some of the follicle cells. Further, such yolk platelets are enclosed by numerous membranous whorls (Figs. 12–14) similar to those described as being associated with the process of yolk demolition in embryonic cells of other species (43, 44). Some regions of the follicle cell cytoplasm contain extensive membranous whorls which probably have resulted from the complete destruction of the yolk platelets (Fig. 12). However, there is no evidence that any of the yolk platelets within the oocyte are undergoing demolition at this stage. Thick Epon sections of the entire oocyte-follicle complex were stained and examined for possible evidence of atresia. As far as could be determined, no evidence of atresia in the egg was evident when yolk platelets were encountered in some of the follicle cells late in oogenesis.

THECA: The outer surface of the follicle cells is generally quite smooth and surrounded by a structureless-appearing matrix in which connective tissue fibers are embedded (Figs. 1–3). This region begins the theca, a layer consisting of numerous fibroblasts which are stellate in shape with numerous, long processes interdigitating with adjacent cells (Fig. 3). The cytoplasm of the theca cells contains mitochondria, free ribosomes, both smooth and rough forms of endoplasmic reticulum, and numerous granules similar in appearance to particulate glycogen (70). Lipid droplets are frequently observed in these cells with the electron microscope and, in suitable histological preparations, some of the cells appear to be filled with lipid droplets (Figs. 5, 6). The space between the cells in the theca is occupied by collagenic fibers, matrix, and capillaries (Figs. 1–3).

SEROSA: The cells comprising the serosa are thin, flattened, and are characterized by numerous and extensive desmosomes (Figs. 1, 3, 17, 18). Large numbers of tonofilaments extend from the desmosomal plaques into the serosal cells and occur throughout much of the cytoplasm (Figs. 18–20). Small, dense granules probably representing particulate glycogen are present early in the development of the serosa cells, and they increase in number during subsequent periods of development (Figs. 18–20). Some overlapping of adjacent serosal cells occurs, and intercellular spaces, sometimes containing a flocculent material, are commonly encountered (Figs. 15–17). In addition to

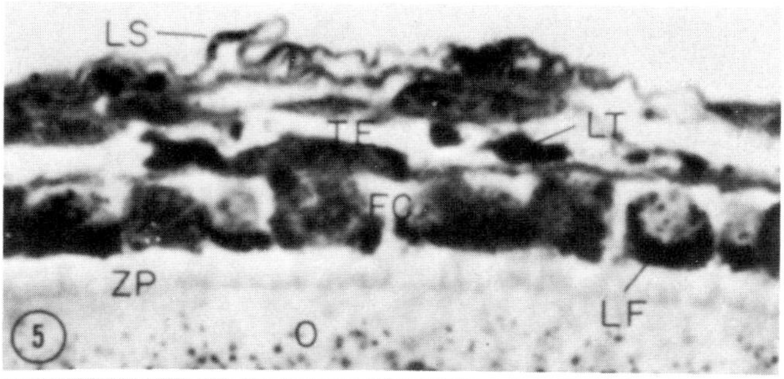

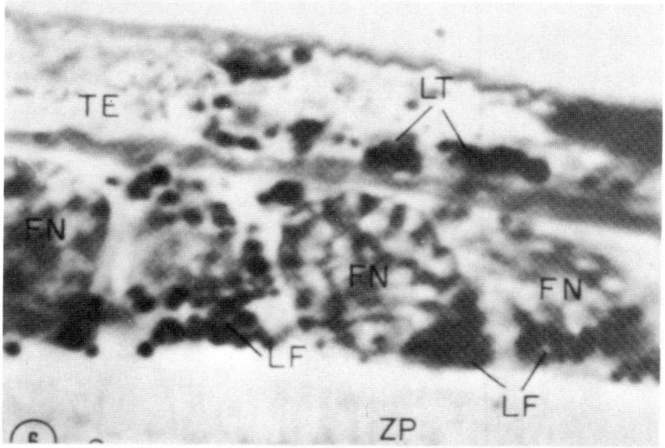

FIGURES 5 and 6 Oocyte-follicle complex ~2 mm in diameter. The photomicrographs illustrate the follicle cell layer (*FC*) and their nuclei (*FN*), the theca (*TE*), serosa (*LS*), zona pellucida (*ZP*), and a portion of the oocyte (*O*). In this preparation, lipid droplets are present in many of the follicle cells (*LF*), theca cells (*LT*), and serosa cells (*LS*). Champy's fixation, Heidenhain's iron hematoxylin. Fig. 5, ×1500; Fig. 6, × 2400.

FIGURES 7–10 Oocyte-follicle complex ~2 mm in diameter. All figures illustrate a portion of the follicle cell cytoplasm which contains extensive arrays of smooth-surfaced endoplasmic reticulum (*AR*), rough-surfaced endoplasmic reticulum (*ER*), and numerous polysomes (*R*). A portion of an extensive intercellular space (*IS*) is included. Pinocytotic vesicles of the coated variety (*CV*) in the process of forming from the plasma membrane are illustrated in Figs. 7–9. Dense, irregularly shaped secretory granules (*SG*) are identified in Figs. 8 and 10. Particulate glycogen indicated at *GY* in Fig. 10. Follicle cell nucleus (*N*). Mitochondria (*M*). Connective tissue fibers of theca (*CT*). Lipid (*L*). Fig. 7, × 23,700; Fig. 8, × 19,000; Figs. 9 and 10, × 50,000.

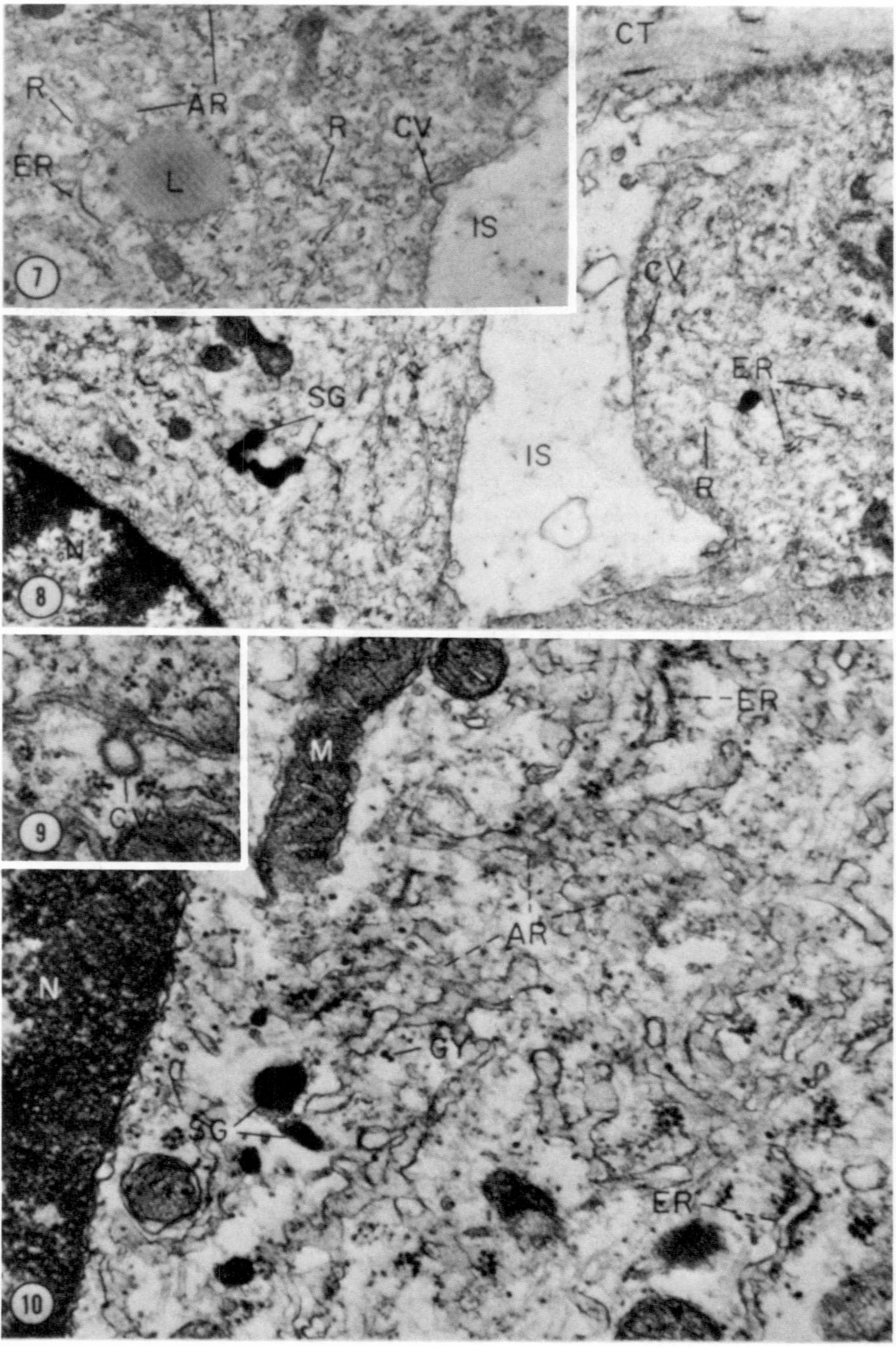
7
R
AR
ER
L
R
CV
IS
CT
8
SG
N
IS
CV
ER
R
9
CV
10
M
ER
AR
N
GY
SG
ER

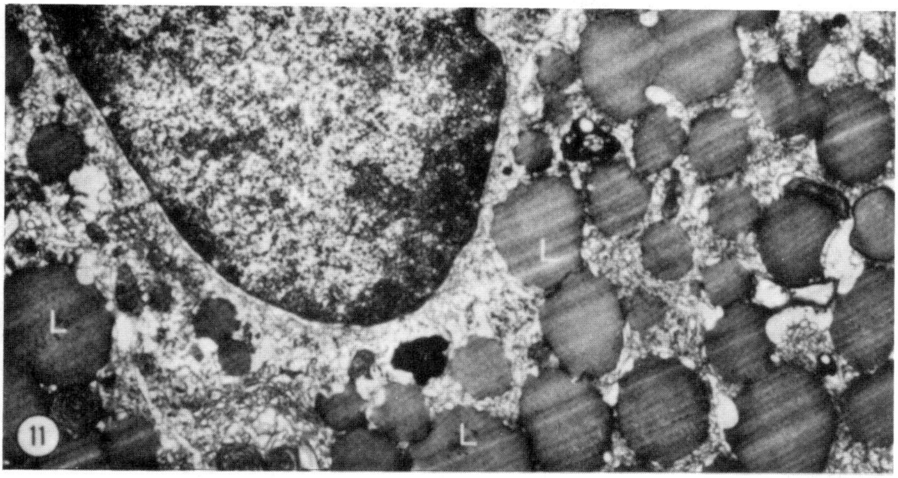

FIGURE 11 Oocyte-follicle complex 3.5–4.0 mm in diameter. The figure illustrates a portion of the nucleus (N) and cytoplasm of a follicle cell. Note the large number of lipid droplets (L) now present at this stage of oogenesis × 10,200.

the extensive array of tonofilaments, mitochondria, particulate glycogen, and pinocytotic vesicles, other membranous structures appear in the cytoplasm of the serosal cells shortly after the formation of this layer. A small amount of rough-surfaced endoplasmic reticulum is present in the cells, as well as a more extensive branching and tubular form of smooth-surfaced endoplasmic reticulum (Figs. 18–20). One or more Golgi complexes are present in each of the serosal cells (Figs. 19, 20). As was the case for the follicle and theca cells, the serosa cells also accumulate lipid droplets during the course of oogenesis (Figs. 5, 16). The number of such lipid droplets, however, varies among serosal cells in the same follicle envelope. In addition to lipid, another type of inclusion product is observed in some of the serosa cells as illustrated in Fig. 19. These inclusions are dense, irregularly shaped granules which seem to have their origin in association with the Golgi complex (Fig. 20) and which in appearance resemble granules also present in the granulosa and theca cells. Microtubules are randomly distributed in all cells of the follicle envelope.

MICROPINOCYTOSIS

SEROSA: Micropinocytosis appears to occur in all cells of the *Necturus* follicle envelope. However, not all cell types in the envelope are equally active, nor does micropinocytosis in these cells occur with constant frequency during the course of oogenesis. The earliest pinocytotic activity is encountered in the serosal cells. From the time the serosa becomes associated with the young oocyte until the oocytes have grown to ~1 mm in diameter, the serosal cells contain numerous vesicles of the nonalveolate or noncoated type which measure 100–130 mμ in diameter (Fig. 15). These vesicles are frequently observed to be continuous with the plasma membrane of the inner, outer, and lateral margins of the cells (Figs. 15, 17, 18). The number of caveolae and isolated micropinocytotic vesicles remains high in the serosal cells as the oocyte grows to a diameter of ~2.5 mm. During subsequent periods of oocyte growth, however, micropinocytotic activity continues but at an apparently reduced rate. The extensive pinocytotic activity in the serosa as evidenced in the electron micrographs suggests that this portion of the follicle envelope is important in the uptake of materials from outside of the follicle and their transport to deeper regions of the follicle envelope.

THECA: Vesicles of the nonalveolate variety similar in size to those seen in the serosa cells are also observed in the theca cells. The vesicles may be isolated in the cytoplasm of these cells or

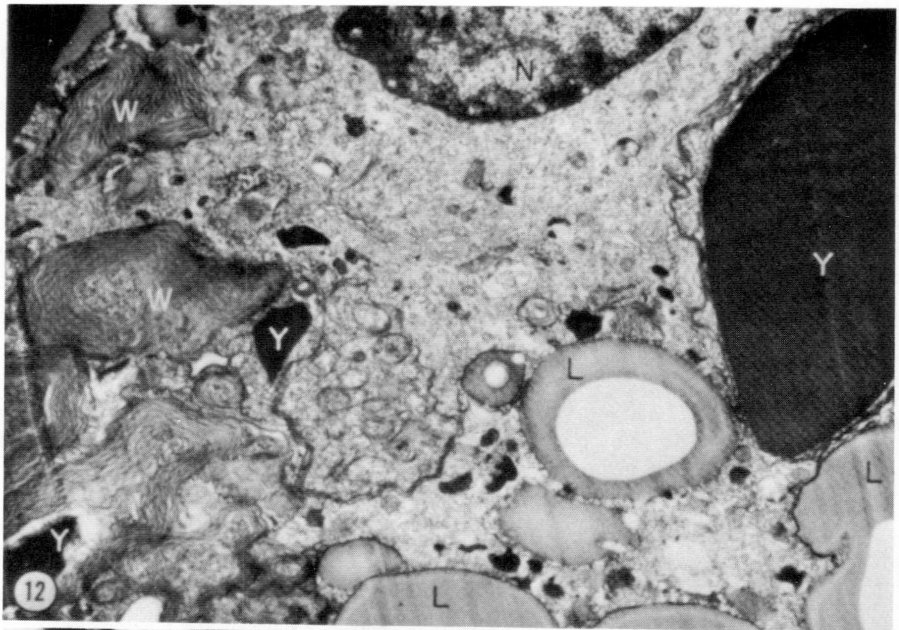

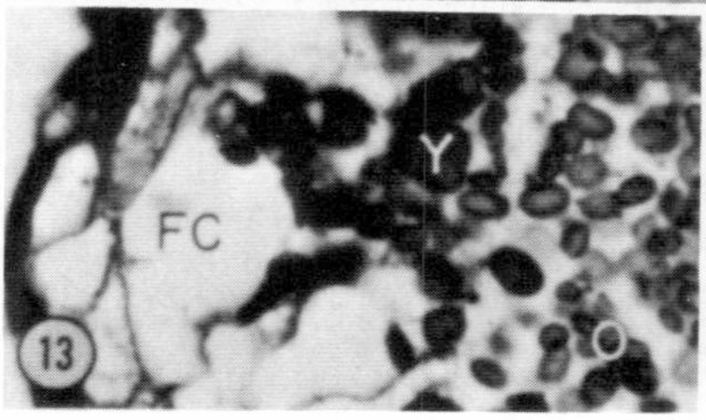

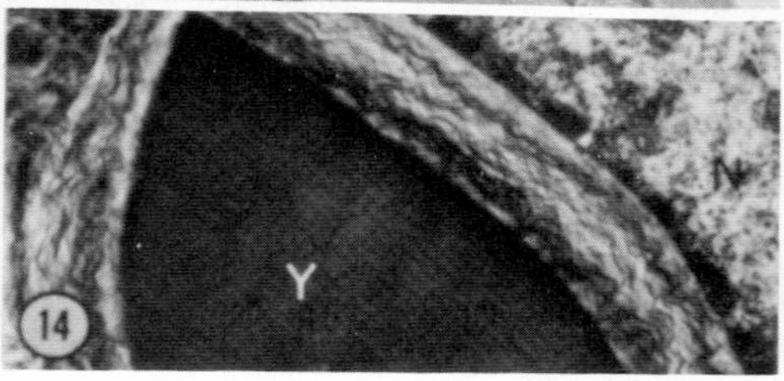

FIGURE 12 Oocyte-follicle complex ~4.5 mm in diameter. The figure illustrates a portion of a follicle cell with nucleus (*N*), lipid droplets (*L*), and yolk platelets (*Y*) surrounded by concentrically arranged membranes. Other regions of the follicle cell cytoplasm contain concentric membranous whorls (*W*) which do not enclose yolk platelets. × 10,200.

FIGURE 13 Oocyte-follicle complex ~5 mm in diameter. The photomicrograph illustrates that some of the follicle cell cytoplasm appears vacuolated (*FC*), while in other cells intensely stained yolk platelets (*Y*) are present and are similar in size and appearance to those in the oocyte (*O*). Bouin's fixation. Mercuric bromphenol blue stain. × 2500.

FIGURE 14 Oocyte-follicle complex ~5 mm in diameter. The figure illustrates a portion of a follicle cell with nucleus (*N*) and a yolk platelet (*Y*) surrounded by concentric membranes. × 19,000.

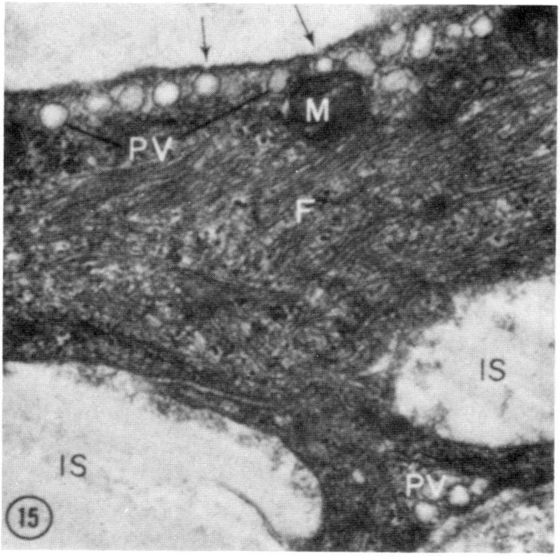

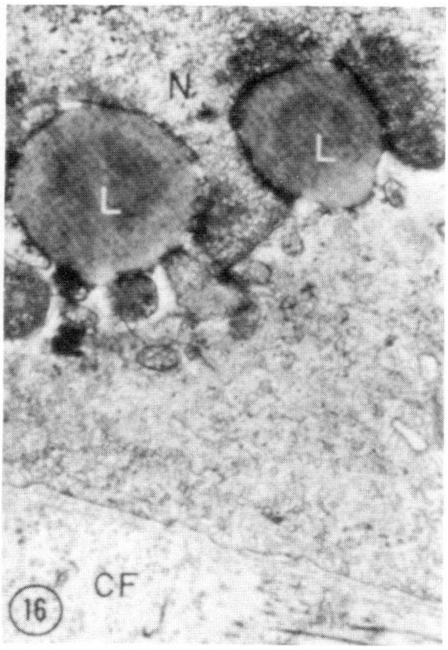

FIGURE 15 Oocyte-follicle complex ~0.5 mm in diameter. Portions of two overlapping serosa cells are illustrated. Intercellular spaces (*IS*) between adjacent cells contain a flocculent material. Note alignment of pinocytotic vesicles along outer and inner surface of the cell (*PV*). Some of the vesicles are continuous with the plasma membrane (arrows). Mitochondria (*M*) are aligned close to the pinocytotic vesicles, and numerous intracytoplasmic filaments are apparent (*F*). × 37,500.

FIGURE 16 Oocyte-follicle complex ~2 mm in diameter. Lipid (*L*) droplets are present in the serosa cell cytoplasm. The nucleus (*N*) of the serosal cell and collagenic fibers of the theca (*CF*) are identified. × 18,900.

continuous at any point with the plasma membrane of the cells. However, the number of such vesicles is not so great in the theca cells as in the serosa. The micropinocytotic vesicles in the theca cells are observed very early in oogenesis and persist throughout all stages studied with the electron microscope.

FOLLICLE CELLS: During the early stages of oogenesis, the follicle cells appear to be the least active of all cells in the follicle envelope. The follicle cells associated with oocytes less than 1 mm in diameter show little or no evidence of pinocytotic activity (Fig. 3). During subsequent stages of oocyte growth, however, occasional micropinocytotic vesicles are found in the follicle cells. These vesicles are similar in size to those described in other cells of the follicle envelope, but those associated with the plasma membrane of the follicle cells are usually of the alveolate (coated) variety (Figs. 7–9).

CAPILLARIES OF THECA: Numerous capillaries traverse the region of the theca. A traversing capillary with an enclosed red blood cell is illustrated in Fig. 1. Perhaps the most extensive micropinycytotic activity occurs in the capillary endothelial cells. The cytoplasm of these cells often appears filled with smooth-surfaced vesicles exhibiting a variety of configurations (Figs. 21–23). Numerous flask-shaped invaginations of the plasma membrane are observed at both the inner and outer surfaces of the endothelial cells, and they are similar in size and shape to intracytoplasmic microvesicles. In addition to the microvesicles, microvesicular rosettes are present in the capillary endothelium (Figs. 21–23). The rosettes consist of a number of microvesicles which sur-

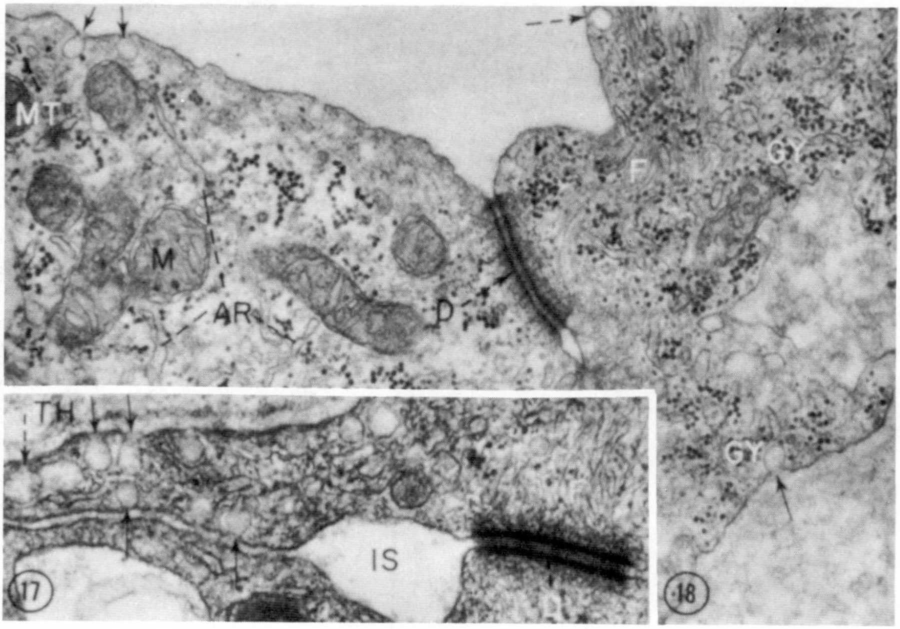

FIGURE 17 Oocyte-follicle complex ~0.5 mm in diameter. Junction of two serosal cells. Desmosome (*D*), intercellular space (*IS*), and tonofilaments (*F*) are indicated. Note caveolae associated with the lateral margins of the serosa cells (upper arrows). Pinocytotic vesicles are also continuous with the plasma membrane of the inner surface of the cell (lower arrows). Basement membrane and connective tissue fibers of the theca (*TH*) are present in upper left of figure. × 50,000.

FIGURE 18 Oocyte-follicle complex ~1.5 mm in diameter. The figure illustrates the junction of two serosa cells. Extensive desmosomes (*D*) are formed at the junction of the two cells, and numerous tonofilaments (*F*) extend from the desmosomal plaques into the cytoplasm. The cytoplasm contains elements of the smooth endoplasmic reticulum (*AR*), mitochondria (*M*), microtubules (*MT*), and glycogen (*GY*). Pinocytotic vesicles are continuous with the plasma membrane at arrows. × 37,500.

round and are continuous with a larger, central vesicle: from one to five small vesicles communicate with the larger, central vesicle (Fig. 23). On occasion, the limiting membrane of the rosette appears to be continuous with the plasma membrane, thus establishing a continuity between the microvesicular rosettes and the extracellular space of the theca (Fig. 21). The capillaries of the theca thus appear to be involved in the transport of large quantities of material from the blood into the follicle envelope. The rate at which such activity occurs does not appear to be constant. If the number of micropinocytotic vesicles present in the capillary endothelium can be used as an index of the amount of material incorporated into the developing follicle, then the activity appears to be greatest in those oocyte-follicle complexes ranging from 1 to 3 mm in diameter.

Postovulatory Follicle

STRUCTURE

Female *Necturus* commonly lay their eggs during the early spring months after an extensive behavioral ritual involving the male. When an egg is ovulated, the follicle envelope becomes converted into a rounded body with a reddish-orange coloration (Fig. 40). Histological preparations of the

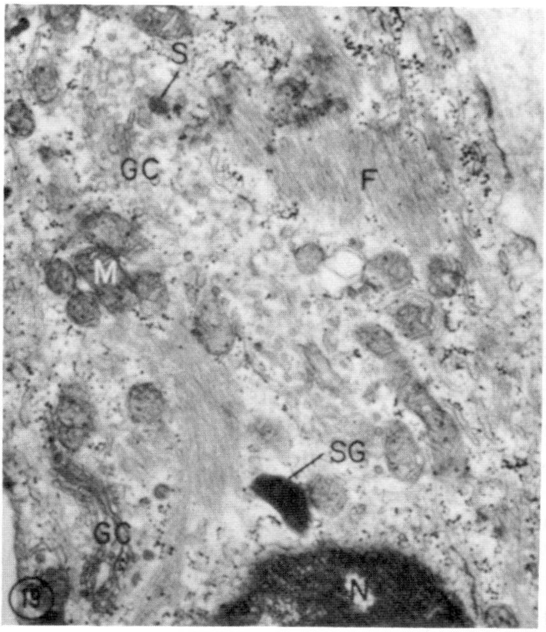

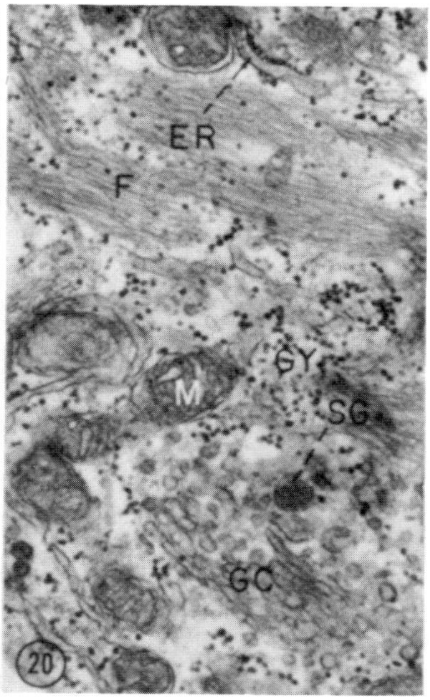

FIGURES 19 and 20 Oocyte-follicle complex ~2 mm in diameter. The Golgi complexes (*GC*) in the serosa cells are illustrated, as are mitochondria (*M*), tonofilaments (*F*), and glycogen (*GY*). The dense, irregularly shaped secretion granule (*SG*) in Fig. 19 appears to have its origin from the Golgi components (Fig. 20, *SG*). The secretory granules are similar in appearance to those present in the follicle cells. Granular endoplasmic reticulum (*ER*). Serosa cell nucleus (*N*). Fig. 19, × 37,500; Fig. 20, × 50,000.

postovulatory follicles show that these follicles are loosely organized, contain connective tissue fibers (probably those of the preovulatory follicle), and are richly vascularized. The maximum diameter of the postovulatory follicles ranges from 1.25 to 1.50 mm. After the egg-laying period in the spring, some of the mature eggs not ovulated appear to undergo degeneration. This activity is accomplished by stellate-shaped cells which invade the oocyte in large numbers (Fig. 24). The postovulatory follicles can be distinguished from atretic follicles on the basis of their coloration and consistency. The postovulatory follicles examined with the electron microscope ranged in size from 1.25 to 1.4 mm. As will become evident, the postovulatory follicle is composed of a number of cell types, each of which is active over a considerable period of time. The description of the fine structure of the postovulatory follicle included here is primarily for the purpose of determining whether this follicle contains cell types similar to those occurring in known steroid-producing organs.

Before the ultrastructure of the cell types encountered in the postovulatory follicle is described, several comments can be made which apply generally to all the cells comprising this follicle. Thus, all the cells have a well developed endoplasmic reticulum (rough and/or smooth) and a large number of lipid droplets and mitochondria. Many of the cells are filled with small, dense granules approximately 250 A in diameter, which appear similar to particulate glycogen, as well as intracytoplasmic filaments.

The most frequently encountered cell type in the postovulatory follicle is characterized by an entensive, centrally located Golgi complex consisting of flattened saccules and both bristle-coated and uncoated vesicles (Fig. 25). Dense granules in

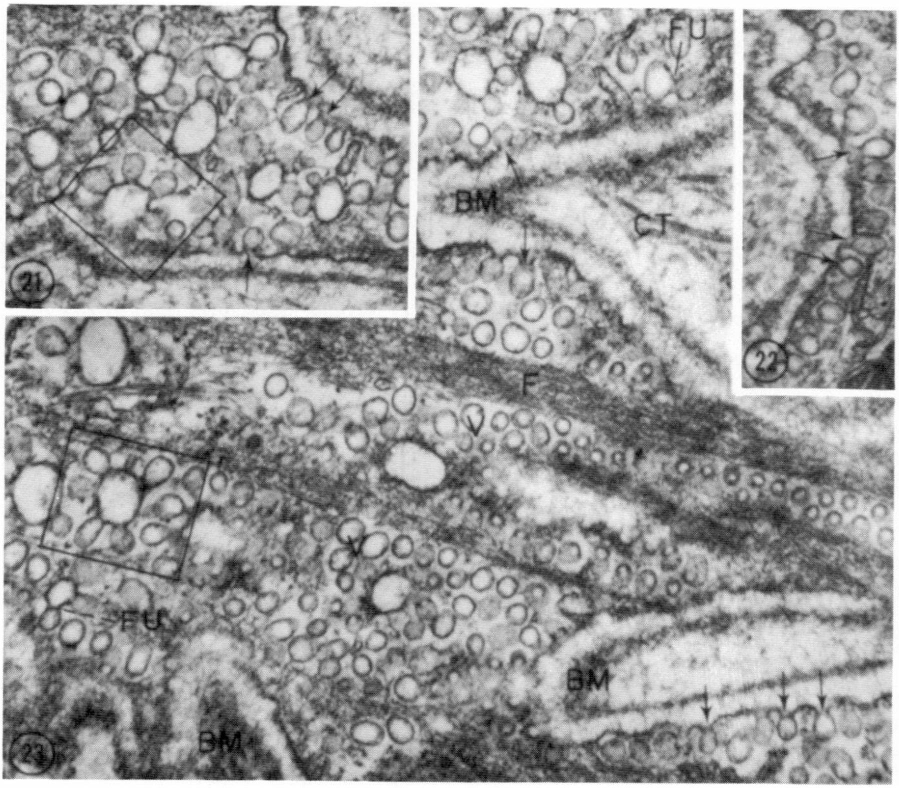

FIGURES 21–23 Portion of capillary endothelial cells in theca. Oocyte-follicle complex ~3 mm in diameter. Note the presence of large numbers of intracytoplasmic microvesicles (*V*). Fusion of microvesicles is indicated at *FU*. Many of the microvesicles appear continuous with the plasma membrane of the endothelial cell (arrows). Microvesicular rosettes (inside squares) are commonly present in the endothelial cells and occasionally are continuous with the plasma membrane (Fig. 21). Intracytoplasmic filaments (*F*), basement membrane (*BM*), and connective tissue fibers of the theca (*CT*) are identified. × 37,500.

a variety of shapes and sizes are closely associated with the Golgi components and, in fact, some of the Golgi cisternae contain a product similar in appearance to the isolated granules (Fig. 25). Elements of the rough-surfaced endoplasmic reticulum and lipid droplets surround the Golgi elements in this cell and occupy most of the remainder of the cell (Fig. 26). In some instances, the cisternae of the rough endoplasmic reticulum are dilated by a finely granular product of medium electron opacity.

Another cell type in the postovulatory follicle contains, in addition to the components already mentioned, a predominantly smooth form of the endoplasmic reticulum (Figs. 27–29). In some regions of this cell, these membranes are arranged into concentric whorls (Fig. 27). In these regions, as well as elsewhere in the cytoplasm, the short cisternae of agranular reticulum are expanded and contain a dense secretory product (Figs. 27, 28). Only a small amount of granular endoplasmic reticulum is present in this cell. What appears to

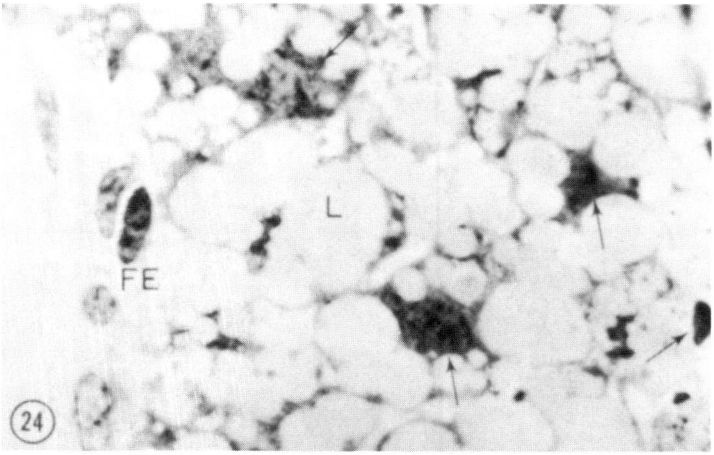

FIGURE 24 Atretic egg. Follicle envelope (*FE*) is present at left in figure. Note that numerous, stellate-shaped cells (arrows) have invade the ooplasm which contains large numbers of lipid droplets (*L*). Epon section. Azure II, methylene blue stain. × 1500.

represent the mature formed product of this cell is illustrated in Fig. 29.

A third cell type in the postovulatory follicle is illustrated in Figs. 30 and 31. These cells contain an extensive Golgi complex and agranular reticulum. They are also characterized by numerous inclusion bodies which become extremely large and dense. The large inclusion bodies appear to result from the fusion of smaller granules (Fig. 31). Lipid and particulate glycogen are also present.

Occasionally, cells are encountered in the postovulatory follicle as illustrated in Fig. 32 which contain predominantly granular endoplasmic reticulum. Dense, homogeneous granules are present within the cisternae of this reticulum. These dense intracisternal granules vary in size and number, but typically are surrounded by a more loosely organized granular material.

Finally, a cell type such as that illustrated in Fig. 33 is occasionally observed. These cells contain large lipid droplets and numerous, small dense bodies some of which are comparable to lysosomes in structure. In this connection, it should be indicated that structures which could be identified as lysosomes were rarely found in the previous cell types during the period of activity described.

It is apparent from the description of the fine structure of the postovulatory follicle that considerable differentiation occurs in the cells of the follicle envelope as they transform into cells of the postovulatory follicle. Furthermore, the cells of the postovulatory follicle are much more active than those of the preovulatory follicle in terms of synthetic activities, as judged from the increase in the extent of their organelle systems and the increase

FIGURES 25 and 26 Postovulatory follicle. The peripheral portion of this cell type contains lipid droplets (*L*) and an extensive development of granular endoplasmic reticulum (*ER*), the cisternae of which are dilated by a homogeneous product. The central portion of the cell contains an extensively developed Golgi complex consisting of Golgi saccules (*GS*) and vesicles (*GV*). The origin of the secretory granules (*SG*) appears to be associated with the Golgi complex since some of the Golgi cisternae are filled and expanded by a dense, homogeneous product (arrows) similar in appearance to the isolated secretory granules. Fig. 25, × 29,700; Fig. 26, × 37,500.

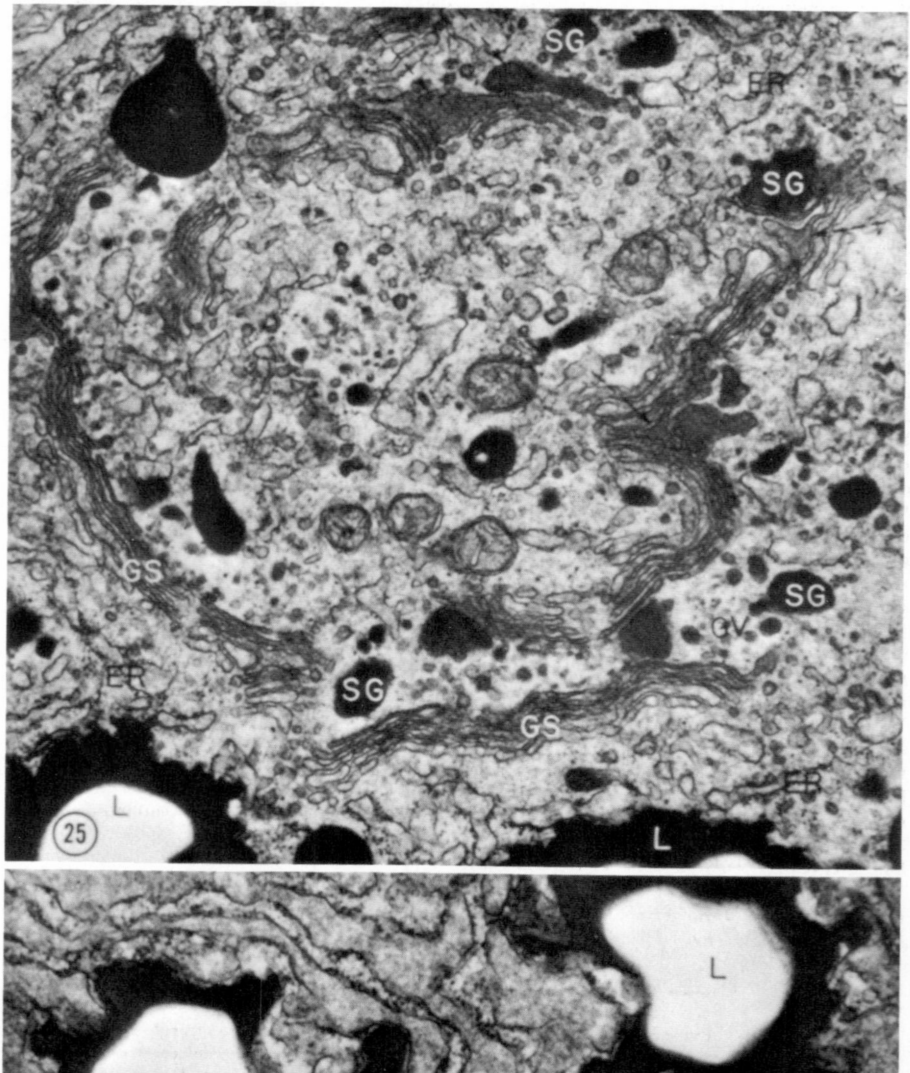
SG
ER
SG
GS
SG
GV
ER
SG
GS
ER
L
25
L

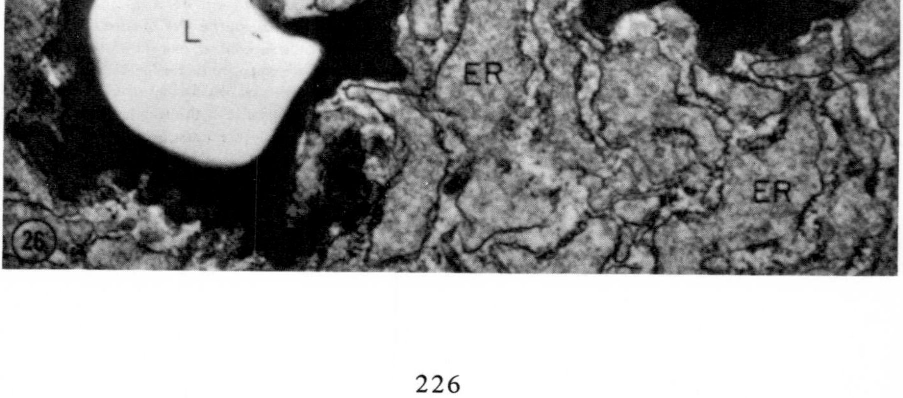
L
L
ER
ER
26

in the amount and kind of formed product observed in the cytoplasm of these cells.

Cytochemistry

That all cell types found in the *Necturus* follicle envelope accumulate lipid droplets during the course of oogenesis was illustrated both in photomicrographs of histological preparations and in electron micrographs. The accumulation of lipid in the follicle envelope is also demonstrated with Nile blue staining. Thus, when follicle envelopes are mechanically isolated from oocytes less than 1 mm in diameter and either the fresh preparations or frozen sections are stained with Nile blue, no red-staining droplets characteristic of neutral lipids are observed in the cells (Fig. 34). However, all follicle envelopes removed from oocytes ranging from 2.5 to 5 mm in diameter showed large numbers of reddish-stained droplets in their cells, these results indicating the accumulation of large quantities of neutral lipid in the follicle (Figs. 35, 36). Follicle envelopes prepared by methods which remove lipid show follicle cells with a highly vacuolated cytoplasm (Fig. 13). Large numbers of red- to pink-staining globules are also present in some of the cells of the postovulatory follicle (Figs. 37, 38). The purple coloration noted in some cells of the postovulatory follicle suggests that acidic lipids are also present in addition to neutral lipids.

Fresh-frozen sections of postovulatory follicles, stained with the Schultz method for the identification and demonstration of cholesterol and cholesteryl esters, contain numerous blue-green colored granules characteristic of a positive reaction (Fig. 39). On the other hand, fresh-frozen sections of follicle envelopes removed from oocytes less than 1.5 mm in diameter did not give a positive reaction for cholesterol. However, similar preparations of follicle envelopes removed from oocytes nearing maturity did contain some blue-green granules characteristic of a positive reaction.

Application of the Gomori method for the demonstration of alkaline phosphatase was made on the follicle envelope at different periods of oogenesis. The results of this technique indicate that large amounts of alkaline phosphatase are localized in the follicle envelope (Fig. 41). While all regions of the envelope appear to contain the enzyme, the reaction is particularly intense in the follicle cells (Fig. 42). The intensity of the reaction appears greater in follicle cells surrounding oocytes <3 mm in diameter than in those greater than 3 mm in diameter.

Soluble Proteins

With the technique of polyacrylamide-gel electrophoresis, an analysis was made of the soluble proteins occurring in the follicle envelope during oogenesis. This analysis was performed on follicle envelopes mechanically isolated from oocytes less than 1 mm in diameter, from oocytes 4–5 mm in diameter, as well as on the postovulatory follicle. Photomicrographs of these three gels are illustrated in Figs. 43–45. With the techniques employed, 22 distinct soluble-protein bands were detected in the original gels of follicle envelopes removed from immature oocytes. In contrast, 14 distinct soluble-protein bands were detected in the original gels of follicle envelopes removed from nearly mature oocytes. In the postovulatory follicle, 18 distinct soluble-protein bands were detected in the original gels. Changes in the migration of many of the soluble proteins in the follicle envelope during oogenesis is evident in Figs. 43–45.

Radioautography

The incorporation of tritium-labeled uridine and amino acids was used as a measure of RNA and protein biosynthesis in the follicle envelope.

Figures 27–29 Postovulatory follicle. Another cell type encountered in the postovulatory follicle contains extensive amounts of a branching and tubular form of smooth-surfaced endoplasmic reticulum (*AR*). A concentric array of smooth membranes is illustrated in Fig. 27. Spherical or oval dense granules representing the formed product of this cell are illustrated in Figs. 27 and 29 (*SG*). The unlabeled arrows direct attention to those regions in which the forming secretory product appears to be located within the cisternae of the smooth reticulum. Small, dense granules (probably glycogen) are numerous (*GY*), as are the mitochondria (*M*). Only a small amount of the rough-surfaced endoplasmic reticulum is present (*ER*). Lipid (*L*). Nucleus (*N*). Intracytoplasmic filaments (*F*). Fig. 27, × 34,000; Figs. 28 and 29, × 50,000.

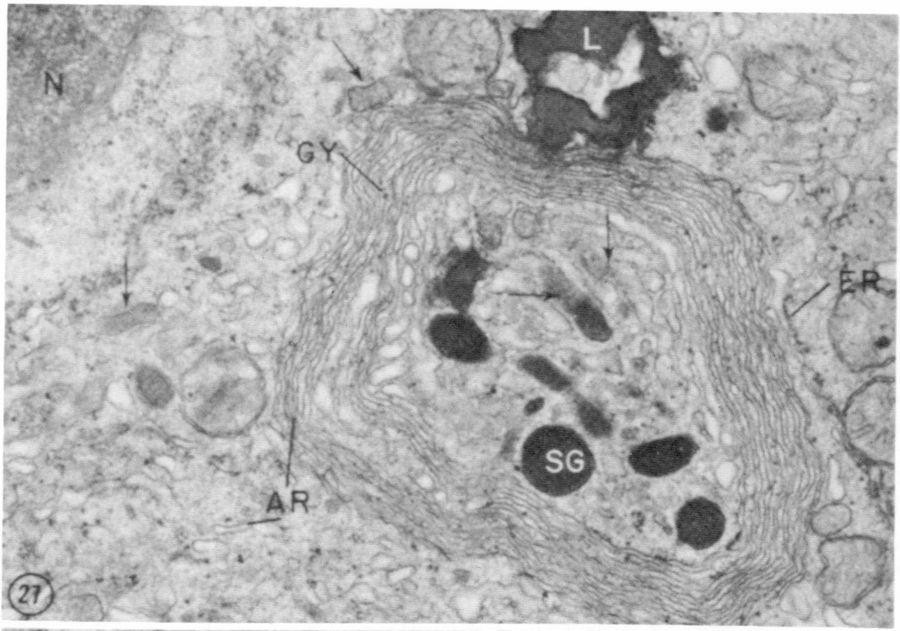
N
L
GY
ER
SG
AR
27

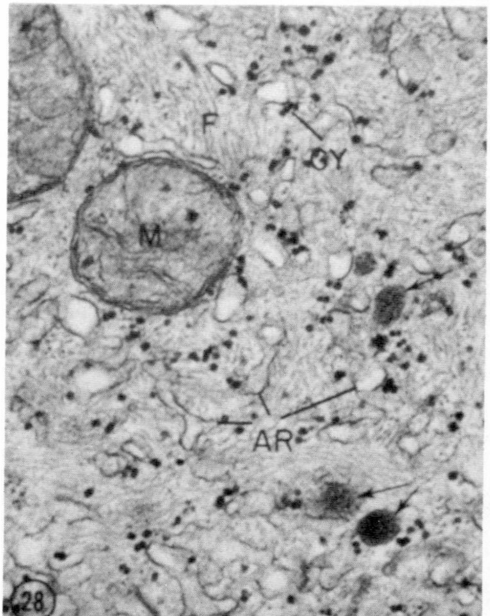
F
GY
M
AR
28

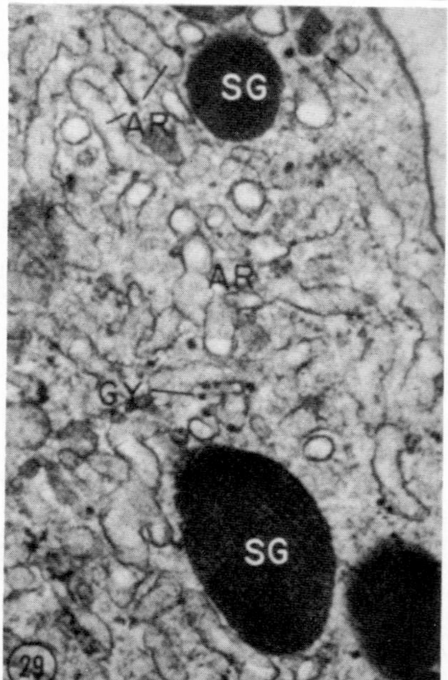
SG
AR
AR
GY
SG
29

Figs. 46–49 illustrate an intense labeling of the follicular envelope with uridine-^{3}H. Fig. 46 illustrates an oogonium as well as two fibroblast nuclei during a very early stage in the formation of the follicle envelope. After 2 hr of in vitro incorporation of uridine-^{3}H, an intense labeling is apparent over the oogonial nucleus and perinuclear cytoplasm. Furthermore, numerous developed silver grains are apparent over the nuclei and cytoplasm of the fibroblast cells. Labeling of the cells of the follicle envelope, especially the follicle cells, with uridine-^{3}H under in vitro conditions remains high as oogenesis proceeds. The oocyte-follicle complexes illustrated in Figs. 47–49 range up to 1 mm in diameter; whereas lipid yolk is already present in such oocytes, protein yolk deposition has not yet begun (49). In Fig. 49, it is apparent that the intensity of the labeling with uridine-^{3}H per unit area over the follicle cell and fibroblast nuclei approaches that apparent over the oocyte nucleus. The cells of the follicle envelope are also labeled with uridine-^{3}H under in vivo conditions. In fact, the cells of the follicle envelope are labeled with uridine-^{3}H under both in vivo and in vitro conditions for a considerable period of oogenesis. In the oocyte-follicle complexes some variation in the intensity of labeling is apparent among oocytes as well as among follicle cells in a single follicle. In some complexes little labeling of the follicle envelope occurs with uridine-^{3}H, whereas in others some of the cells comprising the follicle envelope are more intensely labeled than others. The variation in labeling patterns noted among some of the oocytes may reflect, in part, the fact that not all of the oocytes are in the same state of activity. In general, the labeling of the follicle envelope with uridine-^{3}H is not so significant in those oocyte-follicle complexes nearing the end of their growth period (i.e., 3.5–5 mm in diameter) as was demonstrated during the earlier period of oogenesis. Thus, in oocyte-follicle complexes 4–5 mm in diameter, an occasional follicle cell, theca cell, or serosa cell may have a considerable amount of label associated with it, under both in vitro and in vivo conditions, but the remainder of the cells in the envelope are not heavily labeled.

The cells of the follicle envelope also appear capable of incorporating leucine-^{3}H and lysine-^{3}H under both in vivo and in vitro conditions, but the pattern of labeling differs somewhat from that of uridine-^{3}H. The variations in the pattern of labeling which have been observed with uridine-^{3}H are also apparent with the tritium-labeled amino acids. In general, during the first half of oogenesis (i.e., oocyte-follicle complexes <3 mm), labeling of the follicle envelope, especially the follicle cells, is at a low level as judged from the small number of grains usually located over the envelope. However, as the oocyte-follicle complex increases in size from ~ 3.5 to 5.0 mm in diameter, the number of developed silver grains located over many of the cells in the follicle envelope increases considerably in number under both in vitro and in vivo conditions, as illustrated in Fig. 50. In this radioautograph, a fairly intense labeling of the follicle cells, theca cells, and serosa cells with leucine-^{3}H is apparent under in vitro conditions.

Thin-Layer Chromatography

Tracings of thin-layer chromatograms of lipids extracted from postovulatory follicles are illustrated in Fig. 51. Six different solvent systems were used in an attempt to identify the lipids and to determine whether a steroid material was present in the postovulatory follicle. The possible identification of the materials separated with the different solvents is indicated in Table I. The identification of the compounds is based on the Rf values of the spots as compared to Rf values listed for various materials by Randerath (68) and Stahl (82). It can be noted that a number of the spots obtained with the steroid solvent systems do have Rf values comparable to those of several different steroids. Because of the limitations inherent in the identification of steroids by using Rf values and color reagents applied to thin-layer chromatographic

FIGURES 30 and 31 Postovulatory follicle. Third cell type is characterized by an extensive development of smooth endoplasmic reticulum (*AR*). Golgi complexes are numerous (*GC*). Specific secretory granules (*SG*) vary widely in size probably through fusion of smaller secretory granules (*FG*). Small dense granules (arrows) are concentrated around lipid droplets (*L*), but they are also widely dispersed in the cytoplasm (*GY*). Mitochondria (*M*). Fig. 30, × 19,000; Fig. 31, × 25,200.

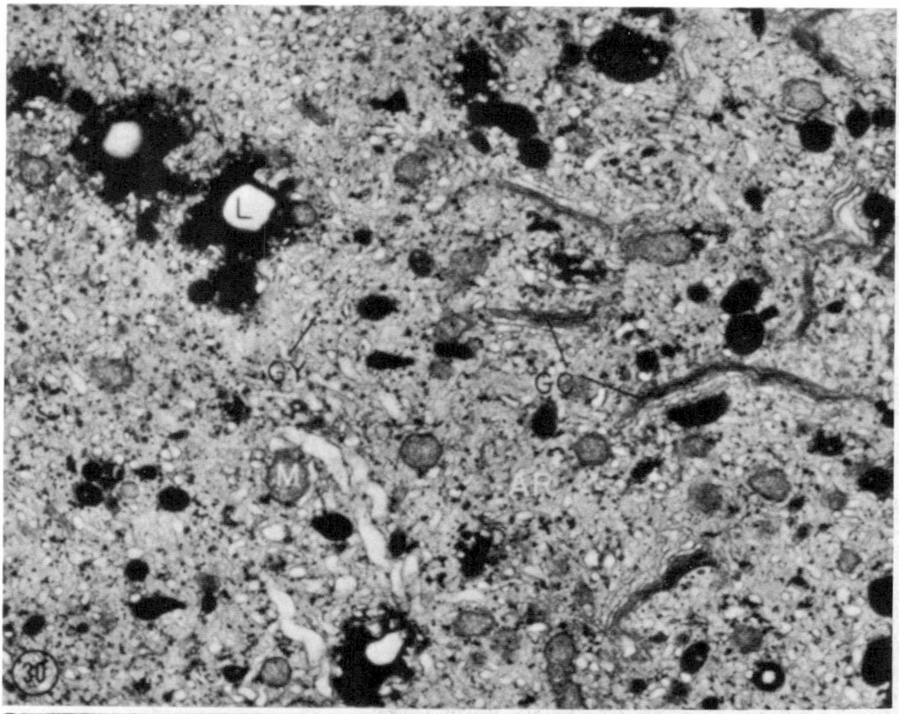
L
GY
GC
M
AR
30

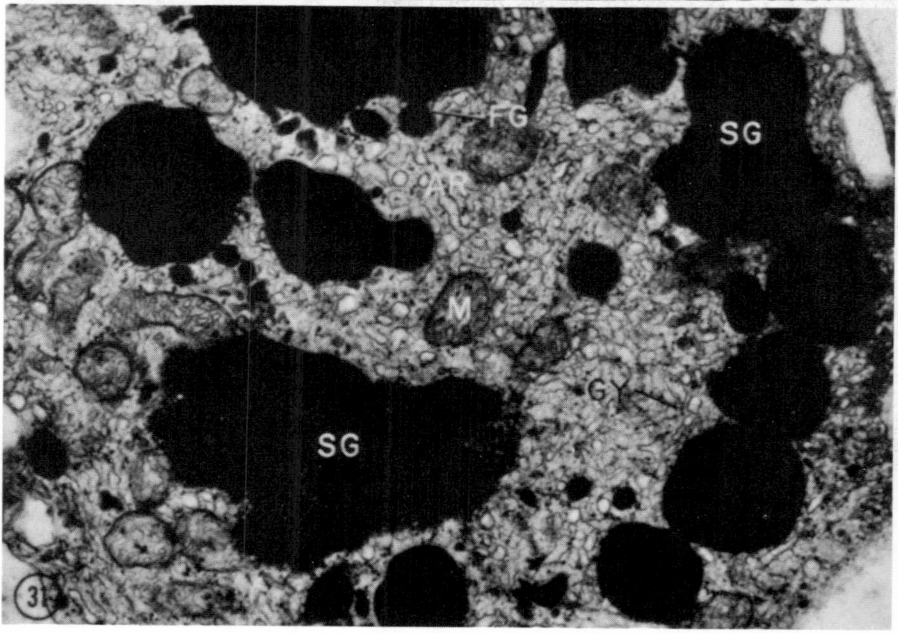
FG
SG
AR
M
GY
SG
31

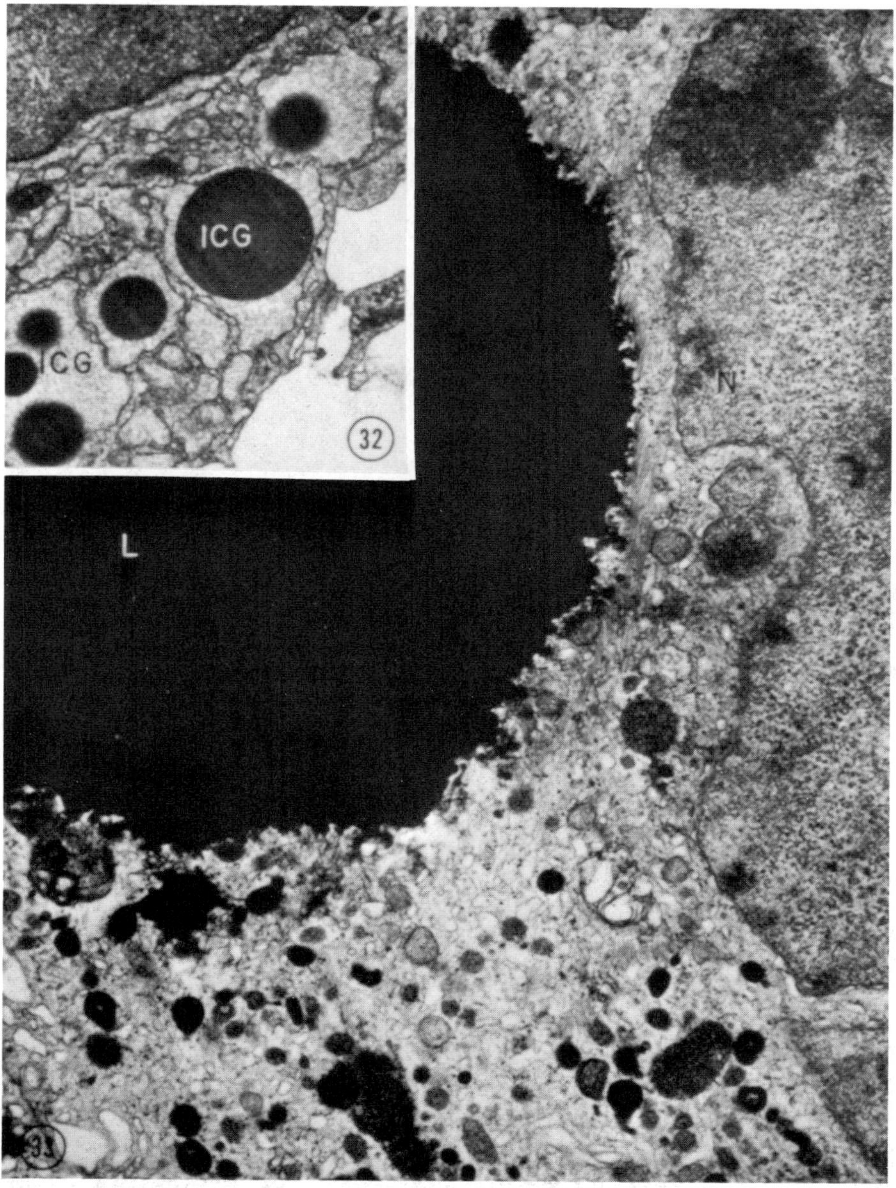

FIGURE 32 Postovulatory follicle. The cell type illustrated here is only occasionally encountered in the postovulatory follicle and contains granular endoplasmic reticulum (*ER*), the cisternae of which are expanded with intracisternal granules (*ICG*). Nucleus (*N*). × 15,750.

FIGURE 33 Postovulatory follicle. This cell type is characterized by extremely large lipid droplets (*L*) and by granules, some of which have the appearance of lysosomes. Nucleus (*N*). × 19,000.

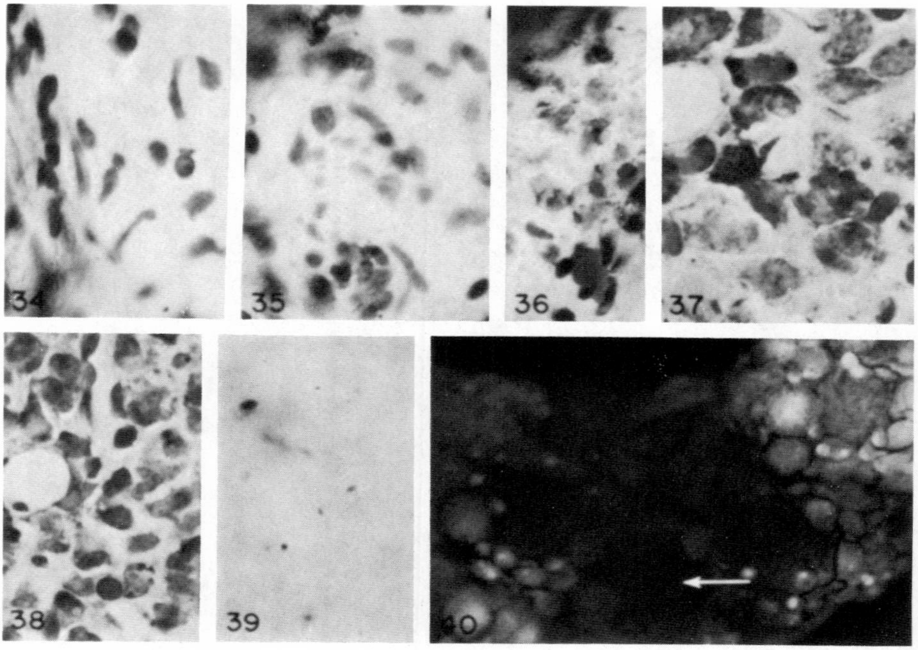

FIGURE 34 Whole mount preparation of follicle envelope removed from oocyte 1 mm in diameter and stained with Nile blue. Note absence of reddish-staining globules and hence of neutral lipid. × 400.

FIGURES 35–36 Whole mount preparations of follicle envelopes removed from oocyte 3–5 mm in diameter and stained with Nile blue. Large number of reddish-staining globules are evident in the cells of the follicle envelope, indicating presence of neutral lipid. × 400.

FIGURES 37–38 Frozen sections of postovulatory follicle stained with Nile blue. × 600.

FIGURE 39 Postovulatory follicle, frozen section, stained for demonstration of cholesterol. Cholesterol is indicated in the bluish-green stained granules. × 400.

FIGURE 40 Gross appearance (unstained) of a *Necturus* ovary. Postovulatory follicles are reddish-orange in color (arrow). × 8.

separations (see 68, 82), the results can be considered as only suggestive evidence for the presence of steroids in the postovulatory follicle.

Steroid Dehydrogenase Activity

The results of the cytochemical localization of 3 β-hydroxysteroid dehydrogenase (3 β-HSD) activity in the postovulatory follicle are illustrated in Figs. 52 and 53. In general, the reaction product appeared finely granular (Fig. 53), but on occasion large purple-staining granules were evident. According to Wattenberg (88), this coloration indicates the most intense activity. In addition, large lipid droplets were commonly encountered which had a number of small, purple-staining granules attached to their surface (Fig. 52). Treatment of frozen sections of postovulatory follicles with acetone prior to incubation in the substrate resulted in the removal of lipids and enzyme activity. In fact, much of the enzyme activity appears to be

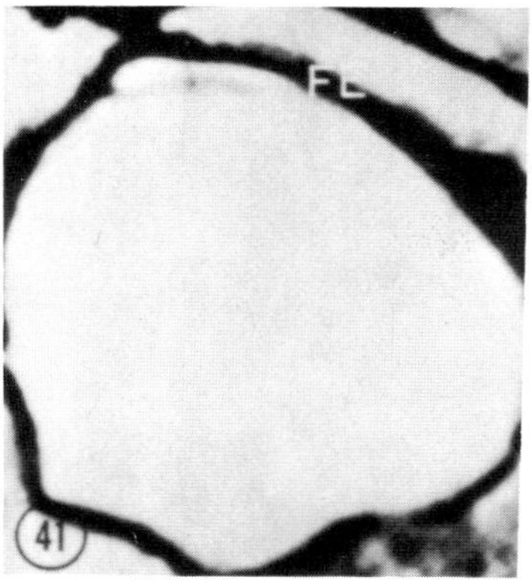

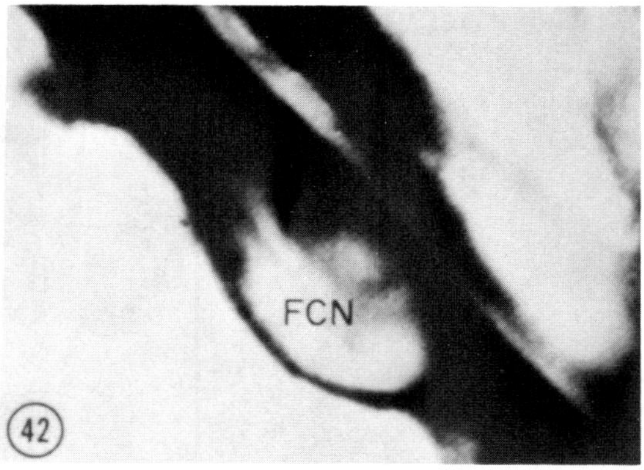

FIGURES 41–42 Frozen sections of follicle envelope from oocyte ~0.75 mm in diameter which have been incubated for alkaline phosphatase reaction. Note intense reaction in entire follicle envelope (*FE*) in Fig. 41. The reaction is particularly intense in follicle cells (Fig. 42). Follicle cell nucleus (*FCN*).

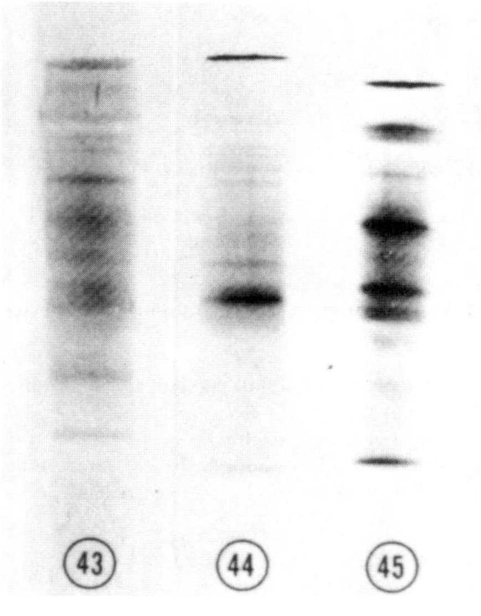

FIGURES 43–45 Polyacrylamide gel electrophoretograms of soluble proteins extracted from follicle envelopes of immature oocytes (<1 mm in diameter) in Fig. 43, from follicle envelopes of nearly mature oocytes in Fig. 44, and in the postovulatory follicle in Fig. 45.

associated with the lipid droplets in the postovulatory follicle, since treatment of the frozen sections with acetone following incubation in the steroid substrate resulted in the disappearance of most of the lipid droplets as well as most of the 3 β-HSD activity. It has previously been shown that a nonspecific deposition of formazan occurs on lipid droplets (52). Although true 3 β-HSD activity is usually stable to acetone, Levy et al. (52) found that 3 β-HSD activity in the theca interna of the rat ovary was acetone labile. Frozen sections of the postovulatory follicle treated for diaphorase activity gave positive results similar to those obtained with 3 β-HSD; however, the reaction for diaphorase was less intense than that obtained for 3 β-HSD.

Attempts were also made to localize more specifically the 3 β-HSD activity in the soluble-protein bands evident after disc electrophoresis on polyacrylamide gels. In the process, the electrophoresed gels were incubated in the steroid substrate (dehydroepiandrosterone). In these preparations, eight narrow bands were apparent in the original gels (Fig. 54). In contrast, only five thin bands were evident in the original electrophoresed gels incubated for diaphorase activity (Fig. 55). Furthermore, in those instances in which the diaphorase- and steroid substrate–reactive soluble-protein bands were identical in position, the intensity of staining of the diaphorase bands was much less (Figs. 54, 55).

In this connection, it should be pointed out that we have recently been concerned with a study of the multiple forms of lactic acid and malic acid dehydrogenases in a variety of vertebrate and

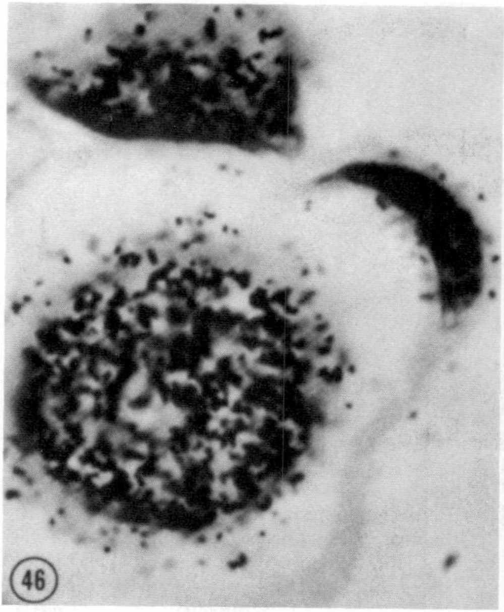

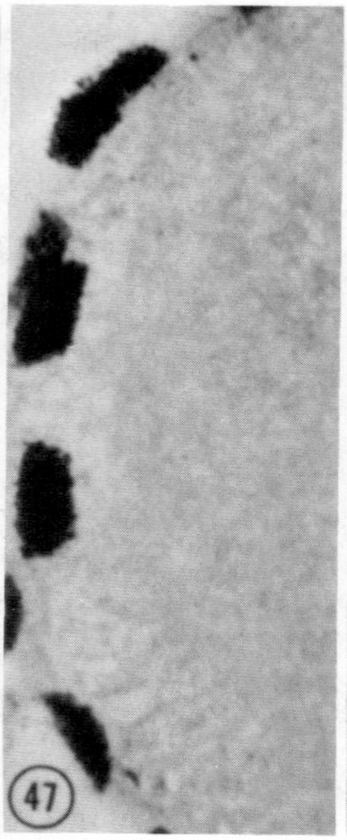

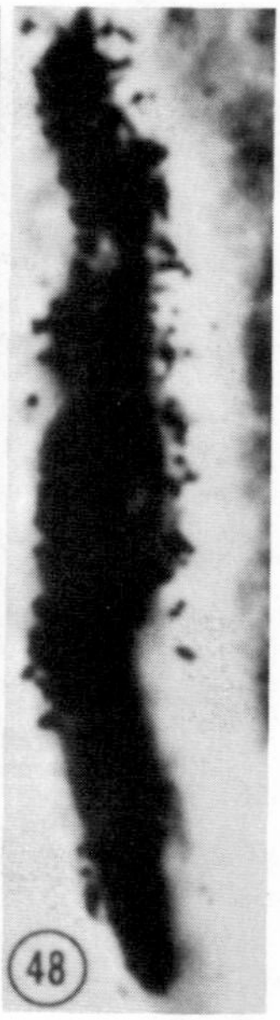

FIGURE 46 Radioautograph of oogonium surrounded by two fibroblast cells during early stage in the formation of follicle envelope. Note intense incorporation of labeled precursor (uridine-^{3}H) into·oogonial nucleus and perinuclear cytoplasm as well as into fibroblast nuclei and cytoplasm. In vitro exposure, 11 μc uridine-^{3}H for 5 hr.

FIGURES 47–48 Follicle cells surrounding oocytes <1 mm in diameter. Follicle cell nuclei are heavily labeled after 22-hr in vitro exposure to 11 μc uridine-^{3}H. Label also appears associated with narrow rim of follicle cell cytoplasm in Fig. 48.

invertebrate tissues. Our preliminary results have suggested that some of the dehydrogenase enzymes may exhibit a lack of substrate specificity and/or that a substrate is not needed for a positive histochemical reaction. For example, treatment of frozen sections of mouse ovaries for LDH and 3 β-HSD activity gave similar results for the localization of both enzymes in the tissue. In addition, electrophoretically separated soluble proteins of *Necturus* and mouse ovaries demonstrate an identical localization for both LDH with a sodium lactate substrate and 3 β-HSD with a dehydroepiandrosterone substrate (cf. 52). Furthermore, treatment of the separated soluble proteins for a diaphorase reaction gave results identical with those obtained in LDH and 3 β-HSD separations, although the intensity of staining of the bands was greatly reduced. The results suggest that, under the conditions employed, the pyridine nucleotide–linked dehydrogenase either is capable of accepting different substrates or is the result of endogenous NAD reacting with the tetrazolium salt to produce a "nothing" dehydrogenase reaction (cf. 95).

DISCUSSION

The presence and distribution of intercellular spaces in the *Necturus* follicle envelope are of considerable interest in connection with the problem of the incorporation of materials into the oocyte from sources external to the oocyte and ovary. Intercellular spaces become evident in both the theca and serosa layers very soon after the completed follicle envelope has been established. Intercellular spaces between follicle cells develop by the time the oocyte-follicle complex has

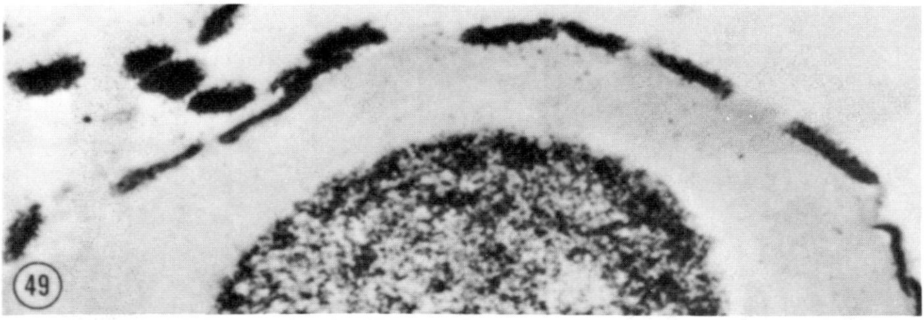

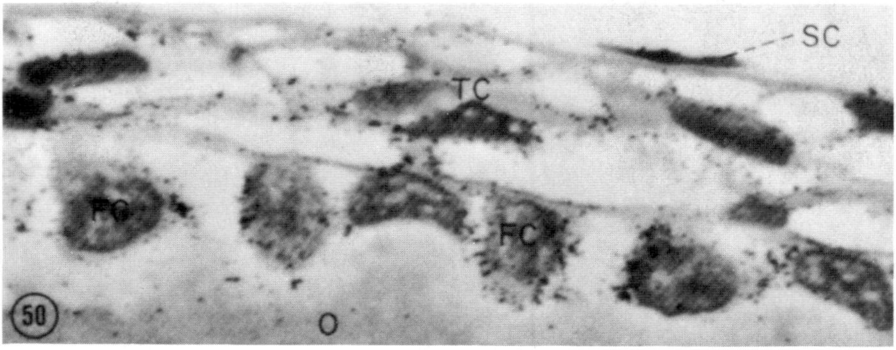

FIGURE 49 Oocyte-follicle complex ~0.75 mm in diameter. Oocyte nucleus, follicle cell nuclei, and ovarian fibroblast nuclei are intensely labeled after 2-hr in vitro exposure to 11 μc uridine-^{3}H. Little labeling is associated with the oocyte cytoplasm.

FIGURE 50 Oocyte-follicle complex ~4 mm in diameter. Note the label associated with the follicle cells (*FC*) after one-hour exposure in vitro to leucine-4,5-^{3}H. Some of the cells in the theca (*TC*) and serosa (*SC*) are also labeled. Oocyte periphery (*O*). × 2400.

grown to ~1.3 mm in diameter, and they persist throughout most of oogenesis. Since such channels exist, materials destined for passage into the growing oocyte need not necessarily pass through the follicle cells. Attention has been called to the presence of intercellular spaces between the follicle cells in *Triturus* oocytes as well (39). The ubiquity of follicle cell processes, oocyte microvilli, and pinocytotic vesicles associated with the oocyte surface in amphibia has inspired the concept that nutritive materials from the blood capillaries in the theca reach the oocyte by diffusion through the follicle cells and their processes (see 1, 16). While such a pathway may be operable in the *Necturus* follicle envelope, it is most probable that the intercellular spaces in the follicle envelope represent a major pathway for the migration of materials from the capillaries in the theca and from those outside of the follicle envelope to the oocyte surface where these materials are incorporated into the oocyte via micropinocytotic vesicles. The flocculent material present in many of the intercellular spaces of the follicle envelope may represent material coagulated by the fixative while in transit through these spaces. There is evidence to suggest that intact blood proteins are capable of passing through the barrier formed by the follicle cell layer and reaching the oocyte (see 30, 59, 77). In this regard, the results of a soluble-protein analysis on the oocyte, follicle envelope, blood, and liver of *Necturus* are of interest (64). For oocytes ranging from 1.0 to 1.5 mm in diameter, the soluble pro-

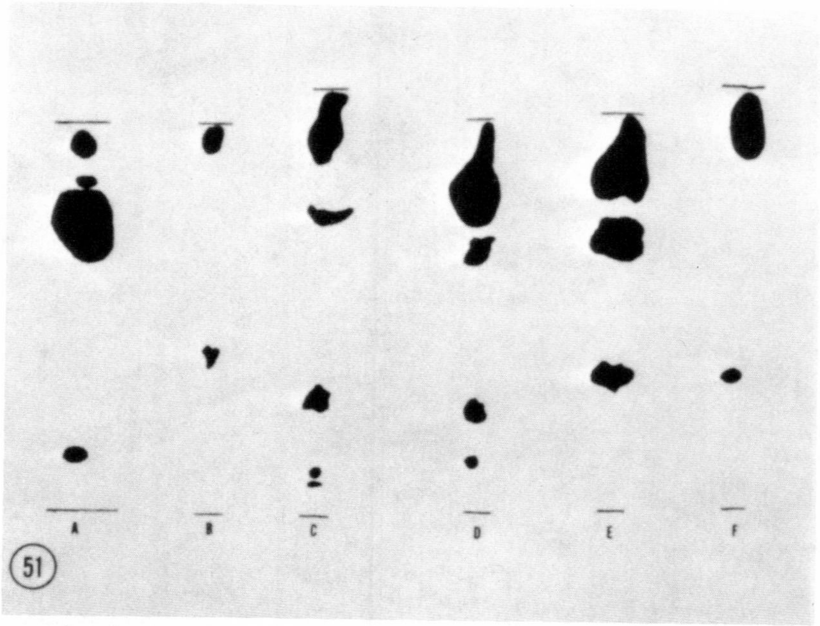

FIGURE 51 Tracings of thin-layer chromatograms of lipids extracted from postovulatory follicles. Solvent for *A*, 90 parts *n*-pentane, 10 parts diethyl ether, and 1 part acetic acid; solvent for *B*, 65 parts chloroform, 34 parts methanol, and 4 parts water; solvent for *C*, 3 parts benzene, 2 parts ethyl acetate; solvent for *D*, 4 parts benzene, 1 part ethyl acetate; solvent for *E*, 4 parts chloroform, 1 part ethyl acetate; solvent for *F*, 9 parts chloroform, 1 part acetone.

teins extracted from either the follicle envelopes or oocytes are present in greatest numbers (64). Furthermore, those soluble proteins isolated from either the follicle envelopes or oocytes during this period of oogenesis are strikingly similar in number and migration to those isolated from the liver or blood (64). Studies on the mechanisms of yolk deposition in the oocyte indicate that this deposition is initiated in oocytes ~1.1 mm in diameter (49). In oocytes ranging from 1.1 to 2.0 mm, the yolk precursor bodies increase considerably in size and number.[1] The results thus suggest that soluble proteins may pass in large amounts from the blood capillaries of the follicle envelope to the developing oocyte at a period of oogenesis in which the initial stages of protein yolk deposition occur. The formation of intercellular spaces in the follicle cell layer at a time closely correlated with the aforementioned activity would appear to be significant in this activity.

That the cells of the follicle envelope are directly and actively engaged in the processes of transport within the envelope is indicated by the large numbers of micropinocytotic vesicles occurring in these cells. Microvesicles described as associated with a transport function have been observed in a number of cells (see, 27, 42, 72, 90). Although some controversy exists regarding the origin, content, and direction of transport by microvesicles, it is likely that these activities vary with the metabolic state of the cell (see 90). Pinocytotic vesicles are most numerous in the serosa cells, theca cells, and capillary endothelial cells of the *Necturus* follicle envelope, and these vesicles are generally of the noncoated or nonalveolate variety. The pinocytotic vesicles associated with the follicle cells, however, are typically of the coated or alveolate variety (see 74), and they are much less numerous.

[1] R. G. Kessel. Unpublished observations.

TABLE I

Chromatography of Postovulatory Follicle

	Solvent	Rf × 100	Possible identification*
	n-Pentane: 90	92	Sterol esters
A	Diethyl ether: 10	83	Triglycerides of nonoxygenated fatty acids
	Acetic acid: 1	71	Triglycerides or free fatty acids
		14	Steroids, cholesterol, or triglycerides containing hydroxy acids
	Chloroform: 65		
B	Methanol: 34	96	Neutral lipids
	Water: 4	40	Phosphatidyl choline or cephalin
	Benzene: 3	88	Neutral lipids
C	Ethyl acetate: 2	69	
		27	
		10	
		8	
	Benzene: 4	83	Neutral lipids and progesterone
D	Ethyl acetate: 1	65	Estrone, and/or androstenedione and/or cholesterol
		25	5-Androstenediol and/or adrenosterone
		13	Pregnanediol
	Chloroform: 4	83	Neutral lipids
E	Ethyl acetate: 1	67	Testosterone propionate and/or androstane-3,17-dione and/or progesterone
		33	17 α-hydroxyprogesterone, and/or androsterone, and/or estradiol
	Chloroform: 9	88	Neutral lipids
F	Acetone: 1	31	Pregnane-3α,20 α-diol, and/or estradiol

* After Stahl (82), for solvents A, B, E, F; after Randerath (68), for solvent D.

Evidence that microvesicles and microvesicular rosettes in fat cells and endothelium may constitute a mechanism for the transport of free fatty acids has been provided by Williamson (90). In their structure and number, the microvesicles and microvesicular rosettes in the capillaries of the *Necturus* follicle envelope are so similar to those described in fat cells that it is tempting to speculate that they may be involved to some extent in the transport of free fatty acids from the blood into the follicle envelope. That such activity appears correlated with the progressive accumulation of lipid in cells of the follicle envelope lends further support to such a viewpoint. However, it is apparent that a number of different molecular species in addition to fatty acids are transported in the capillaries and into the cells of the *Necturus* follicle envelope.

Most of the cytoplasmic differentiation in cells constituting the follicle envelope in *Necturus* takes place as the oocyte-follicle complex grows from ~1.3 to 2.0 mm in diameter. These changes involve, among others, the elaboration of a smooth form of the endoplasmic reticulum. The rough-surfaced endoplasmic reticulum is also present in varying amount in all cells of the follicle envelope throughout oogenesis. The development and presence of both rough and smooth-surfaced forms of endoplasmic reticulum may be related to the biosynthesis of lipid in these cells. All three cell types in the follicle possess Golgi material that appears to be associated with the origin of dense secretory granules which are similar in structure in all cell types comprising the three layers of the follicle envelope. The nature and significance of these secretory granules are unknown. All cells of

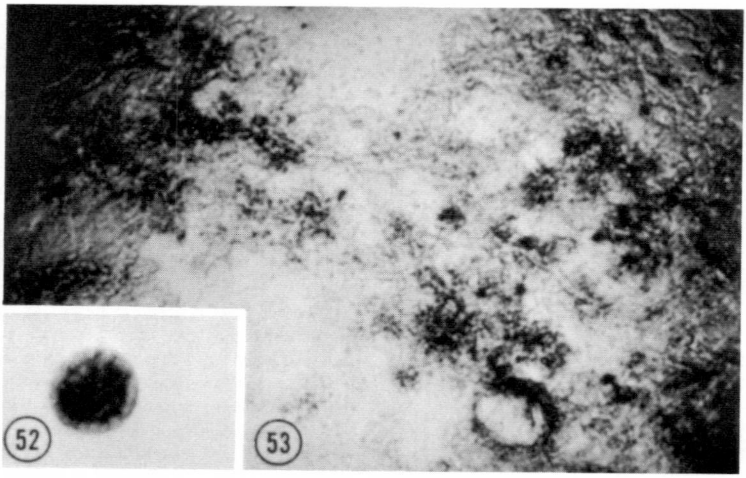

FIGURES 52 and 53 Photomicrographs of frozen sections of postovulatory follicle illustrating reaction product (dark granular material) when assayed for 3 β-HSD activity (Fig. 53). Reaction product is associated with surface of lipid droplet in Fig. 52.

the follicle envelope contain considerable amounts of glycogen, and the accumulation of lipid in all three cell types is particularly active in oocyte-follicle complexes ranging from 1.3 to 2.0 mm in diameter. Thus, the major activities occurring in the preovulatory follicle appear to be the accumulation of lipid and glycogen. These activities proceed during the course of oogenesis so that at its completion most of the follicle cells and many of the serosa and theca cells are filled with lipid droplets. Lipid droplets in generally small amounts have been pointed out in other amphibian follicle cells (see 39). In this connection, it is of interest that in the case of birds the cholesterol content of the follicle epithelium and of the theca interna increases strongly in the last phase of vitellogenesis (56). More recently, it has been demonstrated by Popjak and Tietz (67) that the follicle cells of birds are capable of synthesizing fatty acids and cholesterol from added acetate. Only a small amount of cholesterol was detected in the *Necturus* follicle envelopes at the end of oogenesis. The accumulation of lipid and glycogen in cells of the follicle envelope during oogenesis in *Necturus* appears to be preparation of the cells for their subsequent functional activity as a postovulatory follicle, since cells constituting the postovulatory follicle also contain large amounts of lipid, glycogen, and cholesterol.

Analysis of the number of soluble proteins present in the follicle envelope during oogenesis and in the postovulatory follicle demonstrated a change in the number of soluble proteins in the follicle envelope during development, or a possible change in the solubilities of certain of the proteins. Changes in the patterns of migration of the soluble proteins were also evident. Since soluble proteins have been shown to consist, in part, of enzymes, it is probable that the enzymatic activity of the cells of the follicle envelope may also be changing during the cells' period of development and activity. That this differentiation may, in part, involve loss of the ability to synthesize a pre-existing protein and/or changes in the solubilities of pre-existing proteins has been suggested and discussed by Spiegel (81).

The significance of the intense reaction for alkaline phosphatase obtained in the *Necturus* follicle envelope in general and in the follicle cells in particular may be related to the transport of materials through the follicle envelope. Osawa (62) has reported a strong alkaline phosphatase reaction in the follicle cells of *Triturus*. In a study demonstrating the presence of alkaline phosphatase

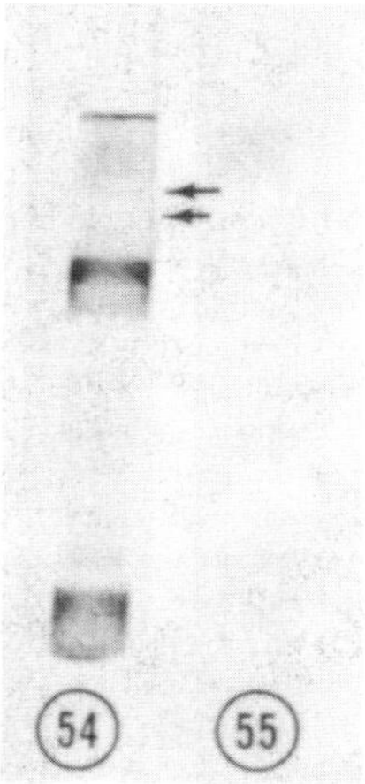

FIGURES 54 and 55 Polyacrylamide gel electrophoretograms of soluble proteins extracted from postovulatory follicle. Gel in Fig. 54 was incubated for 3 β-HSD activity. Gel in Fig. 55 was incubated in same medium but without substrate. Note faint bands (arrows) in Fig. 54.

in the follicle cells of *Chirocephalopsis*, Linder (53) commented that the presence of the enzyme in the follicle is probably related to the follicle's function as a selective membrane controlling the passage of food substances across cell boundaries from the blood to the oocyte.

Based on either cytochemical or radioautographic techniques, it has been suggested that follicle cells in general are rich in RNA (see 69). The radioautographic studies of Ficq (29) indicate the occurrence of DNA synthesis and protein synthesis in follicle cells in other amphibian species. The results of the incorporation of uridine-^{3}H into the *Necturus* follicle envelope suggest that the follicle envelope is capable of the biosynthesis of RNA. That synthesis of RNA rather than accumulation of RNA occurs in the follicle envelope is suggested by the results of the in vitro studies and the treatment of the radioautographs with trichloroacetic acid. Since the radioautographic procedures utilized probably removed the soluble RNA, the incorporation of uridine-^{3}H would appear to occur in ribosomal and perhaps some messenger RNA. The results also indicate that the incorporation of amino acids into proteins occurs in the follicle envelope especially during the later part of oogenesis. The labeling of the follicle cells late in oogenesis may be related to the appearance of yolk platelets in some, but not all, of the follicle cells as demonstrated in the electron micrographs. The radioautographic results thus provide evidence that in *Necturus* the cells of the follicle envelope actively take up uridine, probably into RNA, and later in development are more active in the uptake of protein precursors. Quantitative information regarding the amount of RNA and proteins in isolated amphibian follicle envelopes has only recently become available. Panje and Kessel (64) have recently reported the results of a quantitative analysis of RNA and protein in isolated follicle envelopes of *Necturus* during selected stages of oogenesis. This study showed that in follicle envelopes removed from oocytes 0.7–2.5 mm in diameter there is a rapid rise in the amount of RNA but only a small increase in total protein during this period of oogenesis. In follicle envelopes removed from oocytes ranging from 2.5 to 4.0 mm in diameter, the total RNA continues to increase slowly; but there is, in contrast, a sharp increase in the total protein in the isolated follicle envelopes at these same stages. These quantitative results (64) correlate with the demonstration that the follicle cells intensely incorporate uridine-^{3}H especially during the first half of oogenesis (based on oocyte size) whereas incorporation of leucine-^{3}H and lysine-^{3}H in the follicle cells is particularly high during the latter half of oogenesis.

There is no information regarding the kinds or relative amounts of RNA in the *Necturus* envelope. Thus, it would be of interest to determine whether or not sRNA, mRNA, and rRNA are all present in the follicle cells, what their relative amounts are, and whether or not they vary in amount during the growth period of the oocyte. Nace and Lavin (59) have suggested the possibility that the synthetic machinery (e.g., messenger RNA) may be formed in the amphibian follicle cell nuclei and transported into the oocyte to take part in autosynthetic processes associated with yolk deposition. This activity, they postulate, is consistent with evidence of intense uptake of nucleic acid and protein precursors in the follicle cells (29) which do not appear to accumulate the products of the synthesis implied by such uptake. Although several studies have failed to detect the presence of mRNA in amphibian oocyte nuclei or cytoplasm (11, 12, 19, 32), a more recent study by Davidson et al. (20) reports the presence of mRNA in *Xenopus* oocytes from which the follicle envelopes had been mechanically removed.

During the process of yolk utilization or demolition in various embryological stages of the frog, Karasaki (44) described a condition in which the yolk platelet became progressively surrounded by membranous whorls. As a result of this process, the main body of the yolk platelet was eventually transformed into large concentric membranous structures having the appearance of myelin figures. Karasaki (44) has proposed that these membranes, which are 70 A thick and show a unit membrane structure, are utilized during the course of embryogenesis for the formation of various cytomembranes needed during differentiation. In the present study, when the yolk platelets were observed within the follicle cells, they were surrounded by varying numbers of concentric membranes similar to those suggested as resulting from the demolition of the yolk platelet. Furthermore, the demolition process appears to involve all of the yolk platelets since none were observed in the cells of the postovulatory follicle. In view of the increased differentiation which the cells of the postovulatory follicle undergo compared to their condition in the preovulatory follicle, it is tempting to speculate that the accumulation of yolk in some cells of the follicle envelope late in oogenesis may be associated with their preparation to eventually become highly active in secretion, a process that involves a much more extensive development of cytomembrane systems in those cells of the postovulatory follicle than was the case in the preovulatory follicle cells.

A matter of considerable interest concerns the problem of whether or not the *Necturus* postovulatory follicle is capable of synthesizing steroid and playing a role in the regulation of ovarian activity. Several lines of evidence suggest that the postovulatory follicle may act, among other functions as a steroid-producing organ: (*a*) studies based on thin-layer chromatography of lipids extracted from the postovulatory follicle; (*b*) cytochemical studies for the localization of 3 β-HSD activity and the application of this reaction to soluble proteins separated by polyacrylamide gel electrophoresis; (*c*) the suggested presence of cholesterol (prerequisite for steroid biosynthesis) in the postovulatory follicle based on techniques of thin-layer chromatography and cytochemistry; (*d*) the fine structure of certain cells comprising the postovulatory follicle; and (*e*) the fact that the postovulatory follicle is formed as a result of ovulation and then engages in pronounced cellular activity for a period of time is reminiscent of the activity associated with the mammalian corpus luteum.

The smooth-surfaced endoplasmic reticulum is characteristically a well developed cytomembrane system in known steroid-producing cells. Thus, a tubular form of smooth-surfaced endoplasmic reticulum is encountered in the corpus luteum (5, 25, 26, 93), in the interstitial cells of the testis (17, 18), and in the adrenal cortex (51, 54, 73). In addition, steroid-producing cells are characterized by the presence in the cytoplasm of mitochondria, Golgi complexes, lipid, and lysosomes (see 54). In general, the rough endoplasmic reticulum is not so well developed as the smooth form. The point to be made here is that the postovulatory follicle of *Necturus* does contain cells which, on the basis of kind, number, and disposition of intracellular organelles and formed product, resemble known steroid-producing cells in other animals. In general, no visible accumulation of secretory product occurs in cells known to produce steroid hormones, and the process whereby steroid hormones are secreted is poorly understood. However, Belt et al. (2) have reported electron-opaque granules bounded by a smooth membrane within the cytoplasm of pelican adrenal cells. Granules of similar dimensions and electron opacity but without the membrane were observed between adjacent parenchymal cells and in the subendothelial space. Belt et al. (2) suggested that the dense granules represent hormone which is synthesized within the smooth membranes, carried to the cell surface, and secreted by a process of reverse pinocytosis.

Much of the cholesterol and cholesterol ester in adrenal cortical cells is located in lipid droplets as indicated by the Schultz reaction (see 22). Although cholesterol in certain steroid-producing cells may be stored in solution in the neutral fat droplets, there is also evidence that the extensive agranular reticulum in steroid-producing cells is involved with the storage and metabolism of cholesterol in relation to the biosynthesis of steroid hormones (see 28). Biochemical studies on a variety of cell types indicate that the enzymes required for the synthesis of triglyceride and phospholipids are associated with the rough- and smooth-surfaced elements of the endoplasmic reticulum (see 83, 10, 78, 23, 41). The enzymes catalyzing fatty acid esterification thus appear to be bound to the membranes of the endoplasmic reticulum, so that this activity must occur at, or very near, the membrane (83). The subsequent step, in which the triglyceride molecules aggregate into a visible lipid droplet, has been localized to the intracisternal portion of the endoplasmic reticulum, and the growth of lipid droplets has been suggested to occur by intracisternal accumulation and fusion of smaller droplets (83). The biochemical pathways in the formation of steroids and the subcellular localization of the enzymes involved have been studied by Popjak and Cornforth (66), Olson (61), Halkerston et al. (34), Byer and Samuels (13), Hoffmann (37), Ryan and Engel (75), Sharma et al. (79), and Hayano and Dorfman (35).

Little information is available regarding the significance of the presence of a large Golgi complex in steroid-producing cells. Since recent biochemical studies have emphasized the importance of steroid conjugates, particularly sulfates, in biosynthetic reactions (14), Long and Jones (54) have suggested "that conjugation of steroid hormones occurs in the Golgi apparatus of adrenocortical cells and that the product is secreted as the more water-soluble sulfate or glucuronide."

Some evidence is available to indicate that steroid biosynthesis occurs in the postovulatory follicles of other amphibians. In histochemical studies for the demonstration of 3 β-HSD activity, Botte and Cottino (8) have obtained positive reactions in the ovarian follicle and postovulatory follicles of both *Rana esculenta* and *Triturus cristatus*. On the other hand, such histochemical activity could not be demonstrated in either the ovarian follicle or postovulatory follicles in *Xenopus laevis* and *Bufo bufo* (65). Of particular interest with respect to the capability of steroid biosynthesis by the *Necturus* postovulatory follicle are the results of Callard and Leathem (15) who have recently described studies in which ovarian fragments of *Necturus maculosus* were incubated with labeled pregnenolone, progesterone, or testosterone. Their data suggested that both pregnenolone and progesterone could be used as substrates for in vitro ovarian steroid synthesis in *Necturus*.

The functions of the other cell types constituting the postovulatory follicle are unknown. Some of the cells may function in the continued biosynthesis of lipid and cholesterol. Other cells may be capable of phagocytosis and take part in cleaning up the follicle components after ovulation and after the postovulatory follicle has completed its function, as occurs in other vertebrate species (see 36).

This investigation was supported by research grants (GM-09229; HD-00699) and a Career Development Award (to Dr. Kessel) from the National Institutes of Health, United States Public Health Service. The authors wish to acknowledge the skillful technical assistance of Mrs. Marlene Decker.

REFERENCES

1. Anderson, E., and H. W. Beams. 1960. *J. Ultrastruct. Res.* **3**:432.
2. Belt, W. D., M. N. Sheridan, R. A. Knouff, and F. A. Hartman. 1965. *Z. Zellforsch.* **68**:864.
3. Bier, K. 1962. *Naturwissenschaften.* **49**:332.
4. Bier, K. 1963. *J. Cell Biol.* **16**:436.
5. Blanchette, E. J. 1966. *J. Cell Biol.* **31**:501, 542.
6. Bonhag, P. F. 1955. *J. Morphol.* **96**:381.
7. Bonhag, P. F. 1956. *J. Morphol.* **99**:433.
8. Botte, V., and E. Cottino. 1964. *Boll. Zool.* **31**:491.
9. Brachet, J. 1947. *Experientia.* **3**:329.
10. Brindley, D. N., and G. Hübscher. 1965. *Biochim. Biophys. Acta.* **106**:495.
11. Brown, D. D., and E. Littna. 1964. *J. Mol. Biol.* **8**:669.
12. Brown, D. D., and E. Littna. 1964. *J. Mol. Biol.* **8**:688.
13. Byer, K. F., and L. T. Samuels. 1956. *J. Biol. Chem.* **219**:69.
14. Calvin, H. I., R. L. VandeWiele, and S. Lieberman. 1963. *Biochemistry.* **2**:648.
15. Callard, I. P., and J. H. Leathem. 1966. *Gen. Comp. Endocrinol.* **7**:80.
16. Chiquoine, A. D. 1960. *Am. J. Anat.* **106**:149.
17. Christensen, A. K., and D. W. Fawcett. 1961. *J. Biophys. Biochem. Cytol.* **9**:653.
18. Christensen, A. K., and D. W. Fawcett. 1966. *Am. J. Anat.* **118**:551.
19. Davidson, E. H., V. G. Allfrey, and A. E. Mirsky. 1964. *Proc. Natl. Acad. Sci. U.S.A.* **52**:501.
20. Davidson, E. H., M. Crippa, F. R. Cramer, and A. E. Mirsky. 1966. *Proc. Natl. Acad. Sci. U.S.A.* **56**:856.
21. Davis, B. J. 1964. *Ann. N. Y. Acad. Sci.* **121**:404.
22. Deane, H. W. 1962. *In* Handbuch der Experimentellen Pharmakologie. 1962. O. Eichler and A. Farah, editors. pt. 1. **14**:1. Springer-Verlag, Berlin.

23. DeDuve, C., R. Wattiaux, and P. Baudhuin. 1962. *Advan. Enzymol.* **24**:291.
24. Dollander, A. 1956. *Compt. Rend. Soc. Biol.* **150**: 998.
25. Enders, A. C. 1962. *J. Cell Biol.* **12**:101.
26. Enders, A. C., and W. R. Lyons. 1964. *J. Cell Biol.* **22**:127.
27. Fawcett, D. W. 1959. *In* The Microcirculation. S. R. M. Reynolds and B. W. Zweifach, editors. University of Illinois Press, Urbana. 1.
28. Fawcett, D. W. 1963. *In* Intracellular Membranous Structure. S. Seno and E. V. Cowdry, editors Okayama, Japan. *Symp. Soc. Cellular Chem.*, **14**:15.
29. Ficq, A. 1961. Metabolisme de l'oogenese chez les amphibiens. *In* Symposium on germ cells and development. Instit. Intern. Embryol; 1st. Lombardo, Fondazione A Baselli, Milan. 121.
30. Flickinger, R. A., and D. E. Rounds. 1956. *Biochim. Biophys. Acta.* **22**:38.
31. Folch, J., M. Lees, and G. H. Sloane Stanley. 1957. *J. Biol. Chem.* **226**:497.
32. Gall, J. G. 1966. *In Natl. Cancer Inst. Monograph.* **23**:475.
33. Gomori, G. 1952. Microscopic Histochemistry. The University of Chicago Press, Chicago.
34. Halkerston, I. D. K., J. Eichhorn, and O. Hechter. 1961. *J. Biol. Chem.* **236**:374.
35. Hayano, M., and R. I. Dorfman. 1962. *In* Methods in Enzymology. S. P. Colowick and N. O. Kaplan, editors. Academic Press Inc., New York. 503.
36. Hisaw, F. L., Jr., and F. L. Hisaw. 1959. *Anat. Record.* **135**:269.
37. Hofmann, F. G. 1960. *Biochim. Biophys. Actr.* **37**: 566.
38. Hope, J., A. A. Humphries, Jr., and G. H. Bourne. 1964. *J. Ultrastruct. Res.* **10**:547.
39. Hope, J., A. A. Humphries, Jr., and G. H. Bourne. 1963. *J. Ultrastruct. Res.* **9**:302.
40. Hsu, W. S. 1952. *Quart. J. Microscop. Sci.* **93**:191.
41. Isselbacher, K. J. 1965. *Federation Proc.* **24**:16.
42. Jennings, M. A., V. T. Marchesi, and H. Florey. 1962. *Proc. Roy. Soc.* (*London*), *Ser. B.* **156**:14.
43. Karasaki, S. 1959. *Embryologia.* **4**:247.
44. Karasaki, S. 1963. *J. Ultrastruct. Res.* **9**:225.
45. Kemp, N. E. 1956. *J. Biophys. Biochem. Cytol.* **2**: 281.
46. Kemp, N. E. 1956. *J. Biophys. Biochem. Cytol.* **2**(4, suppl.): 281.
47. Kemp, N. E. 1958. *Anat. Record.* **130**:324.
48. Kemp, N. E. 1961. *J. Appl. Phys.* **32**:1643.
49. Kessel, R. G. 1966. *J. Cell Biol.* **31**:148A. Abstr.
50. Kopriwa, B. M., and C. P. Leblond. 1962. *J. Histochem. Cytochem.* **10**:269.
51. Lever, J. D. 1955. *Am. J. Anat.* **97**:409.
52. Levy, H., H. W. Deane, and B. L. Rubin. 1959. *Endocrinology* **65**:932.
53. Linder, H. J. 1959. *J. Morphol.* **104**:1.
54. Long, J. A., and A. L. Jones. 1967. *Am. J. Anat.* **120**:463.
55. Luft, J. H. 1961. *J. Biophys. Biochem. Cytol.* **9**:409.
56. Marza, V. D., and E. V. Marza. 1935. *Quart. J. Microscop. Sci.* **78**:133.
57. Masui, Y. 1967. *J. Exptl. Zool.* **166**:365.
58. Mazia, D., P. A. Brewer, and M. Alfert. 1953. *Biol. Bull.* **104**:57.
59. Nace, G. W., and L. H. Lavin. 1963. *Am. Zool.* **3**:193.
60. Nandi, J. 1967. *Am. Zool.* **7**:115.
61. Olson, J. A. 1965. *Ergb. Physiol.* **56**:173.
62. Osawa, S. 1951. *Embryologia* **2**:1.
63. Palade, G. E. 1952. *J. Exptl. Med.* **95**:285.
64. Panje, W. R., and R. G. Kessel. 1968. *Exptl. Cell Res.* In press.
65. Pesonen, S., and J. Rapola. 1962. *Gen. Comp. Endocrinol.* **2**:425.
66. Popjak, G., and J. W. Cornforth. 1960. *Advan. Enzymol.* Interscience Publishers, Inc., New York. **22**:281.
67. Popjak, G., and A. Tietz. 1953. *Biochiem. J.* **54**: 590.
68. Randerath, K. 1965. Thin-Layer Chromatography. Academic Press, Inc., New York.
69. Raven, C. P. 1961. Oogenesis: The Storage of Developmental Information. Pergamon Press, New York.
70. Revel, J. P. 1964. *J. Histochem. Cytochem.* **12**:104.
71. Reynolds, E. S. 1963. *J. Cell Biol.* **17**:208.
72. Rhodin, J. A. G. 1962. *Physiol. Rev.* **42**(Suppl. 5, pt. II):48.
73. Ross, M. H., G. D. Pappas, J. T. Lanman, and J. Lind. 1958. *J. Biophys. Biochem. Cytol.* **4**:659.
74. Roth, T. F., and K. R. Porter. 1964. *J. Cell Biol.* **20**:313.
75. Ryan, K. J., and L. L. Engel. 1957. *J. Biol. Chem.* **225**:103.
76. Sabatini, D., K. Bensch, and R. J. Barrnett. 1963. *J. Cell Biol.* **17**:19.
77. Schechtman, A. M. 1947. *J. Exptl. Zool.* **105**:329.
78. Schneider, W. C. 1963. *J. Biol. Chem.* **238**:3572.
79. Sharma, D. E., E. Forchielli, and R. I. Dorfman. 1962. *J. Biol. Chem.* **237**:1495.
80. Sirlin, J. L., and J. Jacob. 1960. *Exptl. Cell Res.* **20**:283.
81. Spiegel, M. 1960. *Biol. Bull.* **118**:451.
82. Stahl, E. 1965. Thin-Layer Chromatography. Academic Press Inc., New York.
83. Stein, O., and Y. Stein, 1967. *J. Cell Biol.* **33**: 319.
84. Thompson, S. W. 1966. Selected Histochemical and Histopathological Methods. Charles C Thomas, Springfield, Illinois.

85. Wartenberg, H. 1962. *Z. Zellforsch.* **58**:427.
86. Wartenberg, H., and W. Gusek. 1960. *Exptl. Cell Res.* **19**:199.
87. Watson, M. L. 1958. *J. Biophys. Biochem. Cytol.* **4**:475.
88. Wattenberg, L. W. 1958. *J. Histochem. Cytochem.* **6**:225.
89. Williams, J. 1969. *In* The Biochemistry of Animal Development R. Weber, editor. Academic Press Inc., New York. **1**:1.
90. Williamson, J. R. 1964. *J. Cell Biol.* **20**:57.
91. Wischnitzer, S. 1966. *In* Advances in Morphogenesis. M. Abercrombie and J. Brachet, editors Academic Press Inc., New York. **5**:131.
92. Witschi, E. 1966. Development of Vertebrates. W. B. Saunders Company, Philadelphia.
93. Yamada, E., and T. M. Ishikawa. 1960. *Kyushu J. Med. Sci.* **11**:235.
94. Zalokar, M. 1960. *Exptl. Cell Res.* **19**:183.
95. Zimmermann, H., and A. G. E. Pearse. 1959. *J. Histochem. Cytochem.* **7**:271.

AUTHOR INDEX

KEY-WORD TITLE INDEX